MW00814588

Springer Series in Advanced Manufacturing

Series Editor

Professor D.T. Pham
Manufacturing Engineering Centre
Cardiff University
Queen's Building
Newport Road
Cardiff CF24 3AA
UK

Other titles in this series

Assembly Line Design
B. Rekiek and A. Delchambre

Advances in Design
H.A. ElMaraghy and W.H. ElMaraghy (Eds.)

Effective Resource Management in Manufacturing Systems: Optimization Algorithms in Production Planning
M. Caramia and P. Dell'Olmo

Condition Monitoring and Control for Intelligent Manufacturing
L. Wang and R.X. Gao (Eds.)

Optimal Production Planning for PCB Assembly
W. Ho and P. Ji

Trends in Supply Chain Design and Management: Technologies and Methodologies
H. Jung, F.F. Chen and B. Jeong (Eds.)

Process Planning and Scheduling for Distributed Manufacturing
L. Wang and W. Shen (Eds.)

Collaborative Product Design and Manufacturing Methodologies and Applications
W.D. Li, S.K. Ong, A.Y.C. Nee and C. McMahon (Eds.)

Decision Making in the Manufacturing Environment
R. Venkata Rao

Reverse Engineering: An Industrial Perspective
V. Raja and K.J. Fernandes (Eds.)

Frontiers in Computing Technologies for Manufacturing Applications
Y. Shimizu, Z. Zhang and R. Batres

Automated Nanohandling by Microrobots
S. Fatikow

A Distributed Coordination Approach to Reconfigurable Process Control
N.N. Chokshi and D.C. McFarlane

ERP Systems and Organisational Change
B. Grabot, A. Mayère and I. Bazet (Eds.)

ANEMONA
V. Botti and A. Giret

Theory and Design of CNC Systems
S.-H. Suh, S.-K. Kang, D.H. Chung and I. Stroud

Machining Dynamics
K. Cheng

Changeable and Reconfigurable Manufacturing Systems
H.A. ElMaraghy

Advanced Design and Manufacturing Based on STEP
X. Xu and A.Y.C. Nee (Eds.)

Artificial Intelligence Techniques for Networked Manufacturing Enterprises Management
L. Benyoucef, B. Grabot (Eds.)

Wei Gao

Precision Nanometrology

Sensors and Measuring Systems for Nanomanufacturing

Springer

Wei Gao, PhD
Tohoku University
School of Engineering
Department of Nanomechanics
Aramaki Aza Aoba 6-6-01
980-8579 Aoba-ku
Sendai
Japan
gaowei@nano.mech.tohoku.ac.jp

ISSN 1860-5168
ISBN 978-1-84996-253-7 ISBN 978-1-84996-254-4 (eBook)
DOI 10.1007/978-1-84996-254-4
Springer London Dordrecht Heidelberg New York

British Library Cataloguing in Publication Data
A catalogue record for this book is available from the British Library

Library of Congress Control Number: 2010929852

Cover design: eStudioCalamar, Figueres/Berlin

Printed on acid-free paper

Springer is part of Springer Science+Business Media (www.springer.com)

To my family

Preface

"When you can measure what you are speaking about, and express it in numbers, you know something about it; but when you cannot express it in numbers, your knowledge is of a meager and unsatisfactory kind; it may be the beginning of knowledge, but you have scarcely in your thoughts advanced to the state of science." – Lord Kelvin

The act of measuring, which is used for determining the size, amount or degree of a parameter by an instrument through comparison with a standard unit, or used indirectly by calculation based on theory, makes science and technology different from imagination. Measurement is also essential in industry, commerce and daily life. If we focus on the manufacturing industry, we can easily find that dimensional metrology plays an increasingly important role not only in the traditional field of manufacturing but also in the advanced field of nanomanufacturing, represented by ultra-precision machining and semiconductor fabrication.

Nanomanufacturing is a process of using precision machines that can generate precision tool motions to fabricate designed surface forms/dimensions with nanometric tolerances. Dimensional measurement of the workpiece and the machine is always an essential process for the purpose of quality control in all kinds of manufacturing. Because accuracy is the most important requirement for nanomanufacturing, the dimensional measurement is a much more crucial process for nanomanufacturing than other kinds of manufacturing. Figure 0.1 shows tolerances with respect to the dimensions of the workpieces and machines in nanomanufacturing. It can be seen that most of the workpieces and machines have sizes ranging from micrometers to meters while the corresponding tolerances range from 100 nm to 0.1 nm. In addition, more and more precision workpieces are required in a shorter amount of time in order to reduce manufacturing costs. Shapes of the precision workpieces are also becoming more and more complex. These factors are bringing greater challenges to the existing measuring technologies of precision metrology and nanometrology.

Precision metrology has a long history tracing back to the inventions of the micrometer (J. Watt 1772), gage block (C. Johansson, 1896), interferometer (A. Michelson 1881), etc. It established the foundation for the Industrial Revolution and contributed greatly to modernize industries based on

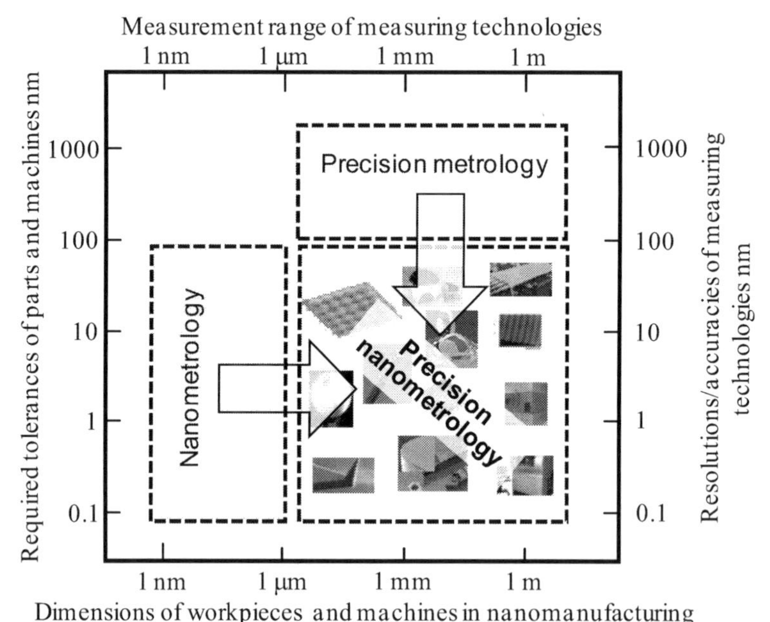

Figure 0.1. From precision metrology and nanometrology to precision nanometrology

interchangeable manufacturing. The term "precision" best reflects the nature of precision metrology, which is often related to the ratio of resolution/accuracy to range. At present, precision metrology is still playing an important role in precision manufacturing, especially due to its ability to make a wide range of measurements. However, it is difficult for precision metrology to reach a measurement resolution/accuracy better than 100 nm (Figure 0.1), which is required by nanomanufacturing. On the other hand, nanometrology, represented by scanning probe microscopy, is a relatively new measuring technology having only been developed since the 1980s. It can reach a high measurement resolution, down to 0.1 nm. As shown in Figure 0.1, however, nanometrology also cannot satisfy the requirement of nanomanufacturing because the measurement range of nanometrology is typically limited to 1 µm (Working Group on Dimensional Metrology of the Consultative Committee for Length, 1998). In addition, most of the commercial SPM instruments are for qualitative imaging use but not precision enough for quantitative measurement.

This book describes a new field of dimensional metrology called precision nanometrology for nanomanufacturing. Precision nanometrology is referred to as the science of dimensional measurement with nanometric accuracy over a broad measurement range from micrometers to meters. It is innovated through improving the measurement accuracy of precision metrology and expanding the measurement range of nanometrology (Figure 0.1). State-of-the-art sensors and measuring systems of precision nanometrology developed by the author's research group at Tohoku University are presented. This book especially focuses on the measurement of surface forms of precision workpieces and stage motions of precision machines, which are important items for nanomanufacturing.

The first half of the book (Chapters 1–4) describes optical sensors used for the measurement of angle and displacement, which are fundamental quantities for precision nanometrology. Technologies for improvement of the sensor sensitivity and bandwidth, reduction of the sensor size as well as development of new multi-axis sensing methods are presented. The methods addressed in this book for detection of the multi-axis positions and angles at a single measuring point are effective for reduction of Abbe errors in various measuring systems.

The second half (Chapters 5–10) presents a number of scanning-type measuring systems for precision nanometrology of surface forms and stage motions. Scanning-type measuring systems have the following advantages: simple structure, flexibility to the size and shape of the specimen, as well as the robustness to the measurement environment. The measuring time can be shortened by increasing the scanning speed. In conventional systems, however, there are still critical issues concerning precision nanometrology that need to be addressed, including reduction of scanning errors, automatic alignment of measuring positions, fast scanning mechanisms, etc. Error separation algorithms and systems for measurement of straightness and roundness, which are the most fundamental geometries treated in nanomanufacturing, are addressed in Chapters 5 and 6. Chapter 7 describes the measurement of micro-aspherics, which requires development in both the scanning mechanism and probing technology. In Chapters 8 and 9, novel systems based on scanning probe microscopy are described for precision nanometrology of micro- and nanostructures in response to new and important challenges from nanomanufacturing. Chapter 10 shows scanning image-sensor systems, which can carry out fast and accurate measurements of micro-dimensions of long structures.

This book is a comprehensive summary of an important part of the research work the author has been involved in over the past ten years. I would like to thank my colleagues and many students in the Nano-Metrology & Control Lab for their marvelous contributions to the technologies addressed in this book. Additionally, a number of students were involved in the preparation of the manuscript. I would also like to acknowledge Mr. Simon Rees, Editorial Assistant at Springer, for helping me to start the writing process and Ms. Claire Protherough, Senior Editorial Assistant, Ms. Katherine Guenthner, Copy Editor, Ms. Sorina Moosdorf, Production Editor, for their dedicated efforts that have made this book possible.

Finally, I wish to express my thanks and dedicate this book to my wife Hong Shen and my daughter Youyang. My wife, a doctor and professor in computer science, has carefully read through and checked the book. This book would never have been completed without their patience, encouragement and assistance. This book is also dedicated to my father who passed away when I was starting my first research project as a graduate student twenty years ago, and to my mother and my parents-in-law for their warm and continuous support.

Sendai, Japan *Wei Gao*
January 2010

Contents

1

Angle Sensor for Measurement of Surface Slope and Tilt Motion

1.1 Introduction

Angle is one of the most fundamental quantities for precision nanometrology. Angle sensors based on the principle of autocollimation, which are conventionally called autocollimators, can accurately measure small tilt angles of a light-reflecting flat surface [1]. Autocollimators have a long history of being used in metrology laboratories for calibration of angle standards, such as polygons, rotary index tables and angle gage blocks. They are also traditionally used in machine shops for surface profile measurements of straightedges, machine tool guideways, precision surface plates, as well as for measurement of tilt error motions of translational stages [2].

In a conventional photoelectric autocollimator, the light rays from a filament lamp are collimated to a parallel light beam with a large beam size, on the order of 30 to 50 mm in diameter [3]. The beam is then projected onto a flat target mirror mounted on the specimen surface. The deviation of the reflected beam with respect to the axis of the incident beam is detected by the autocollimation unit, which is composed of an objective lens and a light position detector placed at the focal plane of the objective lens. Autocollimators using CCD image sensors with image processing can achieve a resolution of up to 0.01 arcsec through employing an objective lens with a long focal length, typically on the order of 300 to 400 mm [3–6]. The large dynamic range of the CCD image sensor also makes the autocollimator have a dynamic range of 60 to 80 dB. However, the requirement of a large target mirror makes it difficult to measure soft specimens such as diamond turned optical surfaces, or thin specimens such as silicon wafers because the target mirror may damage the specimen. The large diameter of the light beam from the filament lamp also limits the lateral resolution for detecting the local slope of a surface. The low bandwidth and large dimension are other disadvantages of conventional autocollimators for the measurement of dynamic tilt error motions of a stage.

This chapter presents angle sensors using different types of photodiodes instead of CCD image sensors for improving the measurement speed and reduction of the sensor size.

1.2 Angle Sensor with Quadrant Photodiode

An angle sensor is a sensor typically for detecting the tilt angle of a surface. Such a sensor with a thin light beam, typically called a surface slope sensor, can also detect surface local slopes. The simplest way to construct an angle sensor is to utilize the method of optical lever. As shown in Figure 1.1, a light beam is projected onto the target surface. The optical spot of the reflected beam on a position photodetector with a distance L from the target will move in the W- and V-directions if the sample tilts about the X- and Y-axes. The two-dimensional components of the tilt angle θ_X and θ_Y can be calculated from the moving distances Δw and Δv of the spot on the photodetector as follows:

$$\theta_X = \frac{\Delta w}{2L}, \tag{1.1}$$

$$\theta_Y = \frac{\Delta v}{2L}. \tag{1.2}$$

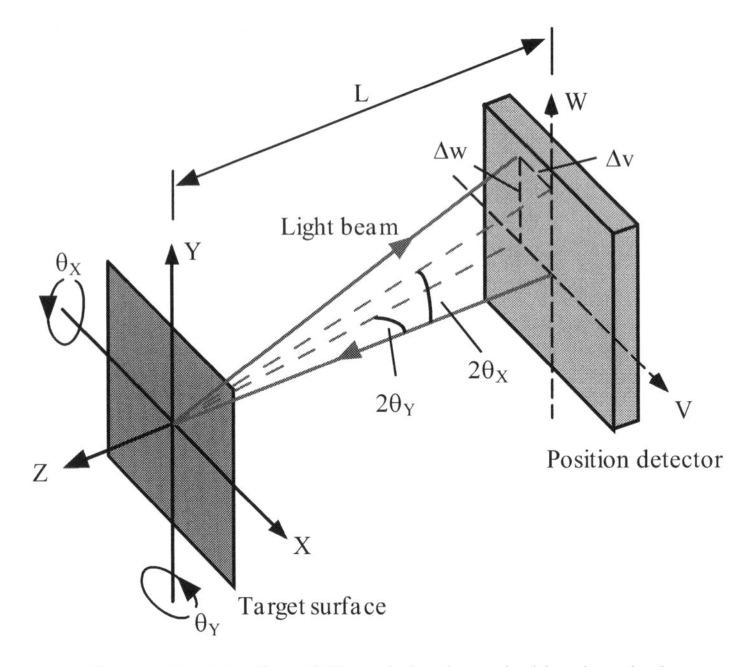

Figure 1.1. Detection of tilt angle by the optical level method

This method is simple but errors arise when distance L changes. This problem can be solved by the technique of autocollimation [7]. As shown in Figure 1.2, an objective lens is placed between the sample and the photodetector. If the photodetector is placed at the focal position of the lens, the relationship between the tilt and the readout of the photodetector becomes:

$$\theta_X = \frac{\Delta w}{2f}, \tag{1.3}$$

$$\theta_Y = \frac{\Delta v}{2f}, \tag{1.4}$$

where f is the focal length of the objective lens. As can be seen in Equations 1.3 and 1.4, the distance between the sample surface and the autocollimation unit composed of the objective lens and the position detector does not affect the angle detection.

In the case of form measurements of precision surfaces and motion measurements of precision stages, the angle of interest is very small and the sensitivity of the angle sensor is required to be very high. The sensitivity of the angle sensor based on autocollimation can be improved by choosing an objective lens with a long focal length. However, this will influence the compactness of the angle sensor. Here, we discuss how to improve the sensitivity of the angle sensor by choosing proper photodetectors without increasing the focal length of the lens.

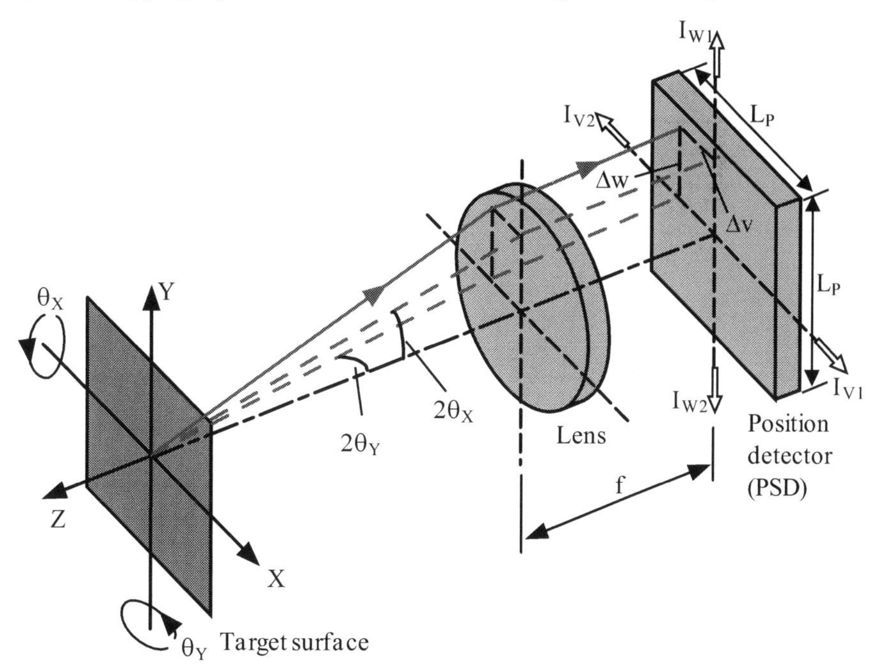

Figure 1.2. Detection of tilt angle by the autocollimation method with a PSD

The linear lateral effect position-sensing device (PSD) is widely used to detect the position of a light spot [8, 9]. PSDs provide continuous position information and have the advantage of good linearity. Position detection is also not affected by the intensity distribution of the light spot. Let the sensitive length of a two-dimensional (2D) PSD be L_P in both X- and Y-directions. The two-dimensional positions Δv and Δw can be obtained from the two-dimensional output $v_{\text{PSD\_out}}$ and $w_{\text{PSD\_out}}$ of the PSD, which are calculated from the photoelectric currents I_{v1}, I_{v2}, I_{w1}, and I_{w2} in Figure 1.2 through the following equations:

$$v_{PSD\_out} = \frac{(I_{v1} - I_{v2})}{(I_{v1} + I_{v2})} \times 100\% = \frac{2}{L_P} \Delta v = \frac{4f}{L_P} \theta_Y , \tag{1.5}$$

$$w_{PSD\_out} = \frac{(I_{w1} - I_{w2})}{(I_{w1} + I_{w2})} \times 100\% = \frac{2}{L_P} \Delta w = \frac{4f}{L_P} \theta_X . \tag{1.6}$$

It can be seen that the sensitivities of a 2D PSD, which are defined as $v_{\text{PSD\_out}}/\Delta v$ (or $x_{\text{PSD\_out}}/\theta_Y$) and $w_{\text{PSD\_out}}/\Delta w$ (or $y_{\text{PSD\_out}}/\theta_X$), respectively, are mainly determined by the sensitive length, and are not adjustable. Since the sensitivities are inversely proportional to the sensitive length, a PSD with a short sensitive length is preferred for obtaining high sensitivity. In Equations 1.5 and 1.6, when an objective lens with a focal length of 40 mm is used, a 0.01 arc-second angle θ_X (or θ_Y) only corresponds to a position change Δw (or Δv) of approximately 4 nm. Assume that the required resolution of the angle sensor is 0.01 arcsec and the dynamic range (measurement range/resolution) is 10,000. The preferred sensitive length of the PSD, which corresponds to the measurement range of the angle sensor, is calculated to be approximately 40 μm. However, commercially available PSDs typically have sensitive lengths of several millimeters, which generate unnecessarily large measurement ranges of angle. Considering the fact that the practical signal to noise ratios (dynamic ranges) of the current/voltage conversion amplifiers used to pick up the photoelectric currents do not easily exceed 10,000, it is difficult to achieve the required resolution of angle detection, which is determined by the measurement range and the dynamic range. Another parameter to determine the resolution of a PSD is the noise current. The noise current level of a 2D PSD is several times larger than that of a one-dimensional (1D) PSD. From this point of view, it is more feasible to use 1D PSDs instead of 2D PSDs. However, two 1D PSDs, with sensitive directions aligned perpendicularly, are necessary for detecting any 2D angle information. This results in a more complicated structure. Misalignment of the sensitive axes of each PSD will also increase the measurement uncertainty. Moreover, just as with a 2D PSD, the resolution of a 1D PSD will not be high enough, where the resolution is basically dominated by the sensitive length and the dynamic range.

Another possible photodetector is the quadrant photodiode (QPD) [10, 11]. As shown in Figure 1.3, a QPD is placed at or slightly apart from the focal point of the objective lens so that a light spot with a width of D_S is generated on the QPD.

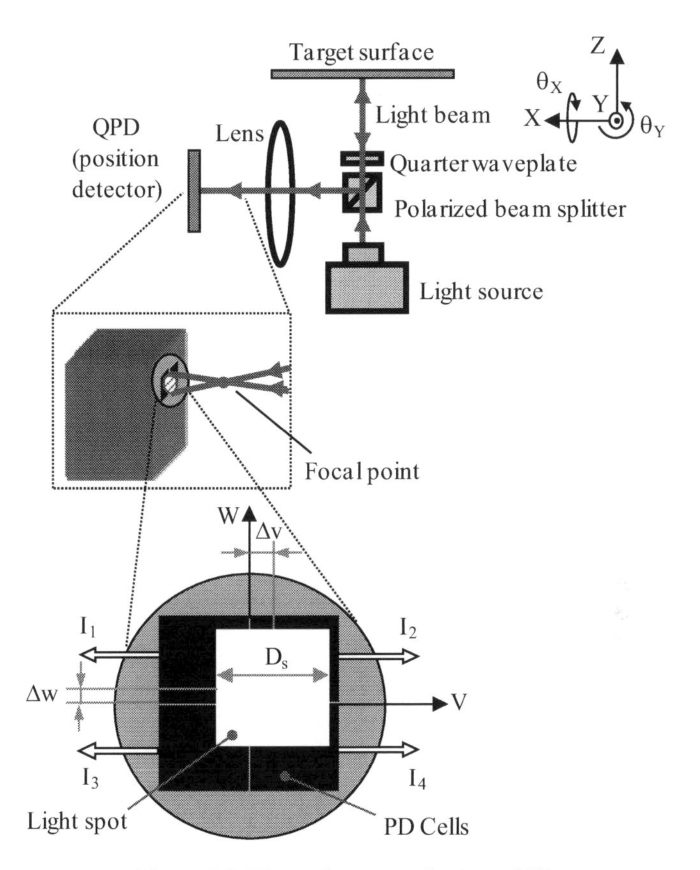

Figure 1.3. The angle sensor of using a QPD

For simplicity, assume the shape of the light spot is rectangular and the intensity distribution of the light spot is uniform. The two-dimensional position of the light spot can be calculated by:

$$v_{QPD\_out} = \frac{(I_1 + I_3) - (I_2 + I_4)}{(I_1 + I_2 + I_3 + I_4)} \times 100\% = \frac{2}{D_s}\Delta v = \frac{4f}{D_s}\theta_Y, \tag{1.7}$$

$$w_{QPD\_out} = \frac{(I_1 + I_2) - (I_3 + I_4)}{(I_1 + I_2 + I_3 + I_4)} \times 100\% = \frac{2}{D_s}\Delta w = \frac{4f}{D_P}\theta_X, \tag{1.8}$$

where I_1, I_2, I_3, I_4 are the photoelectric currents from the QPD cells.

It can be seen that the sensitivity of the QPD for position and/or angle detection is inversely proportional to the width of the light spot on the sensitive window of the QPD. The width of the light spot is a function of the location of the QPD relative to the focal position of the objective lens along the optical axis of the autocollimation unit. A proper measurement range/sensitivity of position and/or angle detection can thus be obtained through adjusting the location of the QPD.

High sensitivity and resolution can be achieved by using this technique. It should be pointed out that if the shape of the light spot were not rectangular but round, the relationships shown in Equations 1.7 and 1.8 would become non-linear. The intensity distribution of the light beam will also influence the linear relationships.

Figure 1.4 shows a schematic view of an angle sensor designed to demonstrate the feasibility of using a QPD as the position photodetector. A laser diode with a wavelength of 780 nm was used as the optical source. The output light from the laser diode unit was a collimated beam with a diameter of 1 mm. An achromatic lens with a short focal length of 40 mm was employed as the objective lens for the sake of compactness. A QPD was used as the photodetector to detect the 2D angle information. Another 1D PSD was also used in the same sensor to detect the 1D angle information about the *Y*-axis so that the sensitivities when using PD and PSD can be compared experimentally. The light beam passing through the objective lens was split into two beams, which were received by the QPD and the PSD, respectively. The PSD has a sensitive length of 2.5 mm, which was the shortest that could be found on the market. The sensor was designed to be within 90 mm (*X*) × 30 mm (*Y*) × 60 mm (*Z*) in size.

Figure 1.5 shows a calibration result of the 2D angle sensor using the QPD. A commercial autocollimator with a resolution of 0.05 arcsec was used as the reference. The target surface in Figure 1.4, which was mounted on a manual tilt stage, can be tilted manually about the *X*-axis and *Y*-axis, respectively. The tilt angle was detected by the angle sensor and the autocollimator simultaneously. The horizontal axes show the applied tilt angles measured by the autocollimator in arcseconds, and the vertical axes show the outputs of the angle sensor in percentages, which are defined in Equations 1.7 and 1.8. As can be seen in Figure 1.4, the angle sensor using a QPD has the ability to detect the two-dimensional tilt angle in a range of approximately 200 arcsec.

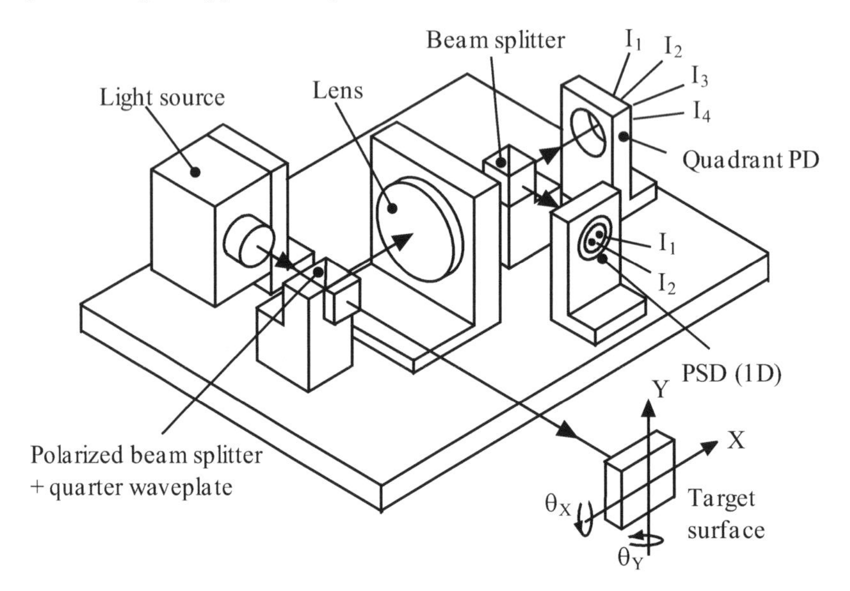

Figure 1.4. An angle sensor with a QPD and a one-dimensional PSD

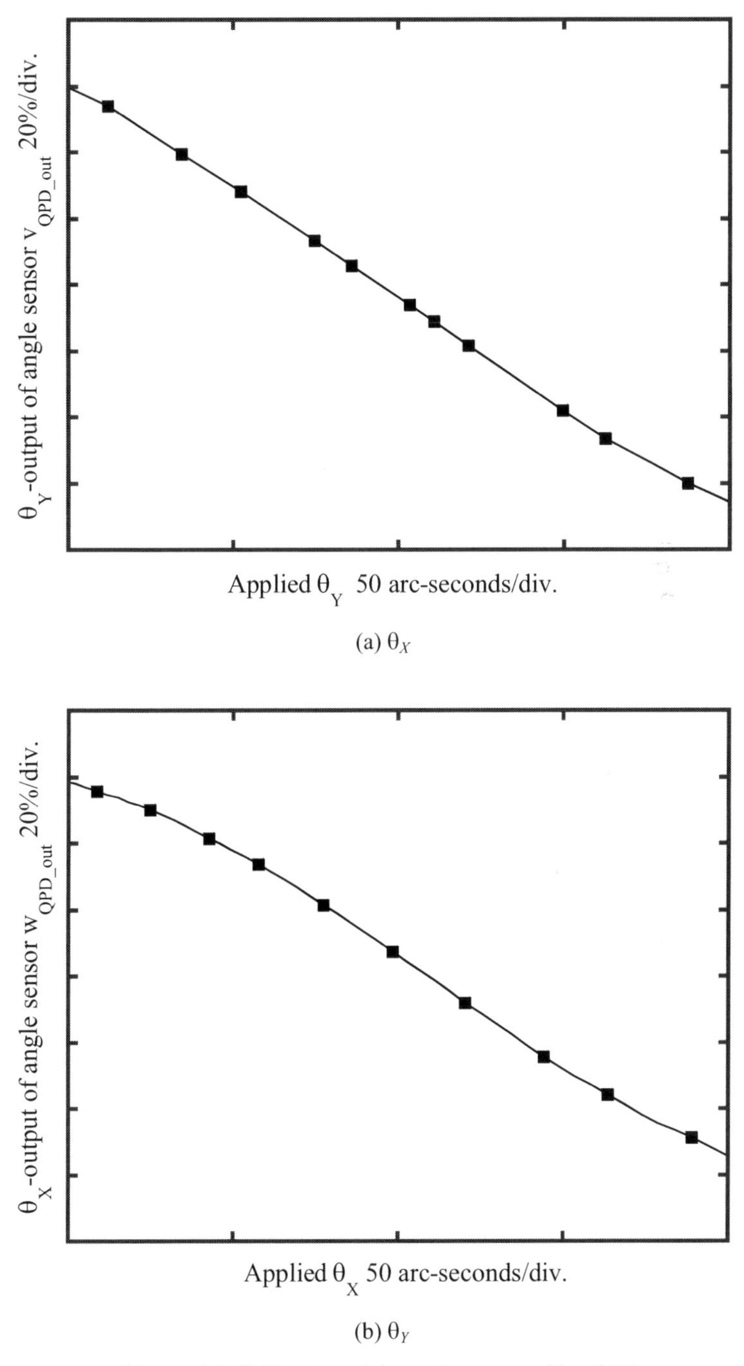

(a) θ_X

(b) θ_Y

Figure 1.5. Calibration of the angle sensor with a QPD

Figure 1.6 shows a comparison of the sensor output when using the QPD and the 1D PSD. Since the 1D PSD can only detect the tilt about the *Y*-axis, the comparison was made only with the *V*-directional output of the QPD. Note that the two vertical scales in the graph are ten times different from each other. It can be seen that the sensitivity when using the QPD is approximately 30 times higher than that when using the PSD.

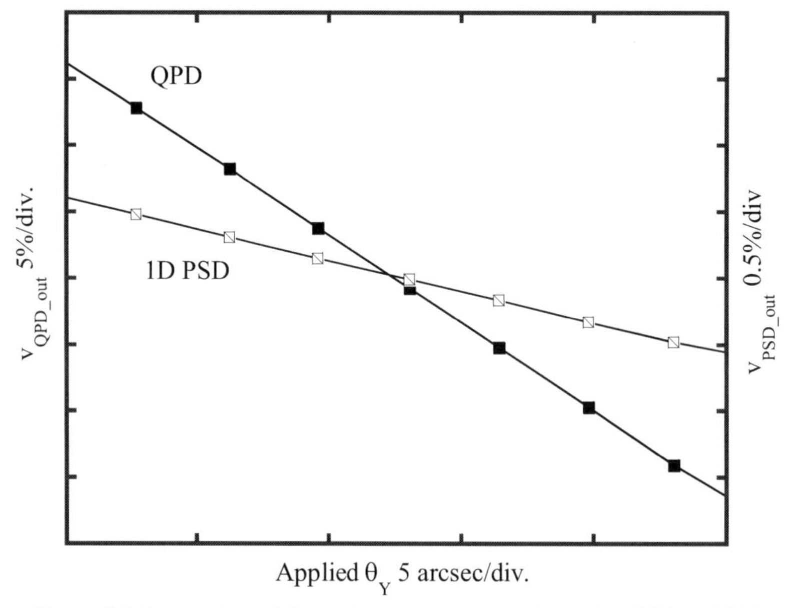

Figure 1.6. Comparison of the angle sensor outputs when using QPD and PSD

The angle sensor with a QPD was employed for measurement of the tilt error motions about the *Y*-axis (yawing) and the *X*-axis (pitching) of an aerostatic bearing linear stage (air-slide) with its axis of motion along the *Z*-direction. Figures 1.7 and 1.8 show the measurement results and the experimental setup, respectively. For comparison, the tilt error motions were also measured by a commercial autocollimator with a resolution of 0.1 arc-second simultaneously. The mirrors for the angle sensor and the autocollimator were mounted on the moving table of the stage. The moving table was first moved 50 mm forward, then kept stationary, then finally moved 50 mm back to the starting point. As can be seen in Figure 1.7, the stage was measured to have a yawing error motion and a pitching error motion of approximately 1 arc-second during the forward motion. The tilt error motions were associated with the acceleration of the moving table of the stage. When the moving table was kept stationary, the tilt error motions of the moving table were on the order of 0.1 arc-second, which were approximated to be one tenth of those during acceleration. When the moving table was moved back, the tilt error motions had the same amplitude as those during the forward motion but with a different tilt direction. This corresponds to the difference of directions between the forward driving force and the backward driving force from the motors. The angle sensor also showed a lower noise level compared with the autocollimator.

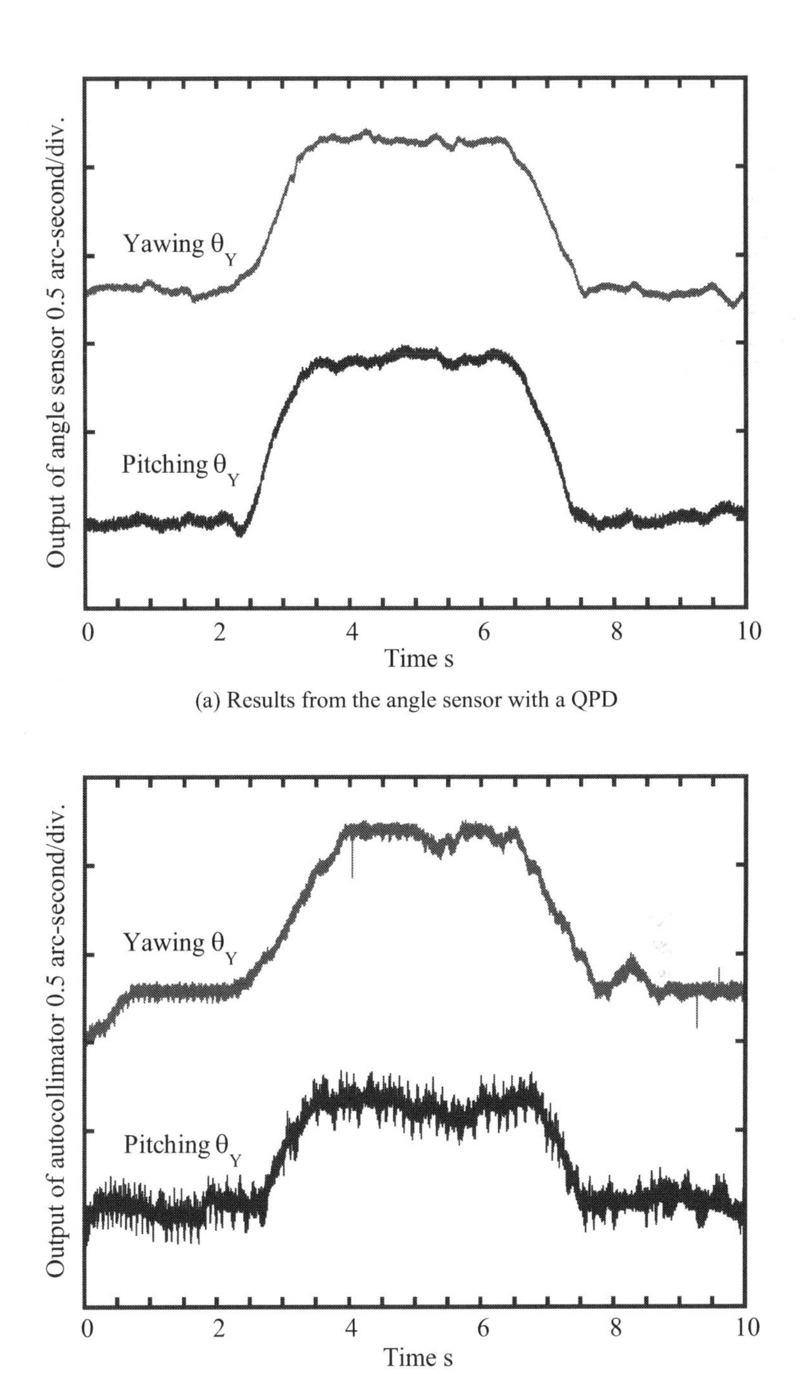

(a) Results from the angle sensor with a QPD

(b) Results from the autocollimator

Figure 1.7. Measurement of tilt error motions by using the angle sensor and a commercial autocollimator

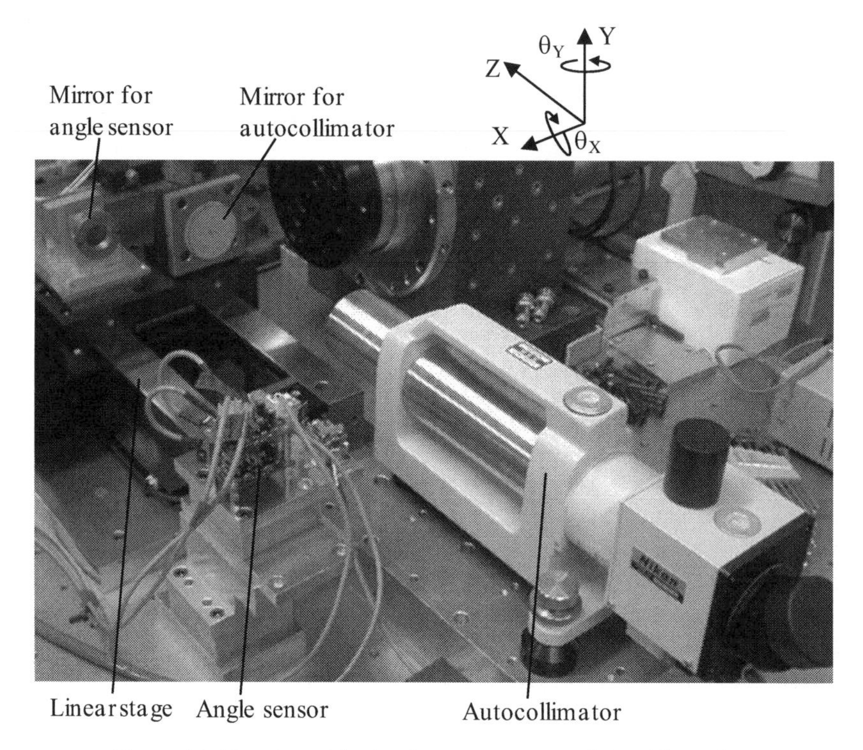

Figure 1.8. Experimental setup for measurement of stage tilt motions

The angle sensor can also be used as a slope sensor for detection of surface local slopes [12–16]. Figure 1.9 shows a schematic of the scanning system for measurement of a silicon wafer substrate by using a slope sensor. The system consists of a wafer spindle, a 2D slope sensor and a linear stage. The spindle axis is along the *Z*-axis and the linear stage moves along the *X*-axis. The slope sensor detects the 2D surface slopes about the *X*- and *Y*-axes.

The coordinates of sampling points on the wafer surface are shown in Figure 1.10. The sampling positions in the *X*-direction are numbered as x_i ($i = 1, \ldots, M$). At each sampling position x_i, the 2D surface local slopes at points along a concentric circle are detected by the slope sensor. The sampling positions along the circle are numbered as θ_j ($j = 1, \ldots, N$). The *Y*-output $\mu_Y(x_i, \theta_j)$ can be expressed by

$$\mu_Y(x_i,\theta_j) = f'_Y(x_i,\theta_j) + e_{CX}(x_i) + e_{SX}(x_i,\theta_j)\,, \tag{1.9}$$

where $e_{CX}(x_i)$ is the roll error of the sensor carriage, and $e_{SX}(x_i, \theta_j)$ is the tilt motion component of the wafer spindle about the *X*-axis. The term $f'_Y(x, \theta)$ is the *Y*-directional local slope of the wafer surface, which is defined as:

$$f'_Y(x,\theta) = \partial f(x,\theta) / \partial y\,. \tag{1.10}$$

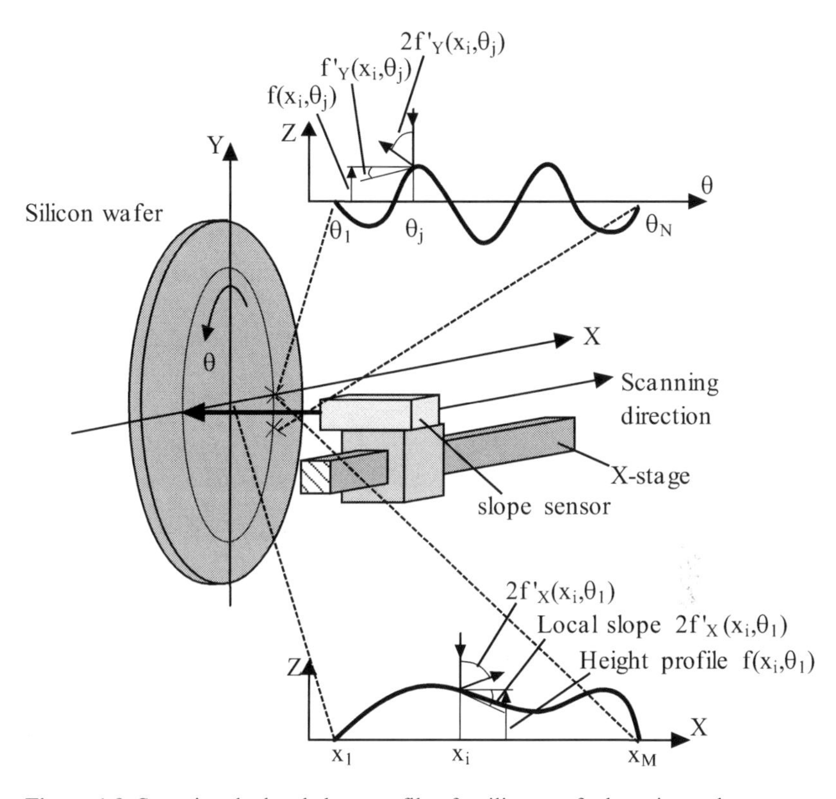

Figure 1.9. Scanning the local slope profile of a silicon wafer by using a slope sensor

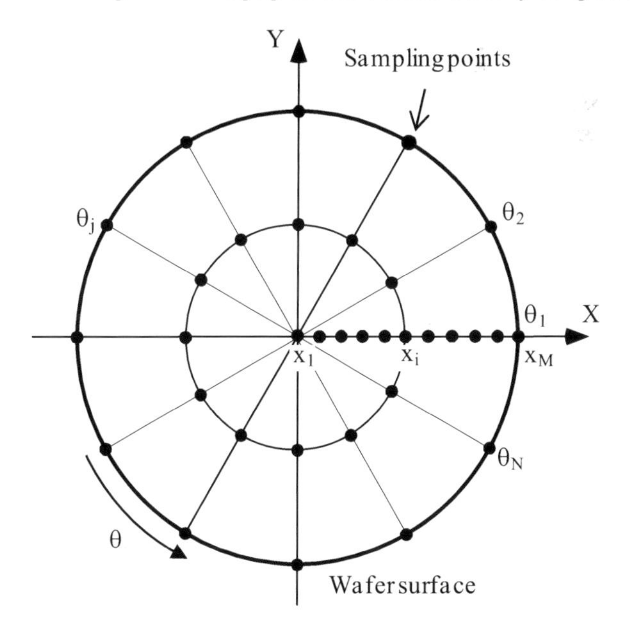

Figure 1.10. Sampling points on the wafer surface

Assuming that the roll error of the sensor stage and the tilt motion of the spindle are small enough or pre-compensated gives

$$\mu_Y(x_i,\theta_j) = f'_Y(x_i,\theta_j) . \tag{1.11}$$

For a fixed x_i ($i = 2, \ldots, M$), the height profile $f(x_i, \theta_j)$ of the wafer section along the ith concentric circle can then be calculated from

$$f(x_i,\theta_j) = \sum_{t=2}^{j} \mu_Y(x_i,\theta_t)(\theta_t - \theta_{t-1})x_i + f(x_i,\theta_1) \, , j = 2, \ldots, N, \tag{1.12}$$

where $f(x_i, \theta_1)$ ($i = 2, \ldots, M$) are the profile heights along the X-axis.

To obtain the entire profile of the wafer surface from profiles $f(x_i, \theta_j)$ along a concentric circle in Equation 1.12, the X-outputs of the slope sensor are used to determine the values of $f(x_i, \theta_1)$ ($i = 1, \ldots, M$). The X-outputs of the slope sensor $\mu_X(x_i, \theta_1)$, which correspond to the tilt about the Y-axis, can be expressed by:

$$\mu_X(x_i,\theta_1) = f'_X(x_i,\theta_1) . \tag{1.13}$$

The yaw error of the sensor carriage is assumed to be small enough or pre-compensated. The term $f'_X(x, \theta)$ is the X-directional local slope of the wafer surface, which is defined as:

$$f'_X(x,\theta) = \partial f(x,\theta) / \partial x . \tag{1.14}$$

The sectional profile $f(x_i, \theta_1)$ along the radial direction can thus be calculated from:

$$f(x_i,\theta_1) = \sum_{t=2}^{i} \mu_X(x_t,\theta_1)(x_t - x_{t-1}) \, , \; f(x_1,\theta_1) = 0 \, , i = 2, \ldots, M. \tag{1.15}$$

The entire surface profile (height map) of the silicon wafer can be obtained by combining the two results in Equations 1.12 and 1.15.

Figure 1.11 shows the experimental system. The wafer was vertically vacuum-chucked on an air-spindle with a rotary encoder and the slope sensor was mounted on a linear motor-driven air-slide with a linear encoder. The spindle and the slide were controlled by a personal computer via RS-232C. The sensor outputs were taken into the PC via a 16-bit AD converter. The sampling positions were determined by the rotary encoder and the linear encoder. A polished silicon wafer substrate with a diameter of 300 mm was used as the specimen. Figure 1.12 shows the outputs of the slope sensor along the circumference direction and the radial direction, which are called the slope maps, respectively. Figure 1.13 shows the evaluated height profile based on Equations 1.12 and 1.15. The out-of-flatness of the wafer was measured to be approximately 10.7 μm.

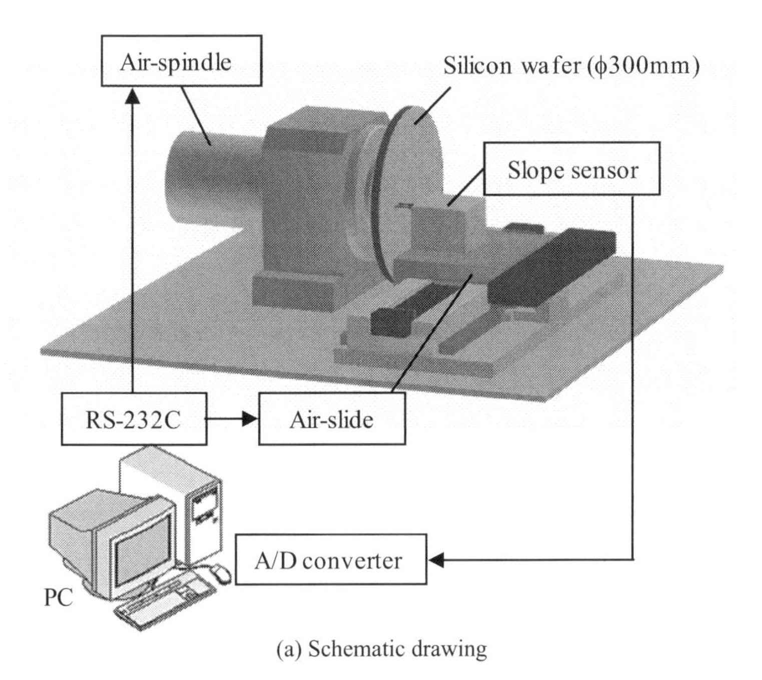

(a) Schematic drawing

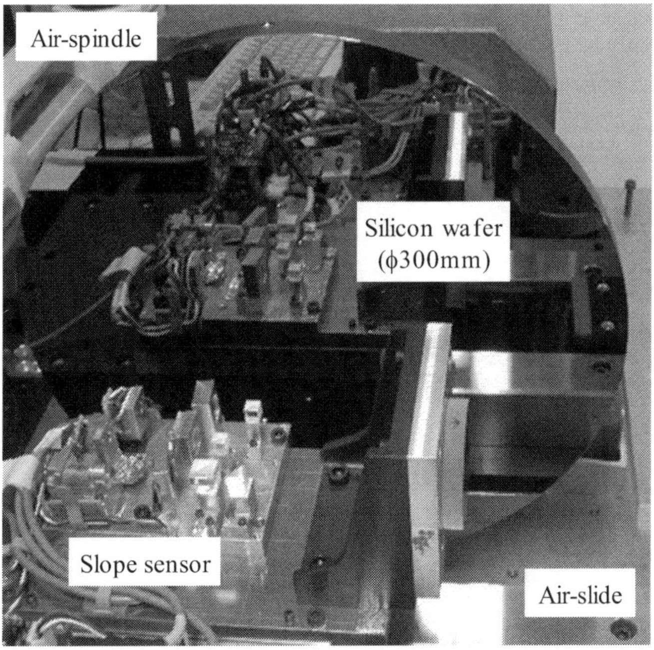

(b) Photograph

Figure 1.11. Experimental system for silicon wafer flatness

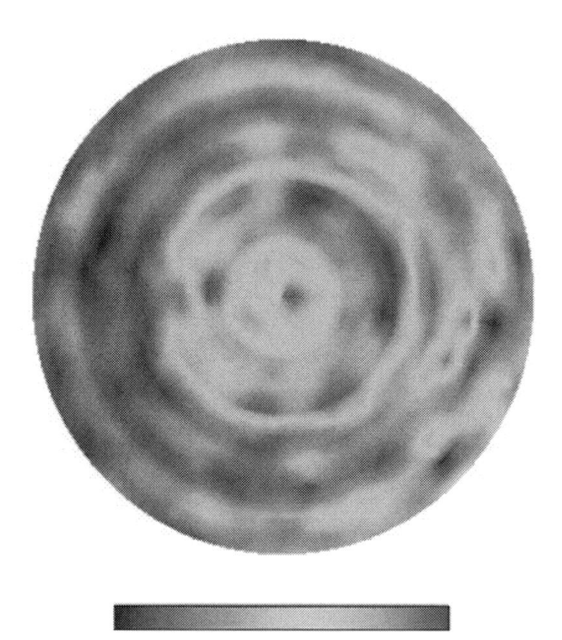

(a) Circumference direction

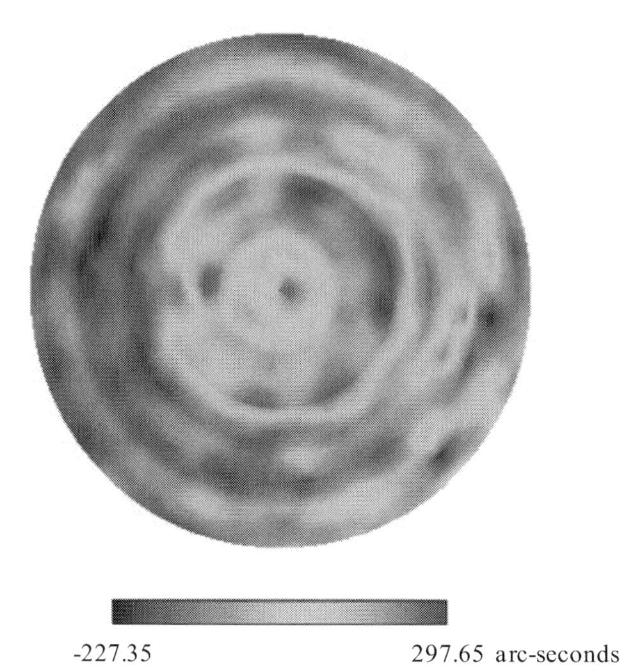

(b) Radial direction

Figure 1.12. Measured slope maps of the wafer surface

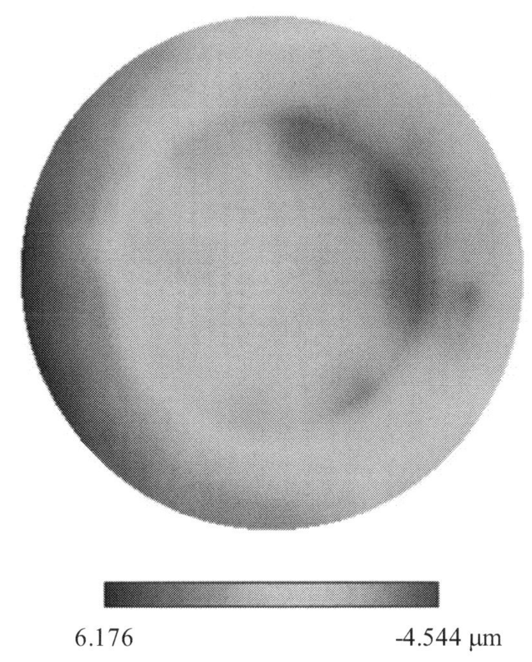

Figure 1.13. Measured height map (out-of-flatness) of the silicon wafer

1.3 Angle Sensor with Photodiode Array

As described in Section 1.2, in the case of using a quadrant photodiode or a bi-cell photodiode, both the resolution and the range of angle measurement are determined by the light spot size on the photodiode. The smaller the spot size, the higher the resolution but the smaller the range. Obtaining a higher resolution means a reduction of range, which is a problem for measurements with large angle variations. This section presents an angle sensor employing a multi-cell photodiode array [17] instead of the bi-cell PD or quadrant PD to achieve a larger measurement range while maintaining the resolution.

Figure 1.14 shows a schematic of the angle sensor with a one-dimensional (1D) multi-cell PD array for the measurement of the tilt angle of θ_Y. The basic structure is the same as that shown in Section 1.2. A multi-cell PD array is placed at the focal position of the objective lens to detect the V-directional displacement Δv of the light spot at the focal plane. Multiple PD cells of the multi-cell PD array are aligned along the V-direction in the VW-plane. The origin of the V-axis as well as the initial position of the light spot is located in the center of the gap between cells 1 and 2.

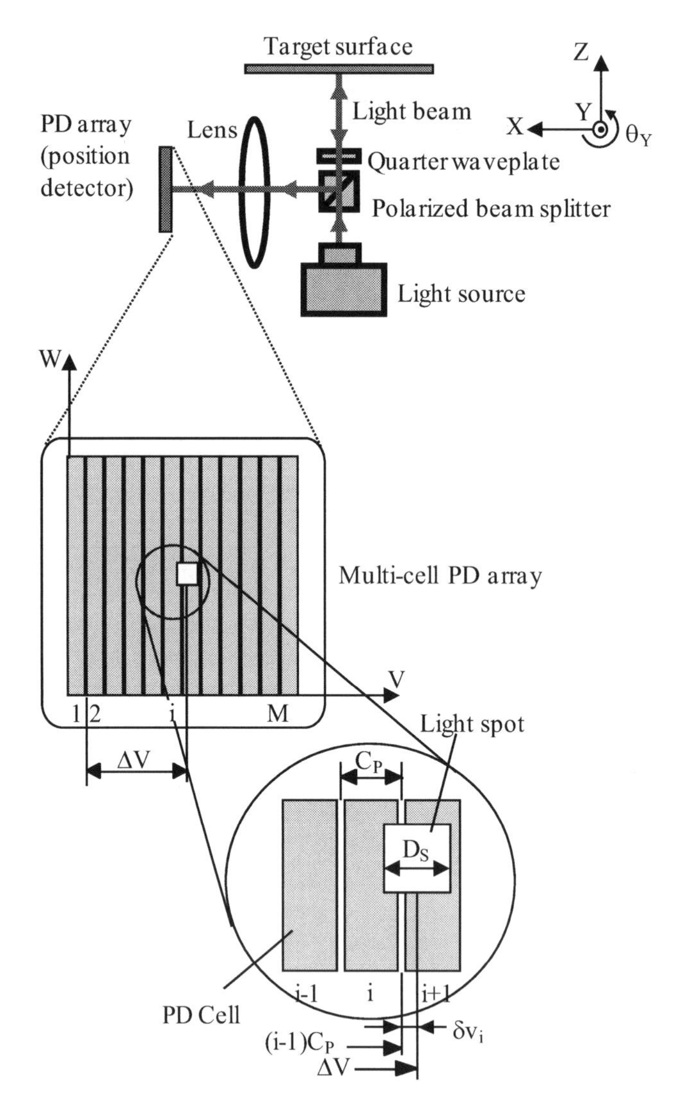

Figure 1.14. The angle sensor using a multi-cell PD array

Assume that the width of each cell is C_P and the total number of cells is M. If the light spot resides on cells i and $i + 1$, the position δv_i of the light spot relative to the center of the gap between cells i and $i + 1$ can be obtained from the following equation, based on the position-sensing principle of a bi-cell PD:

$$\delta v_i = \frac{I_{i+1} - I_i}{I_{i+1} + I_i} \cdot \frac{D_S}{2}, \; i = 1, ..., M - 1, \tag{1.16}$$

where I_i and I_{i+1} are the photoelectric currents of cells i and $i + 1$, respectively, and D_S is the diameter of the light spot. For the sake of clarity, the non-linearity caused by the non-rectangular shape of the light spot is ignored. The width of the gap is also not considered. As can be seen from Equation 1.16, the sensitivity and range of measuring δv_i are determined by the diameter of the light spot. It is necessary to reduce D_S in order to obtain a higher resolution, but this will result in a reduction of range.

To improve both the resolution and the range, the cell width C_P is designed to be equal to or slightly smaller than the diameter D_S. In this case, D_S can be replaced with C_P in Equation 1.16. In addition, when the light spot moves across the multi-cell PD along the V-direction, the light spot will reside on two cells at any position within the sensitive area of the PD so that the position ΔV can be obtained by

$$\Delta V = (i-1)C_P + \delta v_i \,, \tag{1.17}$$

$$\theta_Y = \frac{\Delta V}{2f} = \frac{(i-1)C_P + \delta v_i}{2f} = \frac{(i-1)C_P}{2f} + \Delta\theta_{Yi} \,, \; i = 1, ..., M-1, \tag{1.18}$$

where $\Delta\theta_{Yi}$ is the angle component detected by cells i and $i + 1$, which corresponds to a conventional bi-cell PD.

$$\Delta\theta_{Yi} = \frac{\delta v_i}{2f} = \frac{I_{i+1} - I_i}{I_{i+1} + I_i} \cdot \frac{C_P}{4f} = V_{Array\_out\_i} K_\theta \,, \; i = 1, ..., M-1, \tag{1.19}$$

where $V_{\text{Array\_out\_}i}$ shows the relative variation of the sensor output, and K_θ corresponds to the sensitivity of angle detection. Then,

$$V_{Array\_out\_i} = \frac{I_{i+1} - I_i}{I_{i+1} + I_i} \times 100\% \,, \tag{1.20}$$

$$K_\theta = \frac{C_P}{4f} \,. \tag{1.21}$$

Consequently, the angle measurement range can be improved by M–1 times while maintaining the resolution compared with a bi-cell PD. The constant K_θ can be determined by a calibration process. D_S is determined by the numerical aperture and the aberration of the objective lens. Figure 1.15 shows a photograph of the angle sensor developed for testing the PD array. The diameter of the beam projected onto the target mirror was 0.8 mm and the focal length of the objective lens was 30 mm. The numerical aperture was calculated to be 0.027. Figure 1.16 shows the intensity distribution of the light spot on the focal plane of the objective lens, which was detected by a beam profiler. The light spot had a typical Gaussian distribution. The diameter D_S of the light spot was approximately 50 μm.

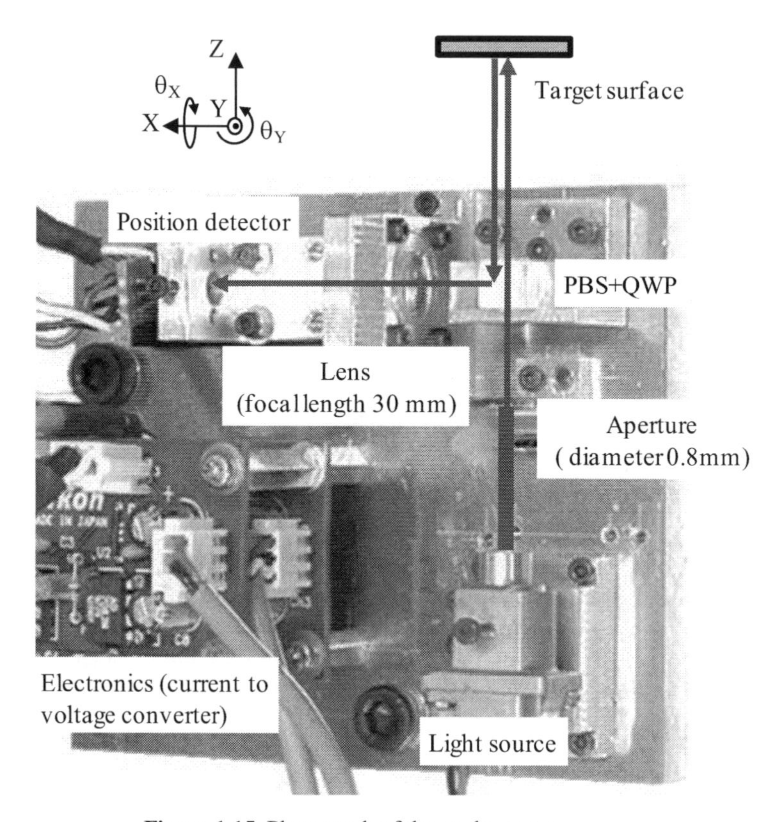

Figure 1.15. Photograph of the angle sensor

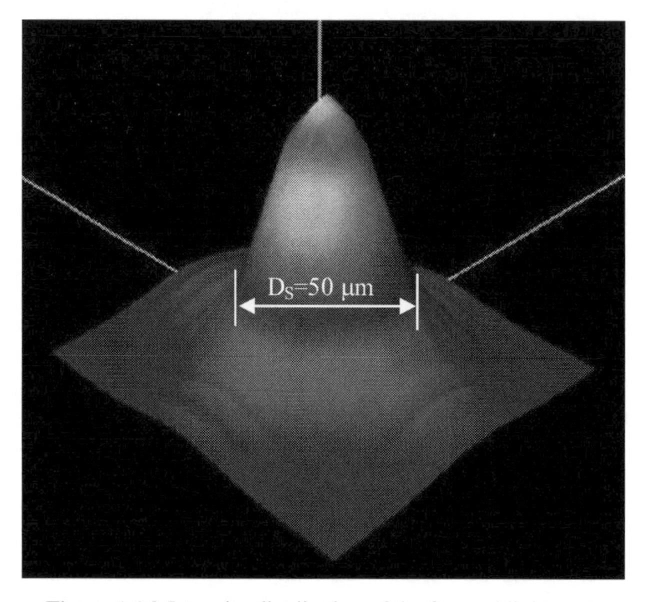

Figure 1.16. Intensity distribution of the focused light spot

On the other hand, however, commercial PD arrays have large cell widths, on the order of 1 mm, and large gaps between cells, on the order of 100 μm, which are not suited for use in the precision angle sensor. A multi-cell PD array was thus specifically designed and fabricated. Figure 1.17 shows the geometry of the sensitive area of the PD array. The pitch between the cells, which is the sum of the gap between cells and the width of each cell, was determined based on the measurement result of the diameter of the focused light spot (D_S) shown in Figure 1.16. The gap between cells was designed to be 5 μm based on the facility capability. The width of each PD cell was designed to be 40 μm so that the sum of the cell width and twice the gap was equal to the spot diameter. The angle range detected by two adjacent cells was calculated to be 156 arcsec. The number of cells was designed to be 16 and the total range of angle detection by the PD array was 2340 arcsec. The cell length, which was not a critical factor, was designed to be 1 mm for ease of alignment. The rectangular contact hole was 20 μm in length. The width of the aluminum electrode and wire was 40 μm, which was the same as the cell width. The wires of two adjacent PD cells were connected to different ends of the cells to avoid contact between the wires.

Figure 1.18 shows the process chart for fabricating the PD array. Figure 1.19 shows the 1D PD array with electrodes. Figures 1.20 and 1.21 show photographs of the sensitive area in the fabricated one-dimensional PD array and the circuit board for mounting the PD array.

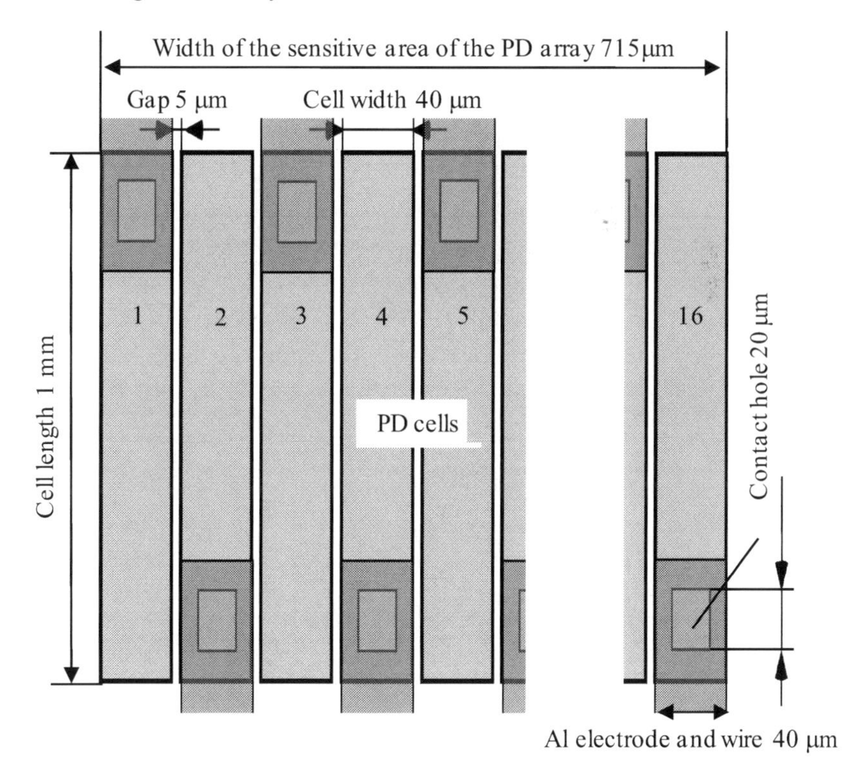

Figure 1.17. Geometric design of the multi-cell PD array

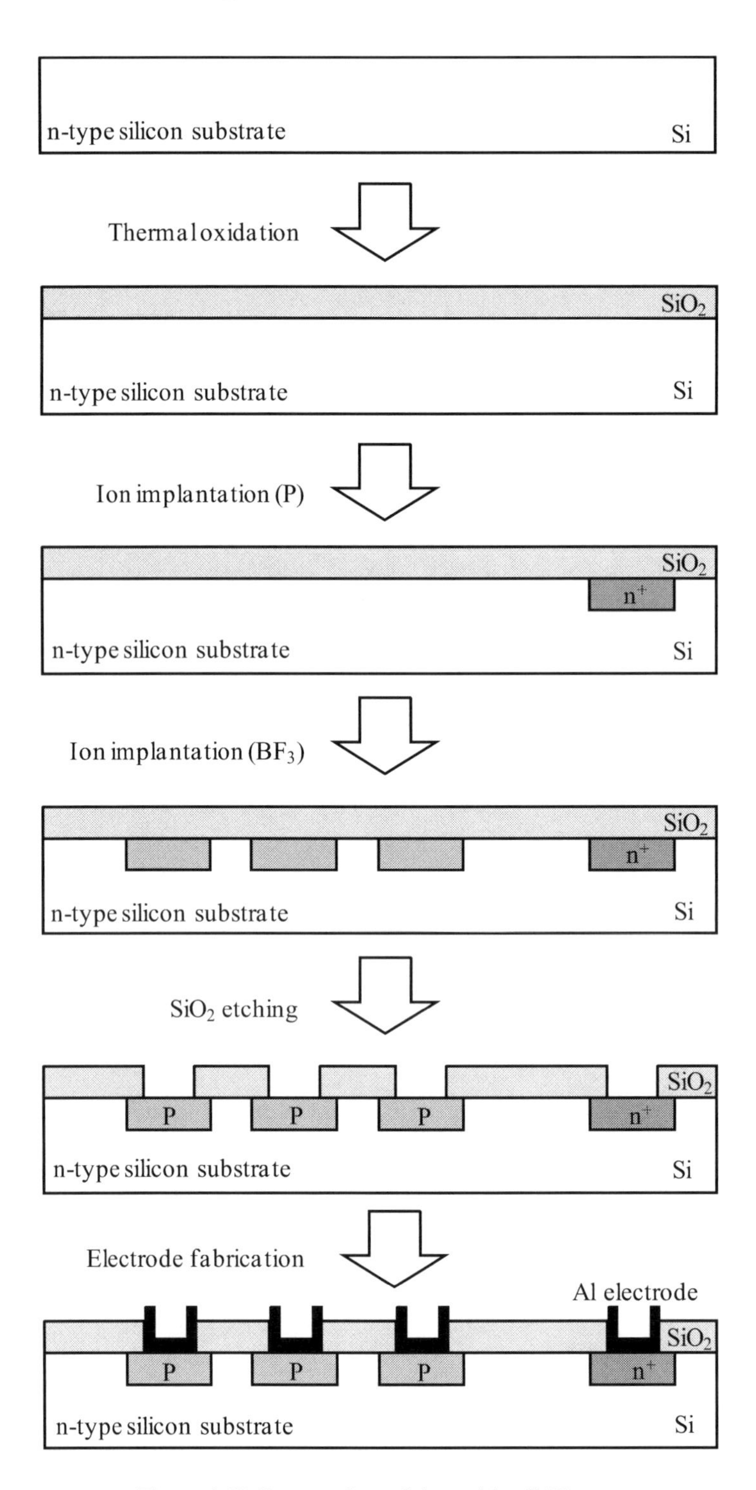

Figure 1.18. Process chart of the multi-cell PD array

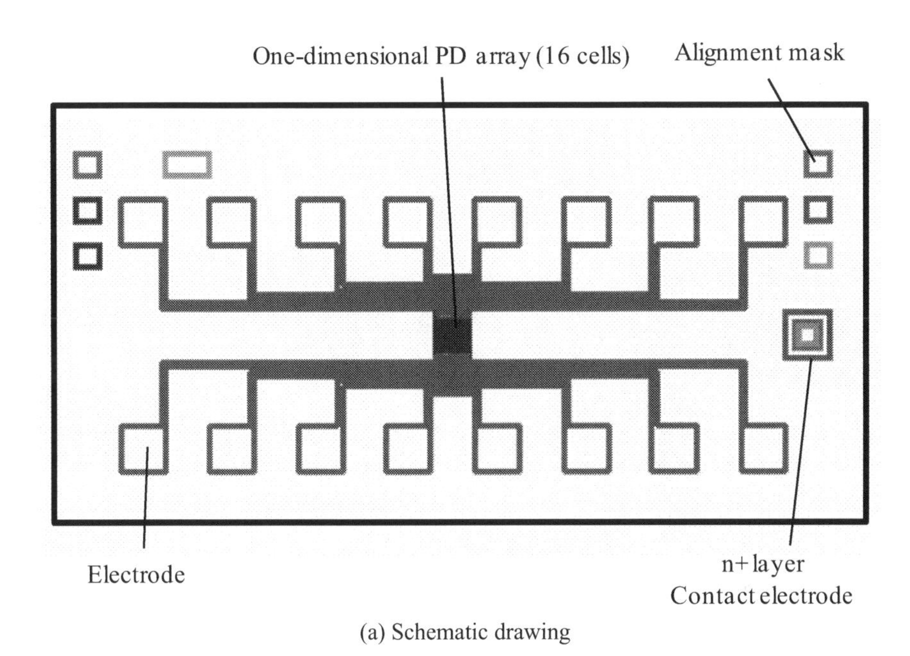

(a) Schematic drawing

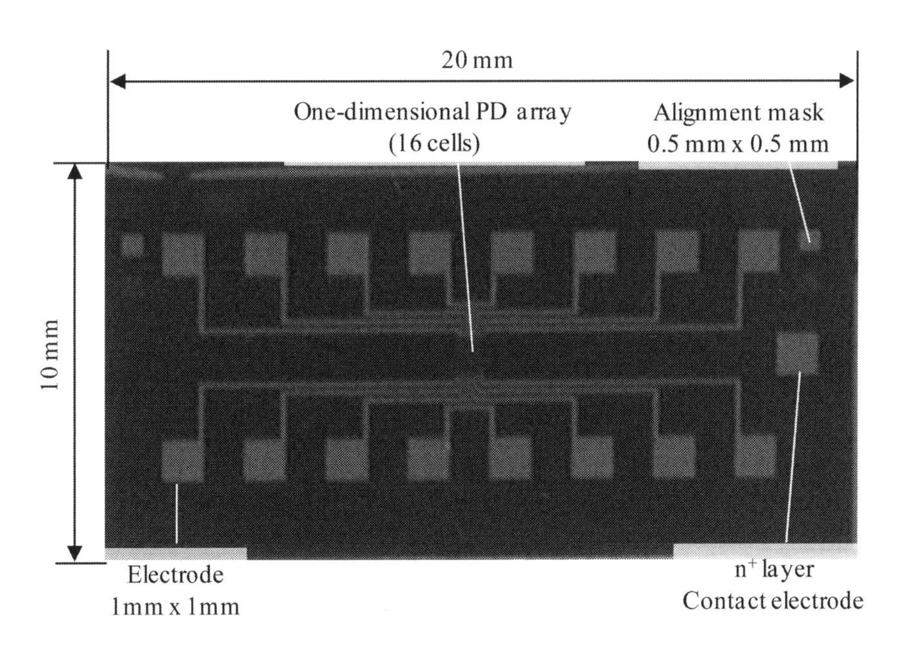

(b) Photograph

Figure 1.19. The 1D PD array with electrodes

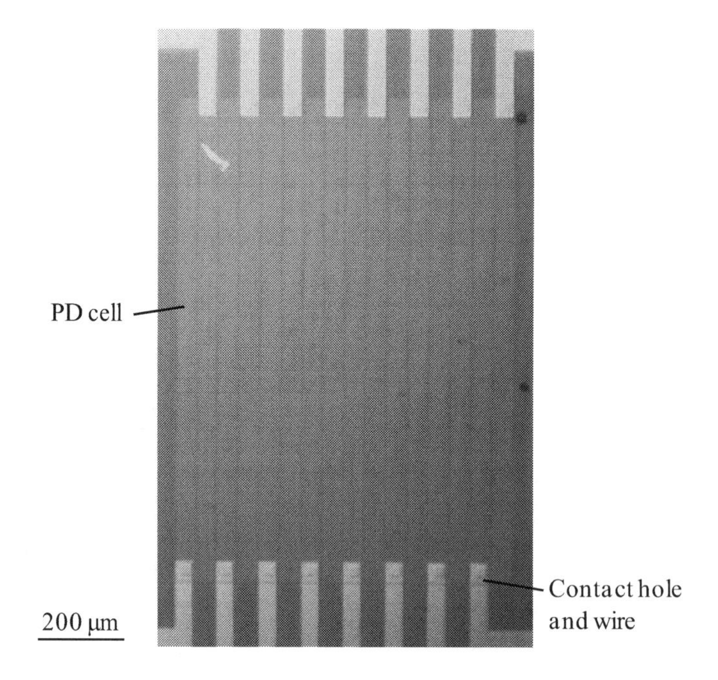

Figure 1.20. Photograph of the sensitive area of the 1D PD array

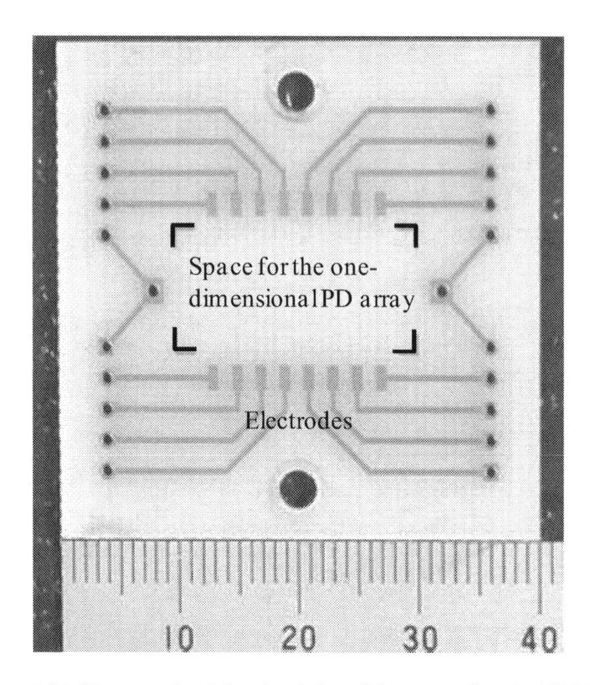

Figure 1.21. Photograph of the circuit board for mounting the 1D PD array

Experiments were conducted to test the angle sensor employing the multi-cell array as the light position detector. As can be seen in Figure 1.22, two mirrors, one for the angle sensor and the other for a commercial autocollimator, were mounted on a tilt stage. The tilt angle θ_Y of the tilt stage was detected by the angle sensor and the autocollimator simultaneously for comparison. The autocollimator had an objective lens with a focal length of 430 mm. The resolution and range were 0.1 and 1200 arcsec, respectively. The dimension was 440 mm (L) × 80 mm (W) × 146 mm (H).

Figures 1.23 and 1.24 show the results of the resolution test where a PZT-driven tilt stage was employed. In the test shown in Figure 1.23, the stage was tilted periodically with an amplitude of approximated 0.3 arcsec. Considering the slow response of the autocollimator, a 0.5 Hz signal was applied to the PZT stage. As can be seen in the figure, the angle sensor responded much better than the autocollimator, indicating that the angle sensor had a higher resolution. The variations in the amplitudes of the sensor output were caused by the instability in the tilt motion of the stage with an open-loop control. Figure 1.24 shows the result when the amplitude was reduced to approximately 0.02 arc-second. The signal frequency was also increased to 50 Hz. The autocollimator could not detect such a small tilt angle but the output of the angle sensor still followed the tilt motion quite well. The resolution of the angle sensor was evaluated to be approximately 0.01 arc-second. The bandwidth of the angle sensor, which was mainly determined by the sensor electronics, was set to be approximately 1.6 kHz.

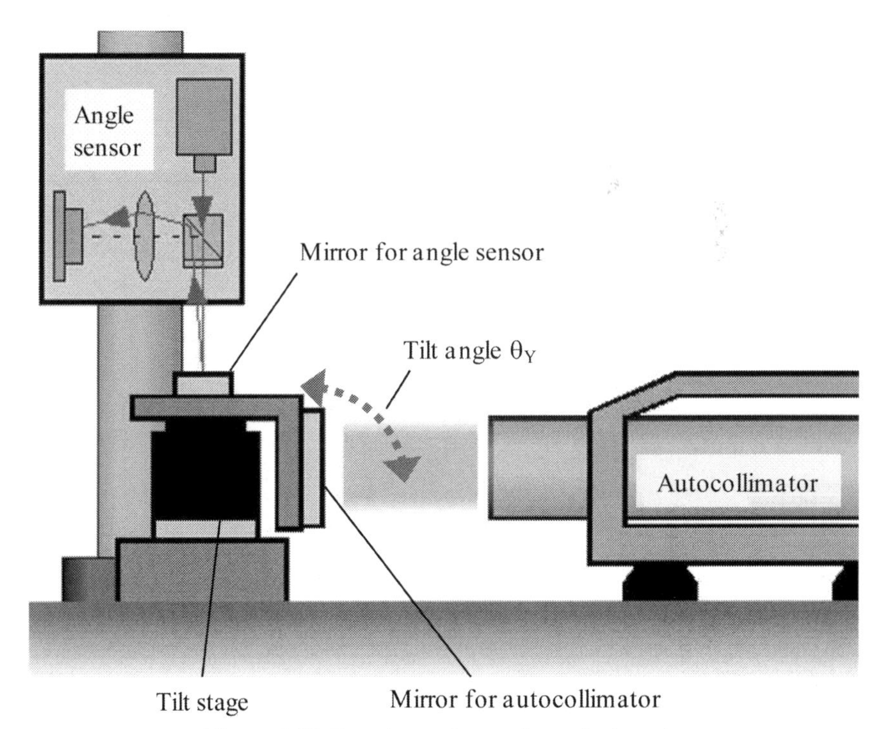

Figure 1.22. Experimental setup for angle detection

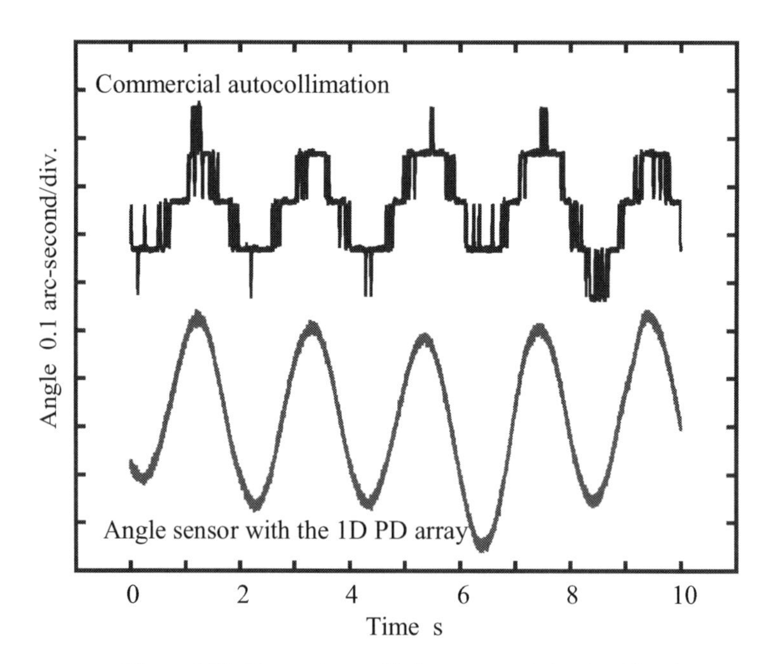

Figure 1.23. Comparison with the commercial autocollimator

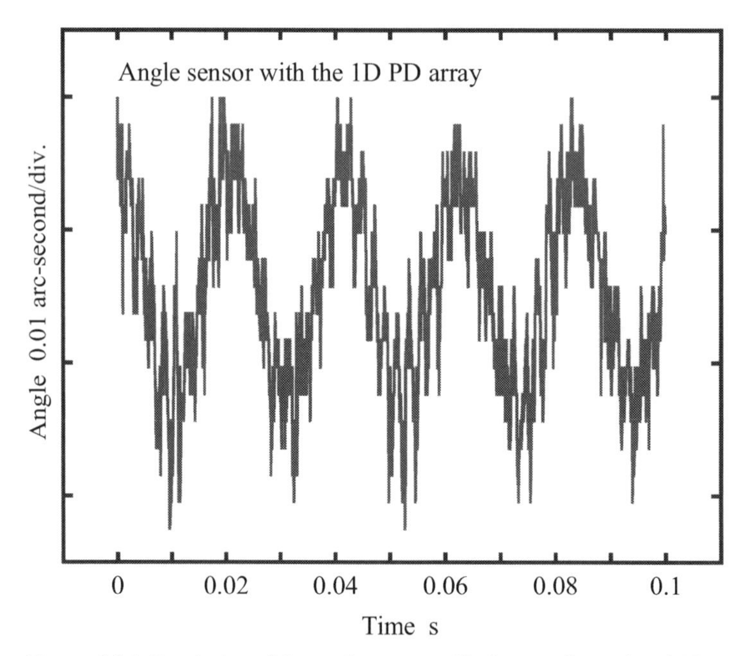

Figure 1.24. Resolution of the angle sensor with the one-dimensional PD array

Figure 1.25 shows the result of testing the measurement range of the angle sensor. A stepping motor-driven tilt stage with a larger tilting range (5°) and a resolution of 1 arc-second was employed instead of the PZT-driven tilt stage shown in Figure 1.22 because the movement range of the PZT-driven tilt stage was smaller than the measurement range of the angle sensor with the PD array. The stage was tilted in such a way that the light spot moved across the PD array from cell 1 to 16. The photoelectric currents of adjacent pairs of PD cells were used to calculate $V_{\text{Array\_out\_}i}$ in Equation 1.20, which was the relative sensor output. As can be seen in Figure 1.25, the tilt angle can be evaluated by each pair of the adjacent PD cells and the measurement range was improved to approximately 2300 arcsec through employing all of the 16 cells.

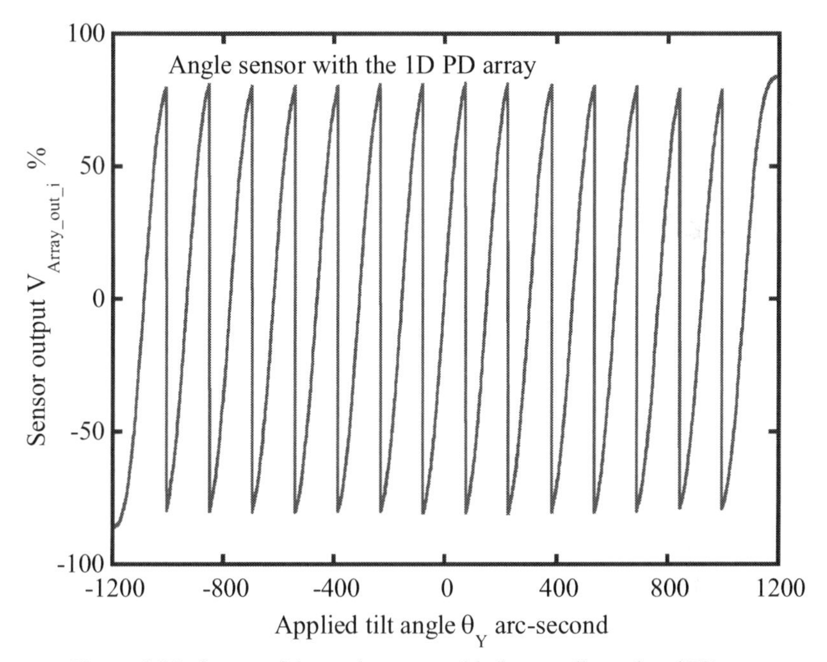

Figure 1.25. Output of the angle sensor with the one-dimensional PD array

The output $V_{\text{Array\_out\_8}}$ from cells 8 and 9 is shown in Figure 1.26. The range of $V_{\text{Array\_out\_8}}$ was 155 arcsec. The non-linearity due to the circular beam cross-section was approximately 27 arcsec. For comparison, a commercial bi-cell PD was also used in the same angle sensor. The cell width was 300 μm, which was much larger than the 50 μm spot diameter. The gap between cells was 10 μm. The sensor output shown in Figure 1.27 corresponds to that defined in Equation 1.7. It can be seen in Figure 1.27 that the effective range was almost the same as that shown in Figure 1.26, indicating that the range is basically determined by the spot size for a bi-cell PD. The non-linearity over an effective range of 160 arcsec was approximately 32 arcsec.

Figure 1.28 shows a fabricated two-dimensional (2D) PD array with 4 × 4 PD cells. Both the tilt angles θ_Y and θ_X can be detected by using the 2D PD array.

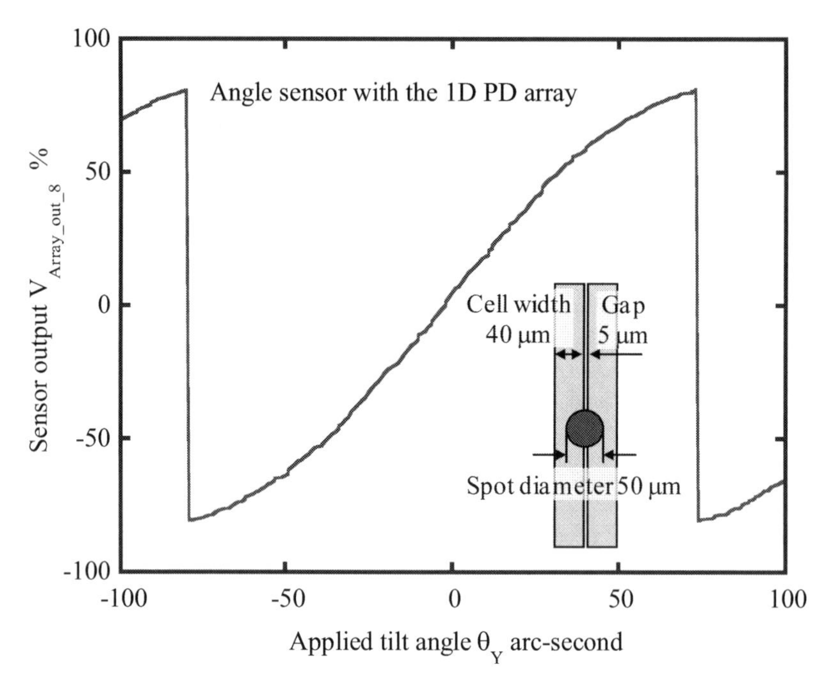

Figure 1.26. Angle sensor output between two adjacent PD cells of the 1D PD array

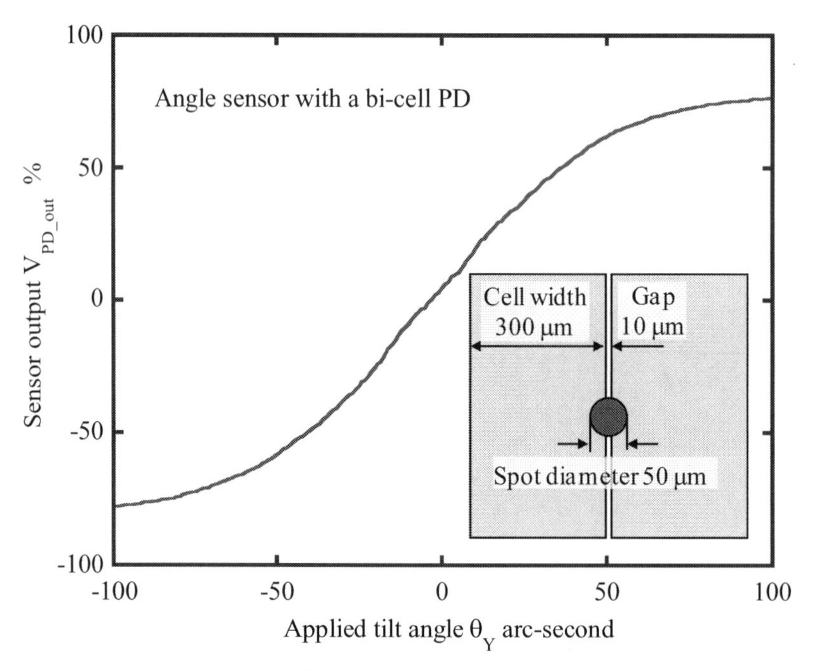

Figure 1.27. Angle sensor output between with a bi-cell PD

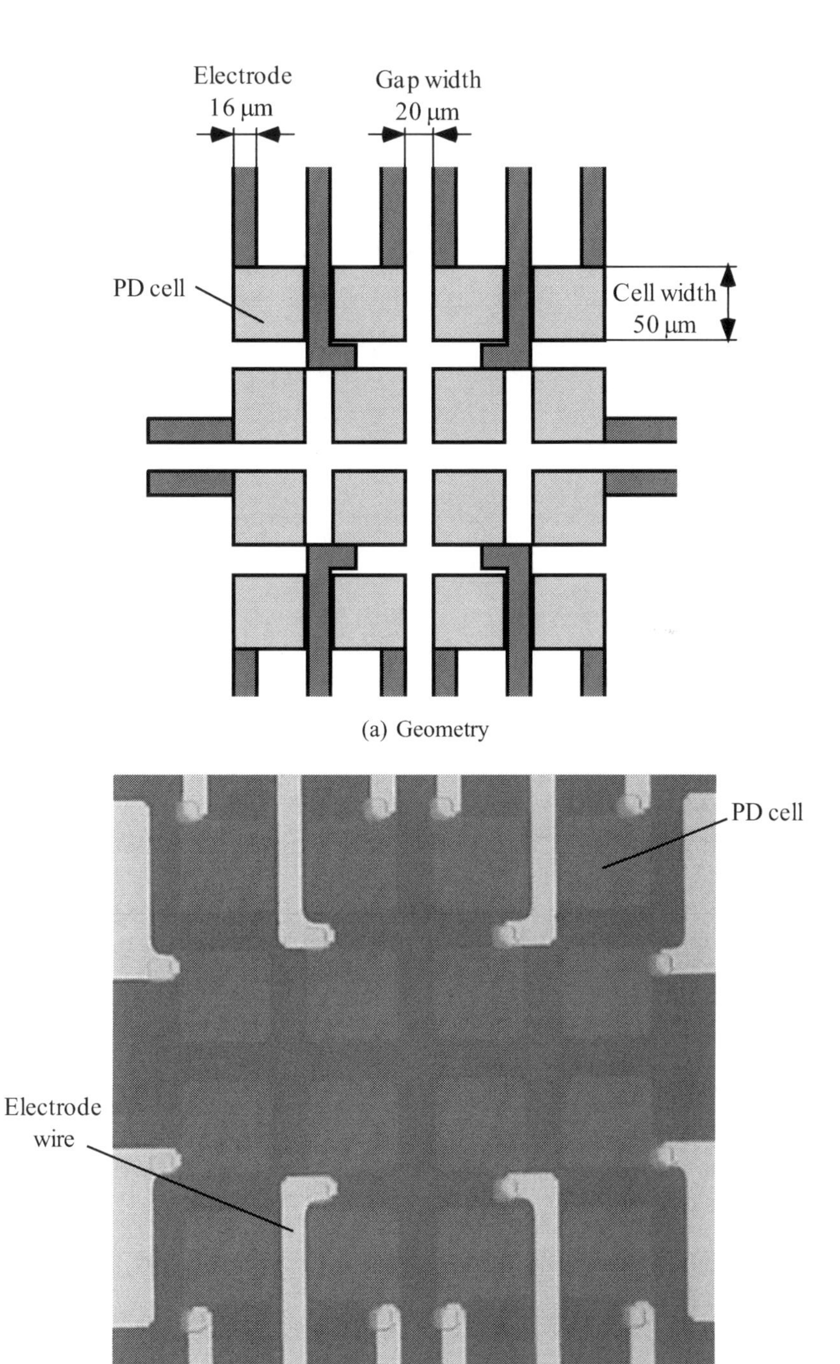

(a) Geometry

(b) Picture

Figure 1.28. A fabricated 2D PD array

1.4 Angle Sensor with Single-cell Photodiode

As shown in the previous sections, in the case of using a quadrant photodiode or a bi-cell photodiode in an angle sensor, the size of the light spot on the QPD is related to the sensitivity of QPD as well as the angle sensor. The smaller the size of the light spot, the higher the sensitivity. It is thus effective to reduce the light spot size for improvement of the sensor sensitivity. However, there are gaps between the PD cells, which are not sensitive to the light spot. The minimum light spot size as well as the sensor sensitivity is thus limited by the gap size between the photodiode cells of the QPD. This section presents an angle sensor using single-cell photodiodes instead of the QPD as the light position detector so that the limitation caused by the insensitive gaps between PD cells of the QPD can be overcome. In the sensor, the light spot size can be reduced to the diffraction limit for realizing a higher sensitivity of angle detection.

Figure 1.29 shows the schematic of the angle sensor using single-cell photodiodes. There are three single-cell photodiodes. PD1 and PD2 are employed for detection of the tile angle components θ_Y and θ_X, respectively. PD3 is employed to monitor the variation of the intensity of the reflected beam. Assume that the beam diameter and the intensity of the light beam received by PD3 are D_0 and INT_0, respectively. The diameter and intensity of the light beams received by PD1 and PD2, which are placed at the focal position of the lens, are D_S and $INT_0/2$, respectively.

As can be seen in Figure 1.29, a tilt motion θ_Y about the Y-axis will cause a displacement of Δv for the light spot on PD1. Similarly, a tilt motion θ_X about the X-axis will cause a displacement of Δw for the light spot on PD2. Assume that the sensitive area of PD0 is large enough so that the light spot on PD0 does not move out of the sensitive area. The output currents of the photodiodes can be expressed by:

$$I_0 \approx C_{on} INT_0, \tag{1.22}$$

$$I_1 \approx C_{on} \frac{INT_0}{2D_s} \Delta v = C_{on} \frac{INT_0}{D_s} f\theta_Y, \tag{1.23}$$

$$I_2 \approx C_{on} \frac{INT_0}{2D_s} \Delta w = C_{on} \frac{INT_0}{D_s} f\theta_X, \tag{1.24}$$

where C_{on} is the coefficient of converting the light intensity to current of the photodiode. The outputs of the angle sensor, in which INT_0 is removed, can be described as:

$$v_{SPD\_out} = \frac{I_1}{I_0} \times 100\% = \frac{f}{D_s} \theta_Y, \tag{1.25}$$

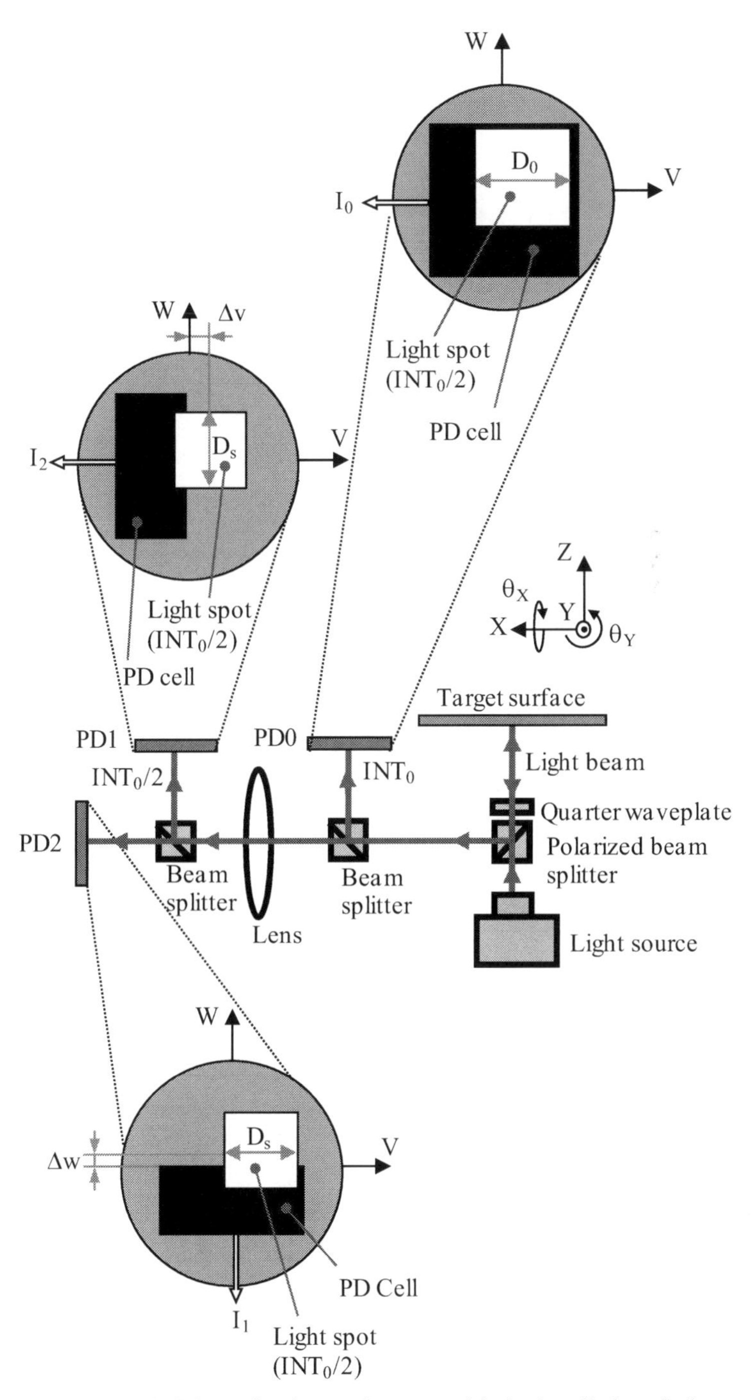

Figure 1.29. Schematic of an angle sensor with single-cell photodiodes

$$w_{SPD\_out} = \frac{I_2}{I_0} \times 100\% = \frac{f}{D_s}\theta_X . \tag{1.26}$$

Rewriting Equations 1.25 and 1.26 gives

$$\theta_Y = K_{S\theta} v_{SPD\_out} , \tag{1.27}$$

$$\theta_X = K_{S\theta} w_{SPD\_out} , \tag{1.28}$$

where

$$K_{S\theta} = \frac{D_s}{f} \tag{1.29}$$

is referred to as the sensitivity for angle detection.

Figure 1.30 shows a photograph of the angle sensor using single-cell photodiodes. In Figure 1.29, PD1 and PD2 share the same lens for simplicity of

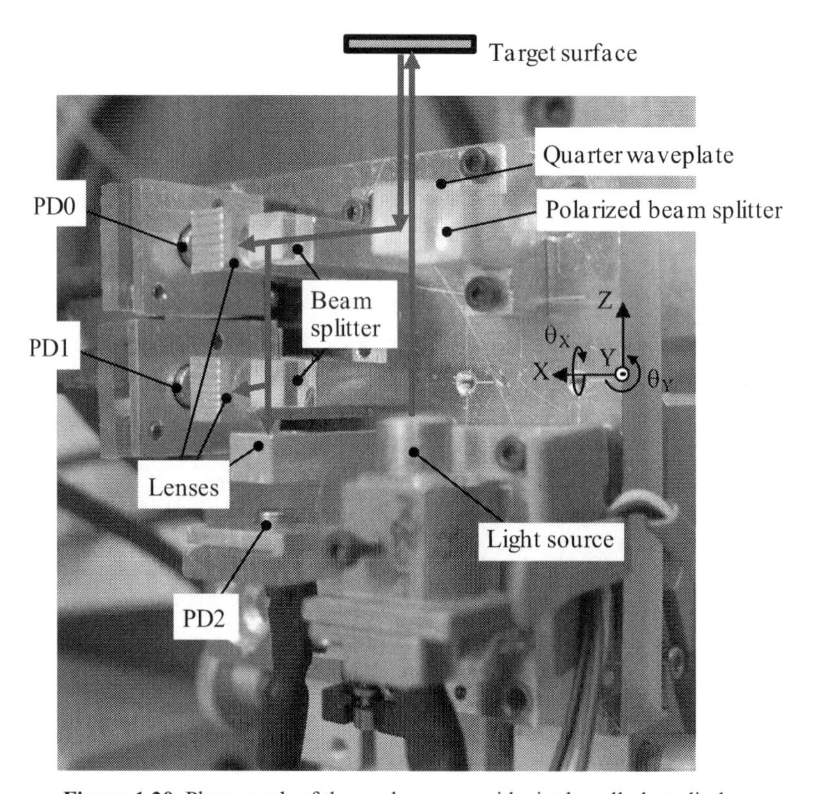

Figure 1.30. Photograph of the angle sensor with single-cell photodiodes

description. Because it is necessary to use a short focal length of lens for generation of a small D_S, two identical lenses were employed for PD1 and PD2 in Figure 1.30. Another lens was placed in front of PD0 for reducing the diameter D_0. The light source was a laser diode with a wavelength of 685 nm. The diameter from the light source was approximately 5 mm. The focal length of the lenses was 7.5 mm. The size of the sensitive area of each photodiode was 1.1 mm × 1.1 mm. The diameter of the light spot D_S was calculated to be approximately 1.25 μm based on the theory of light diffraction. This value was smaller than the 10 μm gap of the QPD used in Section 1.2 and cannot be used in an angle sensor with a QPD.

Figure 1.31 shows the experimental results. The sensitivity $K_{S\theta}$ was approximately 3% / arc-second. It can be seen that the sensor using single-cell photodiodes has a much higher sensitivity than the angle sensor using QPD as shown in Figure 1.5.

1.5 Summary

Angle sensors using different types of photodiodes as light position detectors have been presented. The method of using a quadrant PD was first discussed. It has been confirmed that the diameter of the light spot focused on the QPD is a critical factor to determine the range and sensitivity for angle detection. A smaller light spot can generate a higher sensitivity but reduce the measurement range.

The second method employs a multi-cell PD array as the light position detector. This method can improve the measurement range without reducing the resolution based on the fact that both the range and resolution of an angle sensor, with separate PD elements, are determined by the light spot size on the focal plane of the objective lens where the PD is placed. The cell width of a multi-cell PD array should be designed to be the same or slightly smaller than the spot diameter. A PD array with 16 cells has been designed and fabricated for an angle sensor. A cell gap of 5 μm has been realized. The cell width was designed to be 40 μm based on the gap width and the spot diameter. The angle sensor has a measurement range of approximately 2300 arcsec, which is about 15 times larger than that when employing a conventional bi-cell PD. A two-dimensional PD array has also been designed and fabricated.

The third method of using single-cell photodiodes has been discussed for further improving the sensitivity for angle detection. This method can remove the influence of insensitive gaps between cells in a QPD or a bi-cell PD. The light spot focused on the photodetector can be as small as that determined by the limitation of light diffraction. This method is expected to realize a more sensitive and more compact angle sensor by using this technology.

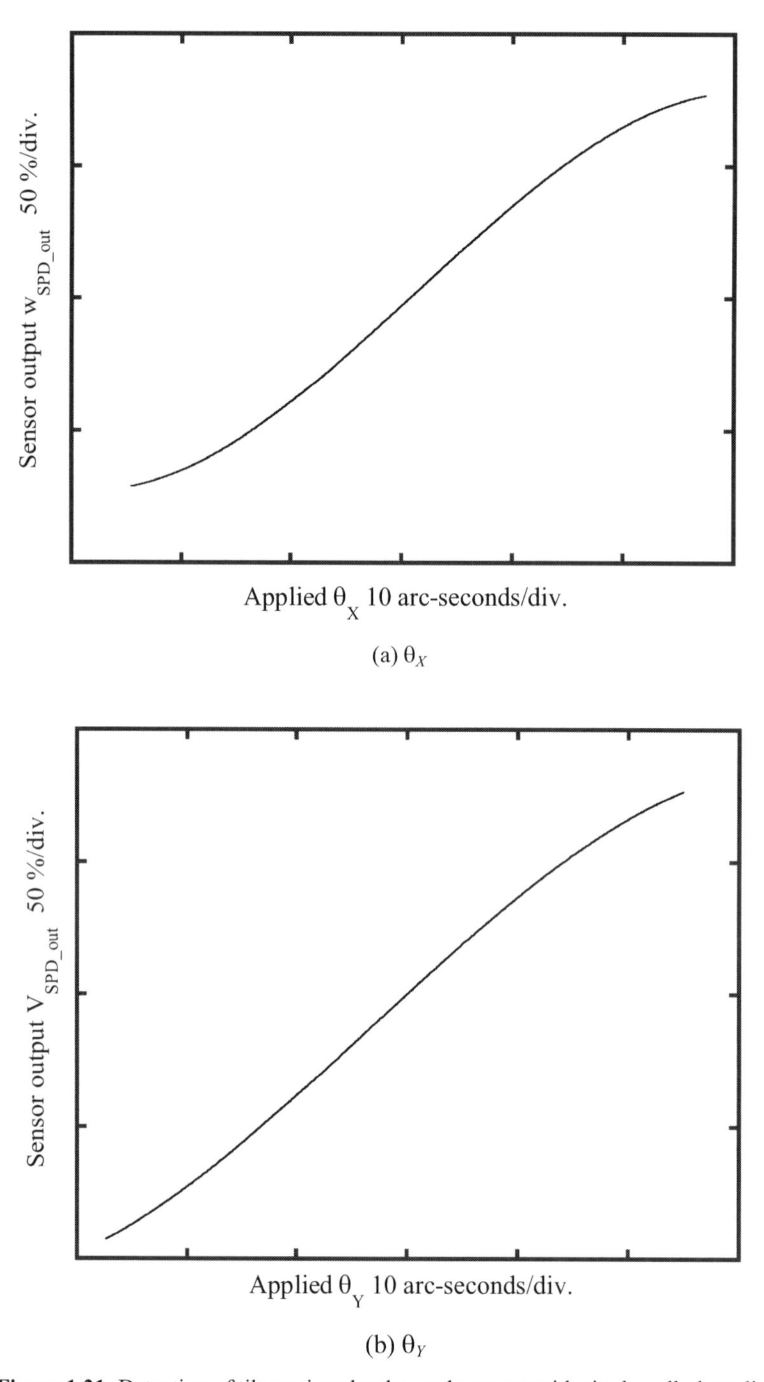

Figure 1.31. Detection of tilt motions by the angle sensor with single-cell photodiodes

References

[1] Estler WT, Queen YH (2000) Angle metrology of dispersion prisms. Ann CIRP 49(1):415–418

[2] Farago FT, Curtis MA (1994) Handbook of dimensional measurement. Industrial Press, New York

[3] Yandayan T, Akgoz SA, Haitjema H (2002) A novel technique for calibration of polygon angles with non-integer subdivision of indexing table. Precis Eng 26(4):412–424

[4] Geckeler RD, Just A, Probst R, Weingartner I (2002) Sub-nm topography measurement using high-accuracy autocollimators. Technisches Messen 69(12):535–541

[5] Vermont Photonics Technologies Corporation (2010) http://www.vermontphotonics.com. Accessed 1 Jan 2010

[6] AMETEK Inc. (2010) http://www.taylor-hobson.com. Accessed 1 Jan 2010

[7] Jenkins FA, White HE (1976) Fundamentals of optics, chap 10. McGraw-Hill, New York

[8] Gao W, Kiyono S (1997) Development of an optical probe for profile measurement of mirror surfaces. Opt Eng 36(12):3360–3366

[9] Hamamatsu Photonics K.K. (2010) http://www.hamamatsu.com. Accessed 1 Jan 2010

[10] Bennett SJ, Gates JWC (1970) The design of detector arrays for laser alignment systems. J Phys E Sci Instrum 3:65–68

[11] OSI Systems Inc. (2010) http://www.osioptoelectronics.com. Accessed 1 Jan 2010

[12] Takacs PZ, Bresloff CJ (1996) Significant improvements in long trace profile measurement performance. SPIE Proc 2856:236–245

[13] Huang PS, Xu XR (1999) Design of an optical probe for surface profile measurement. Opt Eng 38(7):1223–1228

[14] Weingartner I, Schulz M (1999) Ultra-precise scanning technique for surface testing in the nanometric range. In: Proceedings of 9th International Conference on Precision Engineering, Osaka, Japan, pp 311–316

[15] Weingartner I, Schulz M, Elster C (1999) Novel scanning technique for ultra-precise measurement of topography. SPIE Proc 3782:306–317

[16] Gao W, Huang PS, Yamada T, Kiyono S (2002) A compact and sensitive two-dimensional angle probe for flatness measurement of large silicon wafers. Precis Eng 26(4):396–404

[17] Gao W, Ohnuma T, Satoh H, Shimizu H, Kiyono S (2004) A precision angle sensor using a multi-cell photodiode array. Ann CIRP 53(1):425–428

2

Laser Autocollimator for Measurement of Multi-axis Tilt Motion

2.1 Introduction

Precision stages used in nanomanufacturing, including linear stages and rotary stages, have multi-axis tilt error motions. For a linear stage, the tilt error motions are referred to as the pitch, yaw and roll error motions, which cause unexpected Abbe errors [1, 2]. It is necessary to measure the tilt error motions by using angle sensors for evaluation and compensation of Abbe errors. Conventionally, the measurement of pitch and yaw angles is carried out by an autocollimator using a filament lamp as the light source and a CCD image sensor as the light position detector. As described in Chapter 1, conventional autocollimators, which are large in size and slow in measurement speed, are not suited for dynamic measurement of stage tilt error motions. In addition, the conventional autocollimator cannot detect the roll error motion, which is defined as the angular displacement about the normal axis of the target plane reflector, because no light spot displacement can be generated on the light position detector with respect to the roll error motion.

At first, this chapter presents two types of autocollimators based on laser autocollimation [3–5], which are referred to as laser autocollimators, for two-axis measurement of pitch and yaw error motions. Differing from a conventional autocollimator, the laser autocollimator employs a laser as the light source. The laser beam is collimated to a thin parallel beam with a diameter of less than several millimeters so that it is only needed to mount a small target mirror, which will not influence the dynamics of the stage. Because a laser source is used, the sensitivity of angle detection is no longer a function of the focal length of the objective lens, resulting in a compact sensor size. A quadrant photodiode with a high bandwidth is employed as the light position detector, which is important for high-speed measurement.

The laser autocollimator is then improved from two-axis measurement to three-axis measurement by using a diffraction grating as the target reflector. The three-axis angle measurement of the target is carried out by sensing the displacements of the zeroth-order and the positive and negative first-order diffraction light spots on the focal planes of the objective lenses. The methods for calculation of the three-axis angle components are categorized into three types based on behaviors of the zeroth-order and the positive and negative first-order diffraction light spots on the focal planes.

2.2 Two-axis Laser Autocollimator

Figure 2.1 shows the principle of tilt angle detection by the laser autocollimation. The basic principle is the same as the angle sensor described in Chapter 1 although the light source is restricted to a laser. The detection of the tilt angle component θ_X

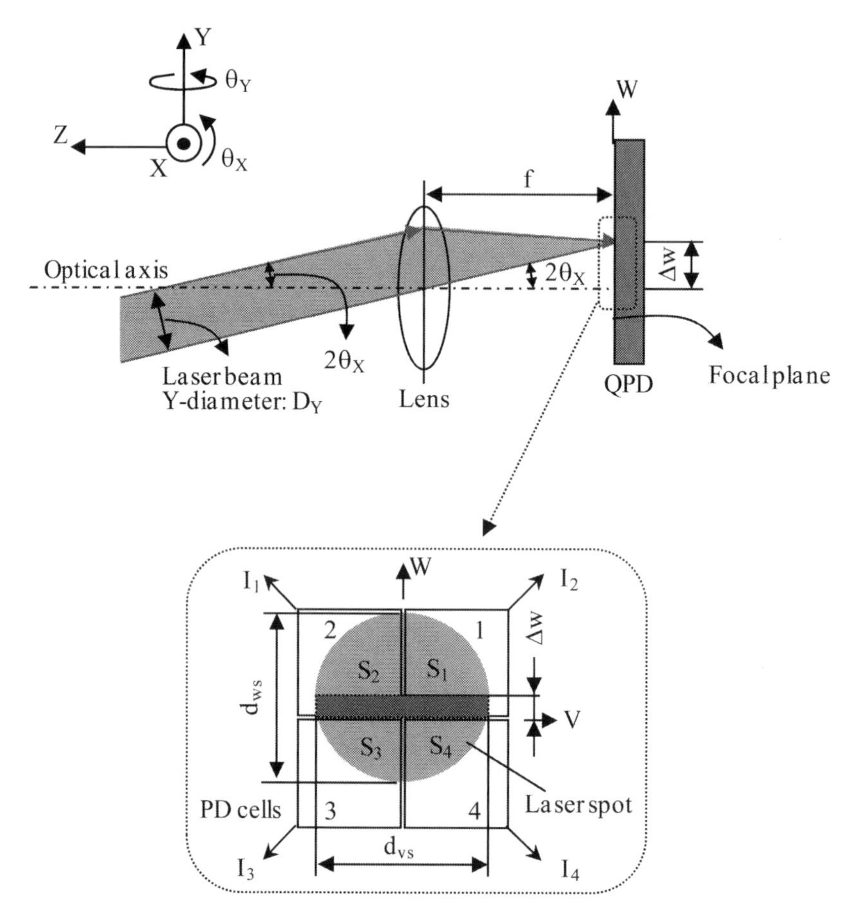

1,2, 3, 4: PD cells

INT_1, INT_2, INT_3, INT_4: optical power on each PD

I_1, I_2, I_3, I_4: output current from each PD

Figure 2.1. Principle of the laser autocollimation (the target mirror is not shown in the figure for clarity)

about the X-axis is shown in the figure. V-directional and W-directional diameters of the focused laser spot on the QPD, which is placed at the focal position of the objective lens, can be expressed as:

$$d_{vs} = \frac{2.44 f\lambda}{D_X}, \tag{2.1}$$

$$d_{ws} = \frac{2.44 f\lambda}{D_Y}, \tag{2.2}$$

where D_X and D_Y are the X-directional and Y-directional diameters of the collimated laser beam entering the objective lens. λ is the wavelength of the laser and f is the focal length of the lens.

As can be seen in Figure 2.1, the laser spot on the QPD will move a displacement of Δw along the W-axis in responding to the tilt angle of θ_X. The output of the QPD can be written as:

$$w_{QPD\_out} = \frac{(I_1 + I_2) - (I_3 + I_4)}{(I_1 + I_2 + I_3 + I_4)} \times 100\%,$$

$$= \frac{(INT_1 + INT_2) - (INT_3 + INT_4)}{(INT_1 + INT_2 + INT_3 + INT_4)} \times 100\%, \tag{2.3}$$

where INT_1, INT_2, INT_3, INT_4 are the optical powers (intensities) received by the PD cells. I_1, I_2, I_3, I_4 are the corresponding current outputs by PD cells.

When θ_X is small, the difference Δs_w between the area of the laser spot within cells 1, 2 and that within cells 3, 4 can be expressed by:

$$\Delta s_w = (S_1 + S_2) - (S_3 + S_4) = 2 d_{vs} \Delta w, \tag{2.4}$$

where S_1, S_2, S_3, S_4 are the corresponding laser spot areas on the PD cells.

The laser spot is assumed to have a uniform intensity distribution and gaps between the PD cells are zero. The intensity received by each PD cell is then proportional to the corresponding light spot area. Equation 2.3 can thus be re-written as:

$$w_{QPD\_out} = \frac{(INT_1 + INT_2) - (INT_3 + INT_4)}{(INT_1 + INT_2 + INT_3 + INT_4)} \times 100\%,$$

$$= \frac{\Delta s_w}{S} \times 100\%, \tag{2.5}$$

where

$$S = S_1 + S_2 + S_3 + S_4 = \frac{1}{4}\pi d_{ws} d_{vs} \quad (2.6)$$

is the area of the light spot on the QPD.

Combining Equations 2.4, 2.5 and 2.6 gives:

$$w_{QPD\_out} = \frac{\Delta s_w}{S} \times 100\% = \frac{8\Delta w}{\pi d_{ws}} \times 100\% . \quad (2.7)$$

On the other hand, based on the principle of autocollimation, θ_X can be related to Δw by

$$\theta_X = \frac{\Delta w}{2f} . \quad (2.8)$$

Substituting Equations 2.2 and 2.8 into Equation 2.7 results in:

$$w_{QPD\_out} = \frac{8\Delta w}{\pi d_{ws}} \times 100\% = k_{\theta x} \theta_X , \quad (2.9)$$

where

$$k_{\theta x} = \frac{8}{1.22\pi} \frac{D_Y}{\lambda} \quad (2.10)$$

is referred to as the sensitivity of detecting θ_X, which is a function of the collimated laser beam diameter D_Y and the laser wavelength λ.

Similarly, the V-directional output for detection of θ_Y can be expressed by:

$$v_{QPD\_out} = k_{\theta y} \theta_Y , \quad (2.11)$$

where

$$k_{\theta y} = \frac{8}{1.22\pi} \frac{D_X}{\lambda} \quad (2.12)$$

is referred to as the sensitivity of detecting θ_Y.

Differing from the angle sensor based conventional autocollimation shown in Figure 1.2 and Equations 1.3 and 1.4, the output of the laser autocollimator is no longer a function of the focal length of the lens. This makes it possible to construct a micro- and sensitive angle sensor by choosing a lens with a short focal length.

Simulation is carried out to identify the characteristics of the laser autocollimator. In the simulation, the laser spot on the QPD is assumed to be round with a diameter of d. Figure 2.2 shows the intensity distribution of the laser spot, which is assumed to have a Gaussian distribution [6] as follows:

$$INT(v,w) = e^{\frac{8(v^2+w^2)}{d^2}}. \tag{2.13}$$

As can be seen in Equations 2.1 and 2.2, the diameter of the focused laser spot on the QPD is proportional to the focal length of the lens. The shorter the lens focal length, the smaller the laser spot diameter. To realize a laser micro-laser autocollimator, it is necessary to employ a lens with a short focal length, which results in a small laser spot diameter on the QPD. On the other hand, there are insensitive gaps between PD cells of the QPD. The PD gap can be a major factor in the sensor design when the laser spot diameter is small. The first simulation is thus carried out to investigate the influence of the gap between the PD cells. In the simulation, the total intensity of the light spot is 1 mW. The focal length of the lens is 10 mm. The diameter D ($D = D_X = D_Y$) of the laser beam entering the lens is 1 mm. The diameter d ($d = d_{ws} = d_{vs}$) of the laser spot focused on the QPD is calculated to be approximately 32 μm based on Equations 2.1 and 2.2. The gap Δg between PD cells, which is illustrated in Figure 2.3, is 10 μm.

Figure 2.4 shows the output of the laser autocollimator. The result when the gap between the PD cells is 0 μm is also shown in the figure for comparison. In the simulation, the tilt angle θ_Y is assumed to be zero and the laser spot moves on the X-axis in responding to θ_X as shown in Figure 2.3. The areas S_1, S_2, S_3, S_4 shown in Figure 2.1 are evaluated by numerical integration to obtain the output of the laser micro-autocollimator shown in Equation 2.5.

It can be seen in Figure 2.4 that the output characteristic of the laser micro-autocollimator is greatly influenced by the PD gap Δg. Over the range of −10 arcsec to 10 arcsec of θ_X, the sensor sensitivity, which is corresponding to the slope of the curve in Figure 2.4, is higher when Δg = 10 μm than when Δg = 0 μm. Figure 2.5 shows the optical powers of the laser spot received by the PD cells. Because the intensity of the focused laser spot has a Gaussian distribution as shown in Figure 2.2, the center area of the laser spot occupies a large part of the optical power of the laser spot. For this reason, not only does the optical power received by each of the PD cell but also the total optical power received by all the four PD cells of the QPD when Δg = 10 μm as shown in Figure 2.5 (b) become much smaller than those when Δg = 0 μm, as shown in Figure 2.5 (a). On the other hand, the total optical power, which is the denominator in Equation 2.5 for calculation of the sensor output, remains almost constant when θ_X changes, as shown in Figure 2.5 (a). However, the total optical power decreases when θ_X approaches zero if there is an insensitive gap between the PD cells as shown in Figure 2.5 (b). The reduction of the denominator in Equation 2.5 makes the sensor output-applied tilt angle curve have a larger slope. Similar phenomena can be observed when the focal length of the lens is changed as shown in Figure 2.6.

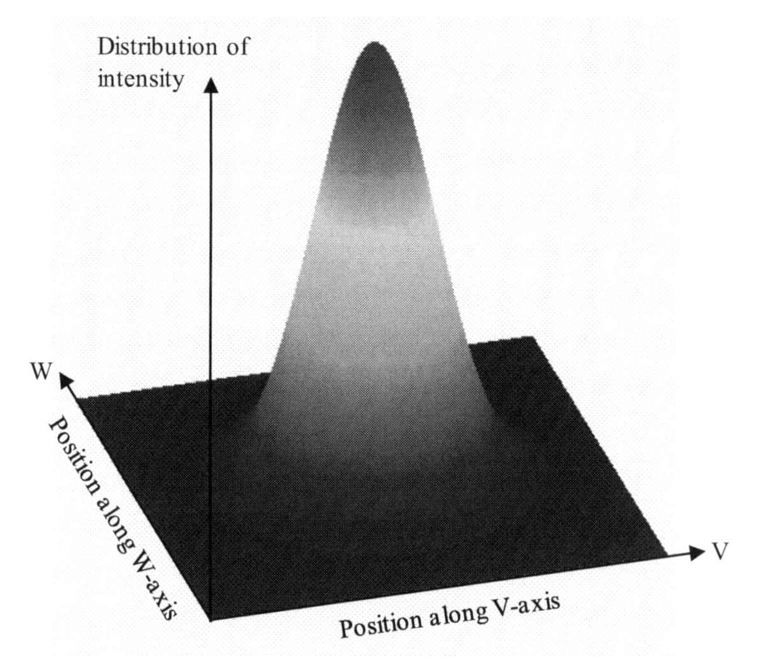

(a) 3D expression

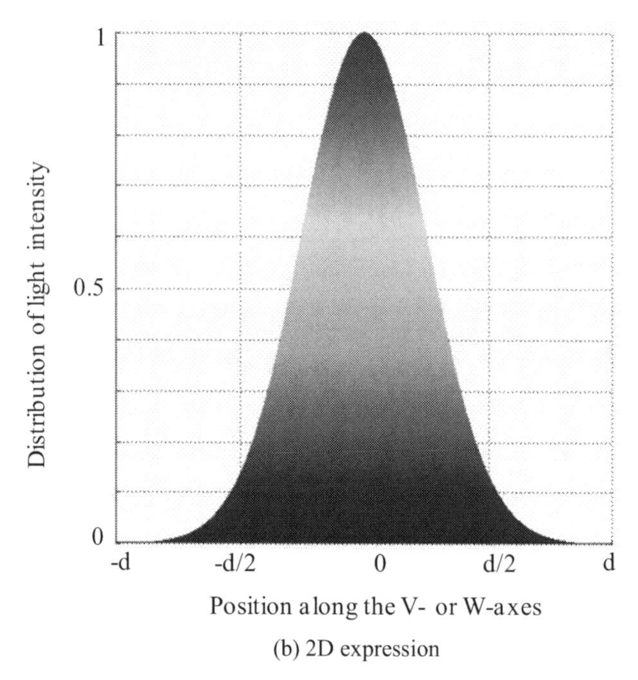

(b) 2D expression

Figure 2.2. Intensity distribution of the focused Gaussian laser spot

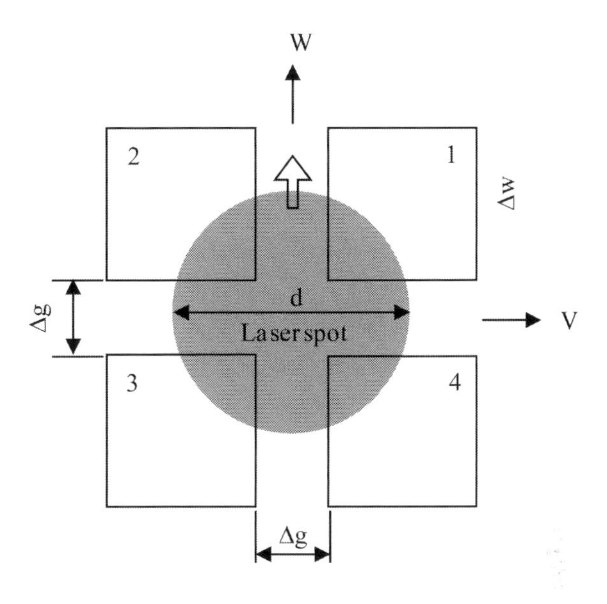

Figure 2.3. Gaps between the PD cells of a QPD

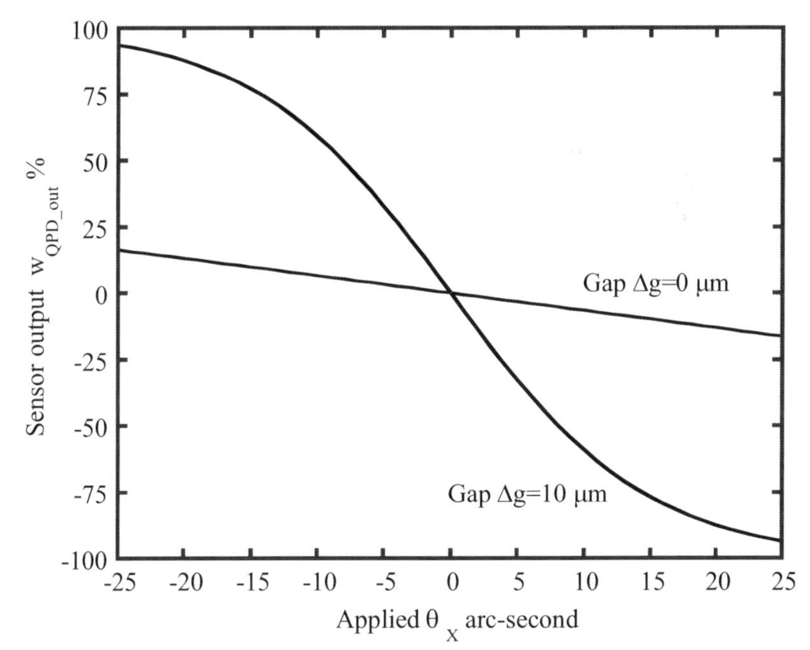

Figure 2.4. Simulation results of sensor output for different PD gaps (QPD is located at the focal plane of the lens)

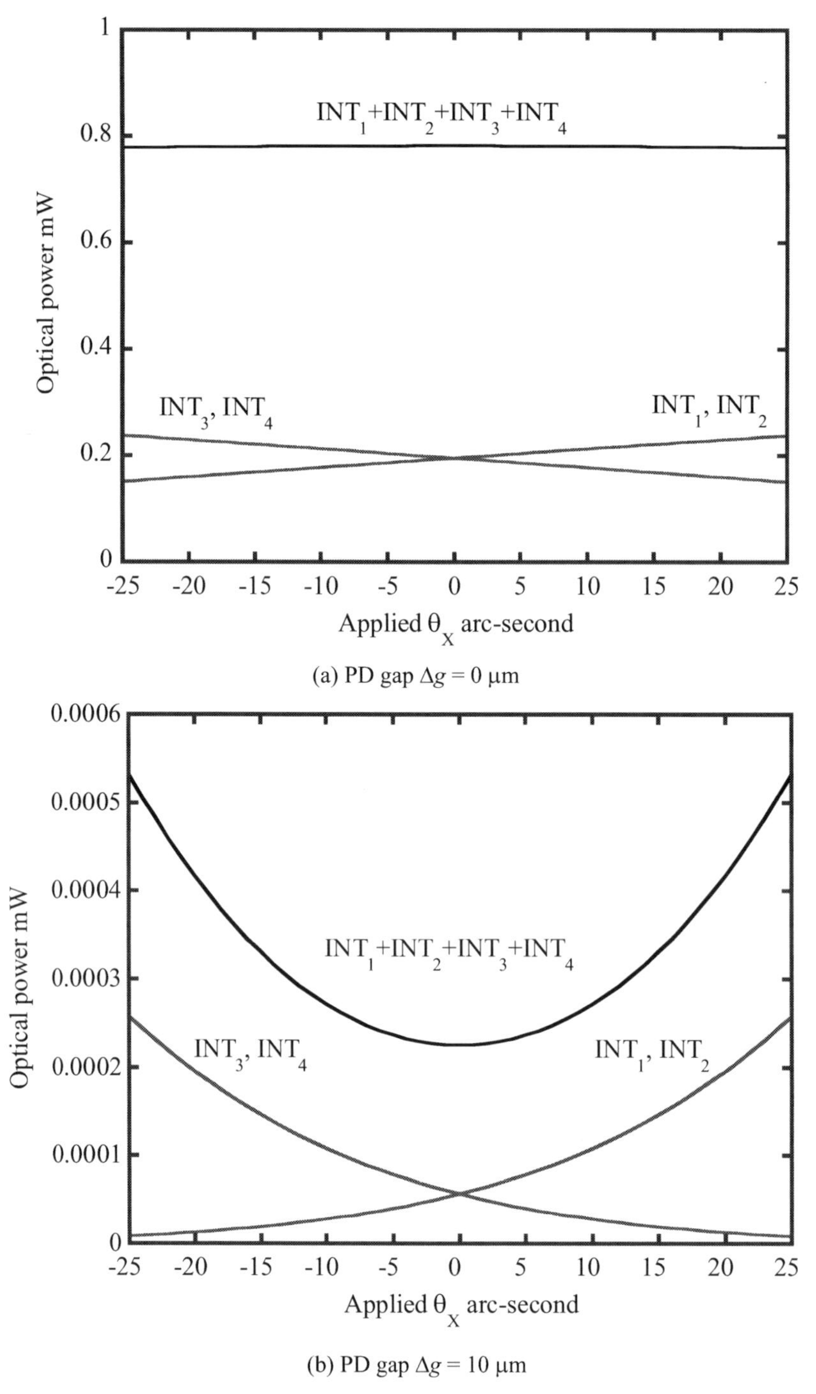

(a) PD gap $\Delta g = 0$ μm

(b) PD gap $\Delta g = 10$ μm

Figure 2.5. Simulation results of optical powers of QPD for different PD gaps (QPD is located at the focal plane of the lens)

Figure 2.6 shows the simulation results for a focal length of 20 mm. Other parameters used in the simulation are the same above. The result for the focal length of 10 mm is also shown in the figure for comparison. As can be seen in Figure 2.6 (a), the output characteristic of the sensor changes with the focal length, which is different from the principle of laser autocollimation. The sensitivity of the sensor output is also different from that shown in Equations 2.11 and 2.12. The longer the focal length, the lower the sensor sensitivity. This is different from that of conventional autocollimation in which the sensor sensitivity is proportional to the focal length. These phenomena can be explained from Figure 2.6 (b) showing the light intensities received by the PD cells. The focused laser spot becomes larger in size when the focal length increases as indicated in Equations 2.1 and 2.2. Consequently, the intensities in Figure 2.6 (b) are larger than those in Figure 2.5 (b). This makes the slope of the sensor output-applied tilt angle curve smaller in the center part because of the division processing shown in Equation 2.3.

Although the PD gap seems to have the effect of making the laser micro-autocollimator have a higher sensitivity for detection of the tilt angle, the small current outputs of the PD cells will reduce the signal to noise ratio of the sensor, making the sensor vulnerable to electronic noises. To solve this problem, it is desirable to expand the diameter of the focused laser spot on the QPD without changing the laser wavelength, the diameter of the collimated laser beam and the lens focal length. This can be achieved by adjusting the position of the QPD along the optical axis of the lens as shown in Figure 2.7. Assuming that the offset of the QPD from the focal plane of the lens is Δz, the diameter d of the laser spot on the QPD can be expressed by:

$$d(\Delta z) = d_0 \left[1 + \left(\frac{4\lambda\Delta z}{\pi d_0^{\ 2}} \right)^2 \right]^{\frac{1}{2}}, \tag{2.14}$$

where d_0 is the diameter of the laser spot at the beam waist, which is defined in Equations 2.1 and 2.2. It can be seen that d is a function of Δz and d can be expanded by increasing Δz.

Figure 2.8 shows the simulation result when Δz is 200 μm. The other parameters are the same as above. The result obtained when Δz is 0 μm is also shown in the figure for comparison. The sensor sensitivity when Δz is 200 μm is lower than that when Δz is 0 μm. The sensor output-applied tilt angle curve is similar to that when the gap size between PD cells is zero as shown in Figure 2.4. Figure 2.9 shows the optical powers received by the PD cells. The optical powers have been increased significantly compared with that when Δz is 0 μm, as shown in Figure 2.5 (b), indicating the effectiveness of the method shown in Figure 2.7.

Computer simulation is also carried out to check the cross-talk error between detections of θ_X and θ_Y. Figure 2.10 shows the results of the sensor output of θ_X detection at different θ_Y. In the simulation, the PD gap size Δg and the QPD offset Δz are set to be 10 μm and 200 μm, respectively. It can be seen that there are no cross-talk errors in the sensor output.

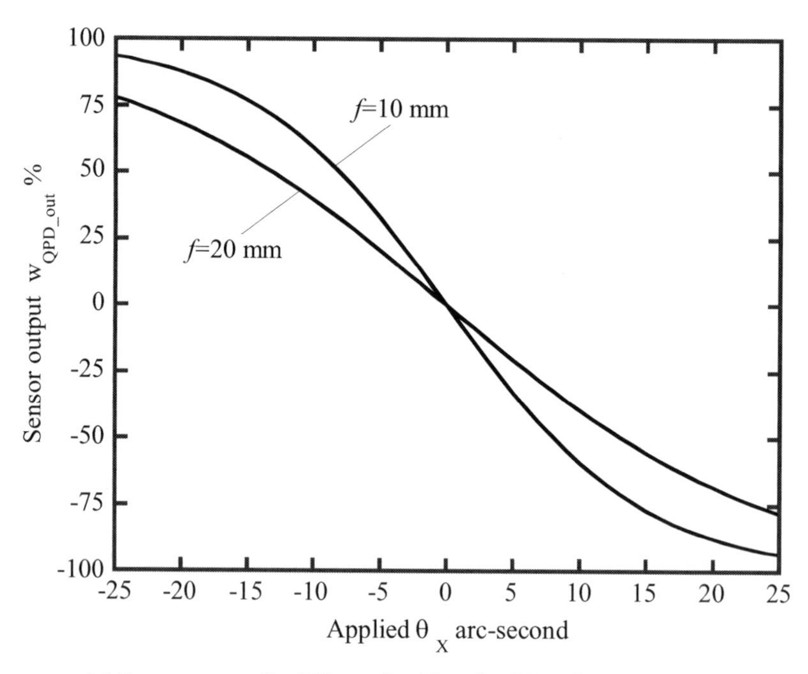

(a) Sensor outputs for different focal lengths (PD cell gap Δg = 10 μm)

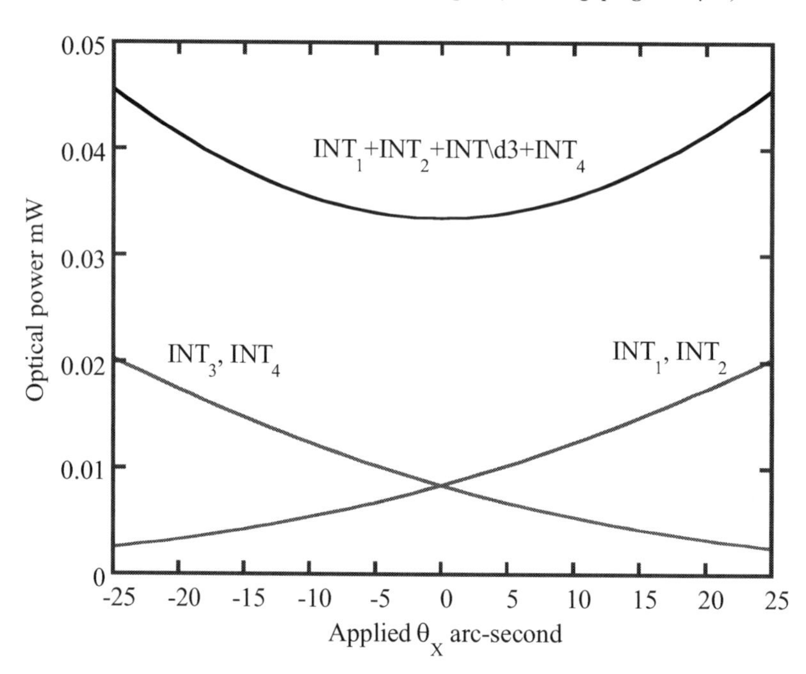

(b) Optical powers received by the PD cells
(Lens focal length f = 20 mm, PD cell gap Δg = 10 μm)

Figure 2.6. Simulation results of the sensor output for different focal lengths of lens

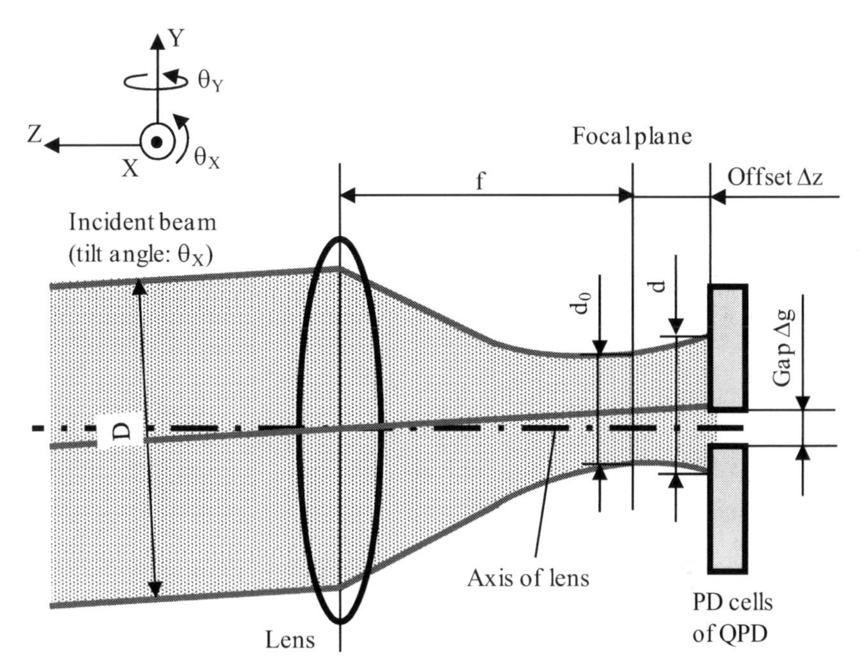

Figure 2.7. Geometric model for applying an offset to the QPD

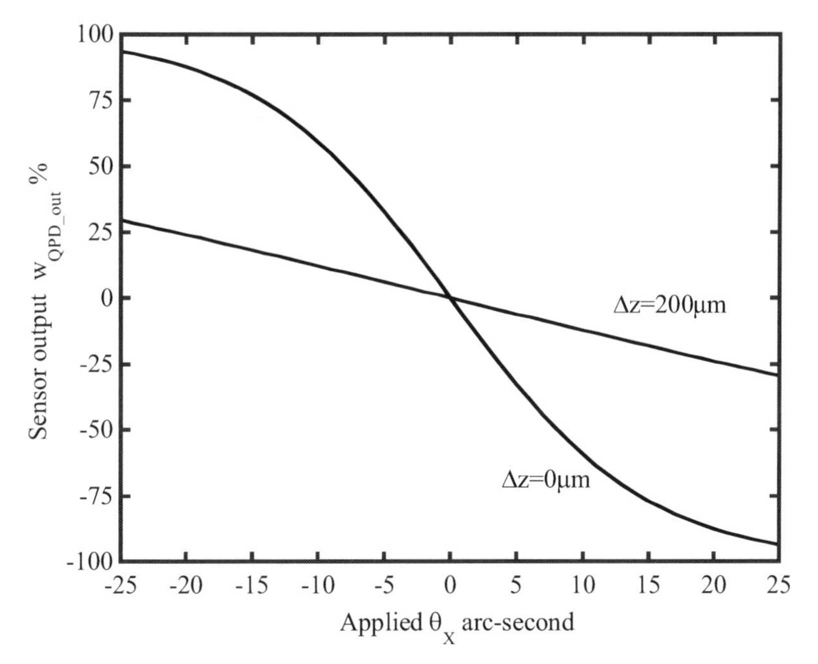

Figure 2.8. Simulation results of sensor output for different QPD offsets (PD cell gap $\Delta g = 10$ μm)

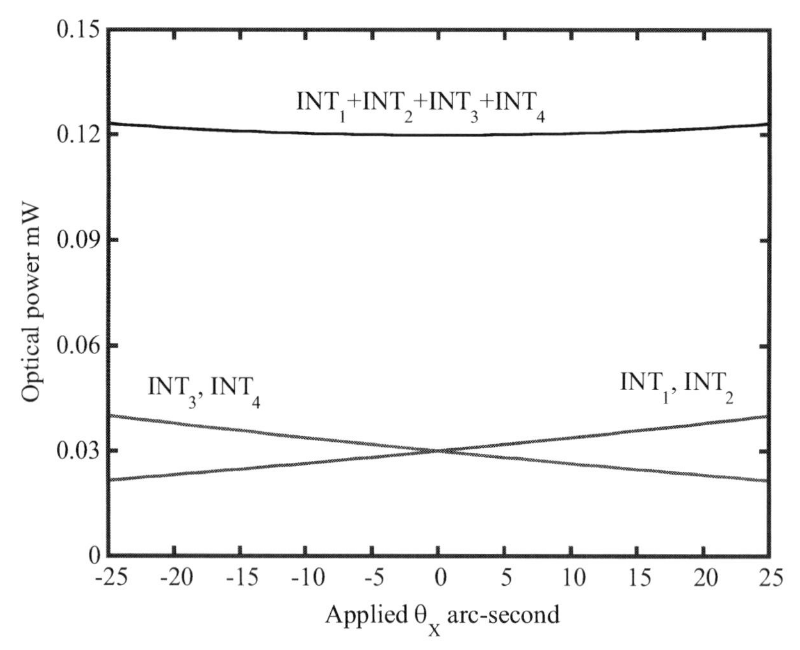

Figure 2.9. Simulation results of optical powers received by PD cells with an offset of Δz = 200 μm (PD cell gap Δg = 10 μm)

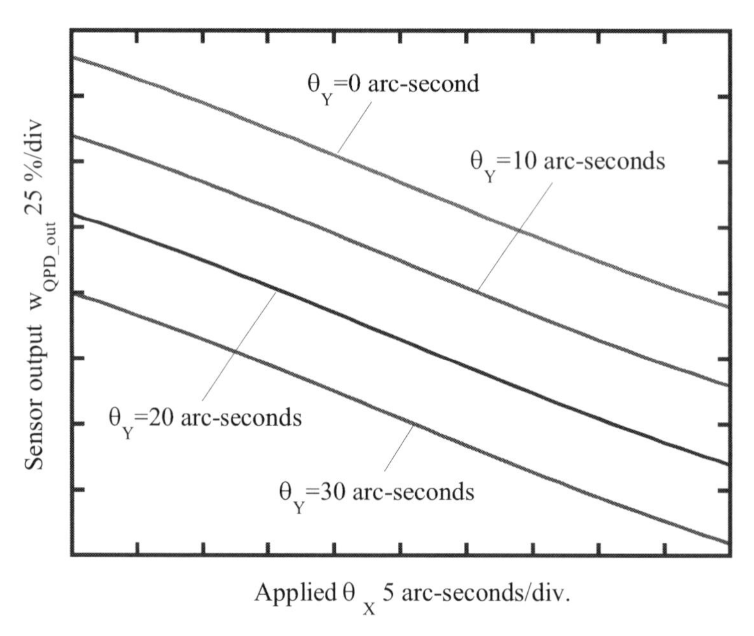

Figure 2.10. Simulation results of cross-talk between detections of θ_X and θ_Y (QPD offset Δz = 200 μm, PD cell gap Δg = 10 μm)

To make the sensor compact, it is necessary to employ a laser diode as the laser source. Most of the laser diodes, however, output ellipsoidal beams with different beam diameters D_X and D_Y, which make the sensor sensitivities $k_{\theta x}$ and $k_{\theta y}$ different. Computer simulation is carried out to investigate the relationship between the QPD offset Δz and the sensor sensitivities when the cell gap Δg is set to be 10 μm. Figure 2.11 shows the results, in which D_X and D_Y are set to be 4 and 1 mm, respectively. According to Equations 2.10 and 2.12, $k_{\theta x}$ and $k_{\theta y}$ are proportional to D_Y and D_X, respectively. The term $k_{\theta x}$ is smaller than $k_{\theta y}$ because D_X is larger than D_Y. As can be seen in Figure 2.11, $k_{\theta x}$ and $k_{\theta y}$ also vary with Δz. Although both $k_{\theta x}$ and $k_{\theta y}$ decrease when Δz increases, $k_{\theta y}$ decreases faster than $k_{\theta x}$. $k_{\theta y}$ becomes the same as $k_{\theta x}$ when Δz is approximately 20 μm. $k_{\theta y}$ then becomes smaller than $k_{\theta x}$ when Δz is larger than 20 μm. From a practical point of view, it is desirable to obtain similar sensor sensitivities in θ_X- and θ_Y-directions. The phenomenon shown in Figure 2.11 can be utilized to match $k_{\theta x}$ and $k_{\theta y}$ when a laser diode with an ellipsoidal beam is employed. In most cases, however, it is difficult to make the PD cells able to receive enough optical powers while matching $k_{\theta x}$ and $k_{\theta y}$ only by adjusting the Δz. In the following laser micro-autocollimator, a laser diode with a circular beam is employed as the laser source.

Figure 2.12 shows a schematic of the two-axis laser micro-autocollimator. The wavelength and the maximum output power of the laser diode are 635 nm and 5 mW, respectively. The diameter of the collimated laser beam is approximately 1 mm. A polarized beam splitter (PBS) and a quarter waveplate (QWP) are employed to bend the reflected beam from the target mirror and to isolate the reflected beam from returning to the laser diode. The PBS and the QWP are glued to each other. The size of the combination of PBS and QWP is approximately 5.5 mm × 5 mm × 5 mm. The lens is an aspherical one with a focal length of 10 mm and a diameter of 6.25 mm. The cell gap of the QPD is 10 μm. The size of the sensor is 26 mm (L) × 12 mm (W) × 14 mm (H). Photographs of the constructed laser micro-autocollimator are shown in Figure 2.13. The mount for the optical components of the sensor has a monolithic structure for reduction of the sensor size. The monolithic structure with a small size is also helpful to improve the thermal stability of the micro-autocollimator. The QPD offset Δz is adjusted in such a way that the output levels of the PD currents are larger enough than the electronic noise.

Figure 2.14 shows the outputs of the two-axis laser micro-autocollimator with the circular beam. For comparison, the laser source is also replaced by a laser diode with an ellipsoidal beam (D_X = 4 mm, D_Y = 1 mm). The outputs are shown in Figure 2.15. The applied tilt angles are measured by a commercial autocollimator. The commercial autocollimator employs a filament light as the light source and a CCD as a detector. The focal length of lens is 380 mm. It can be seen from Figure 2.14 that consistent sensor sensitivities in θ_X- and θ_Y-directions can be achieved by using the laser diode with a circular beam. On the other hand, the sensor sensitivity in the θ_X-direction is approximately 2.5 times larger than that in the θ_Y-direction in Figure 2.15. Figure 2.16 shows the results of testing the resolution of the two-axis micro-autocollimator. The micro-autocollimator has resolutions better than 0.1 arcsec in both the θ_X- and θ_Y-directions. The resolution is also better than that of the conventional autocollimator with a large size.

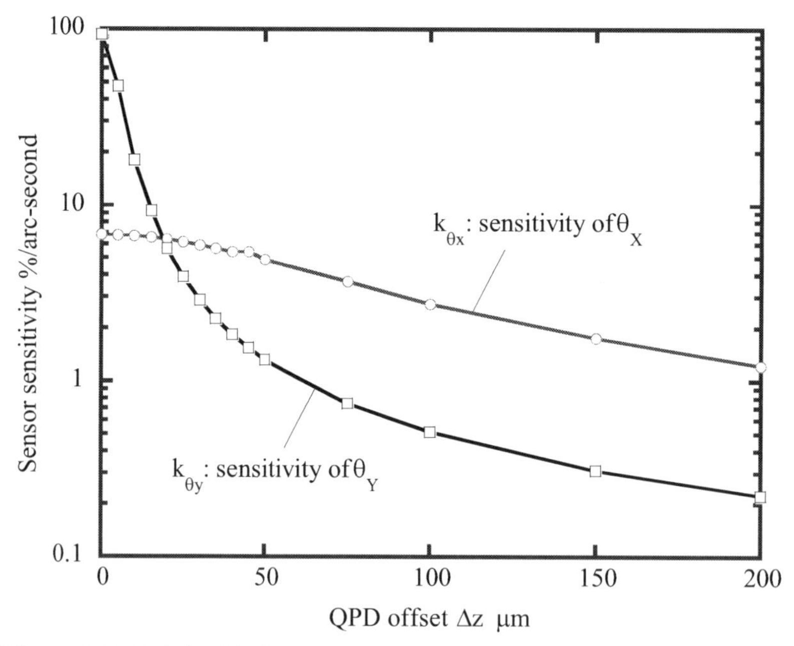

Figure 2.11. Relationship between the sensor sensitivities and the QPD offset for a ellipsoidal laser beam ($D_X = 1$ mm, $D_Y = 4$ mm, PD cell gap $\Delta g = 10$ mm)

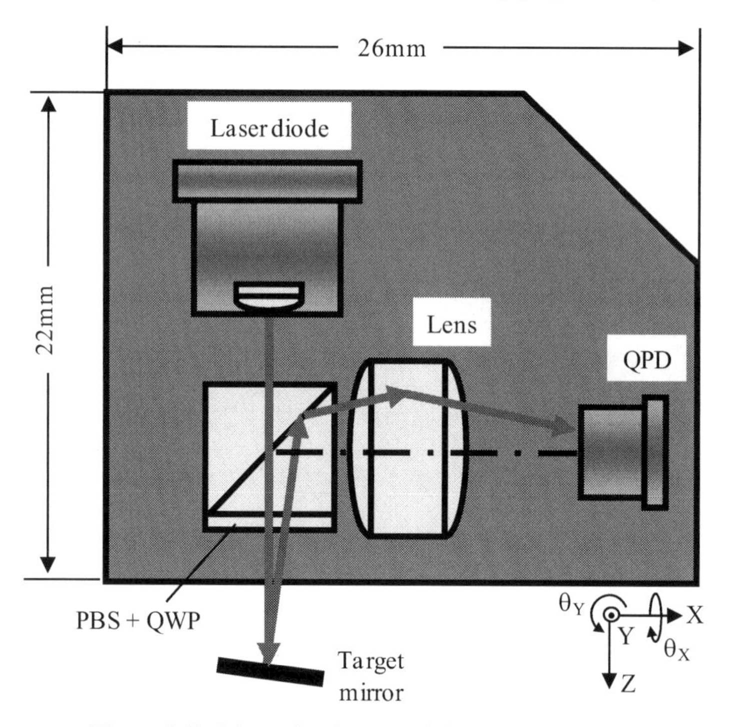

Figure 2.12. Schematic of a two-axis laser micro-autocollimator

(a) The laser micro-autocollimator and a 500 JPY coin with a 26.5 mm diameter

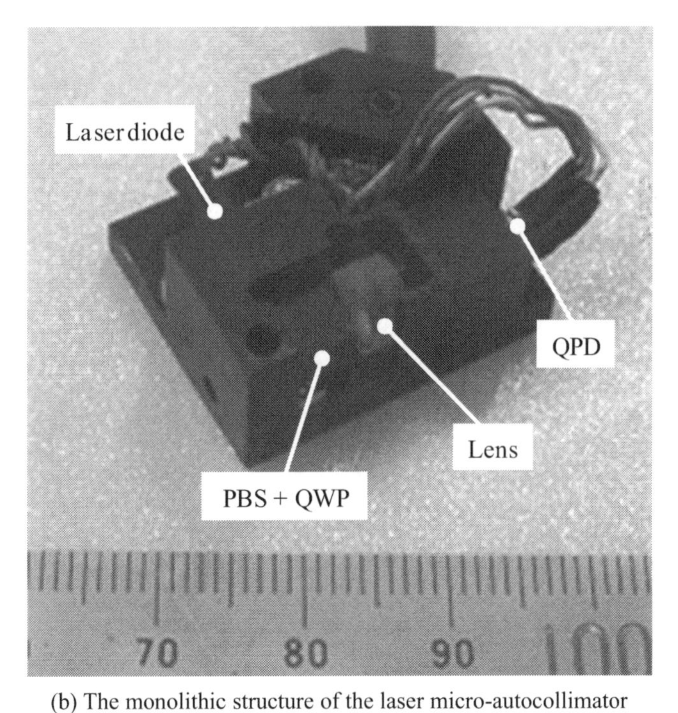

(b) The monolithic structure of the laser micro-autocollimator

Figure 2.13. Photographs of the two-axis laser micro-autocollimator

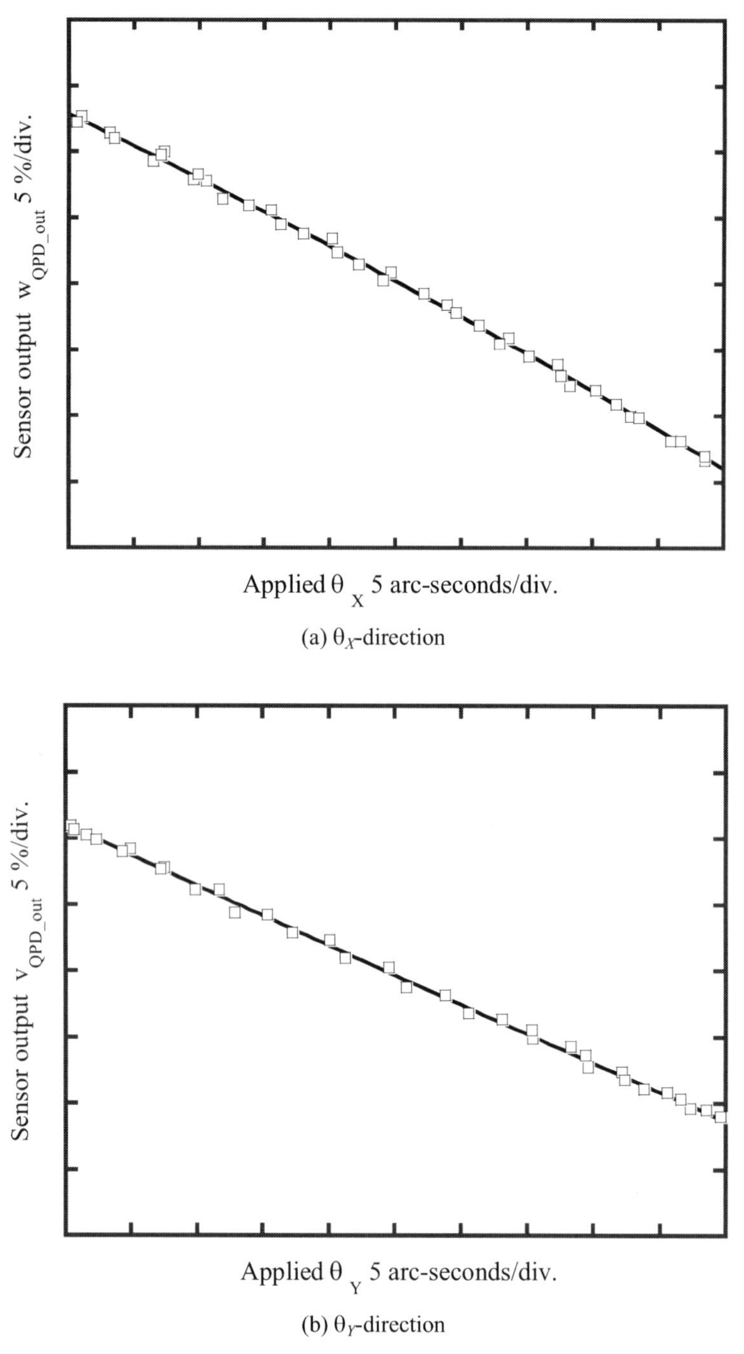

Figure 2.14. Outputs of the two-axis laser micro-autocollimator by using a laser diode with a circular beam (D = 1 mm)

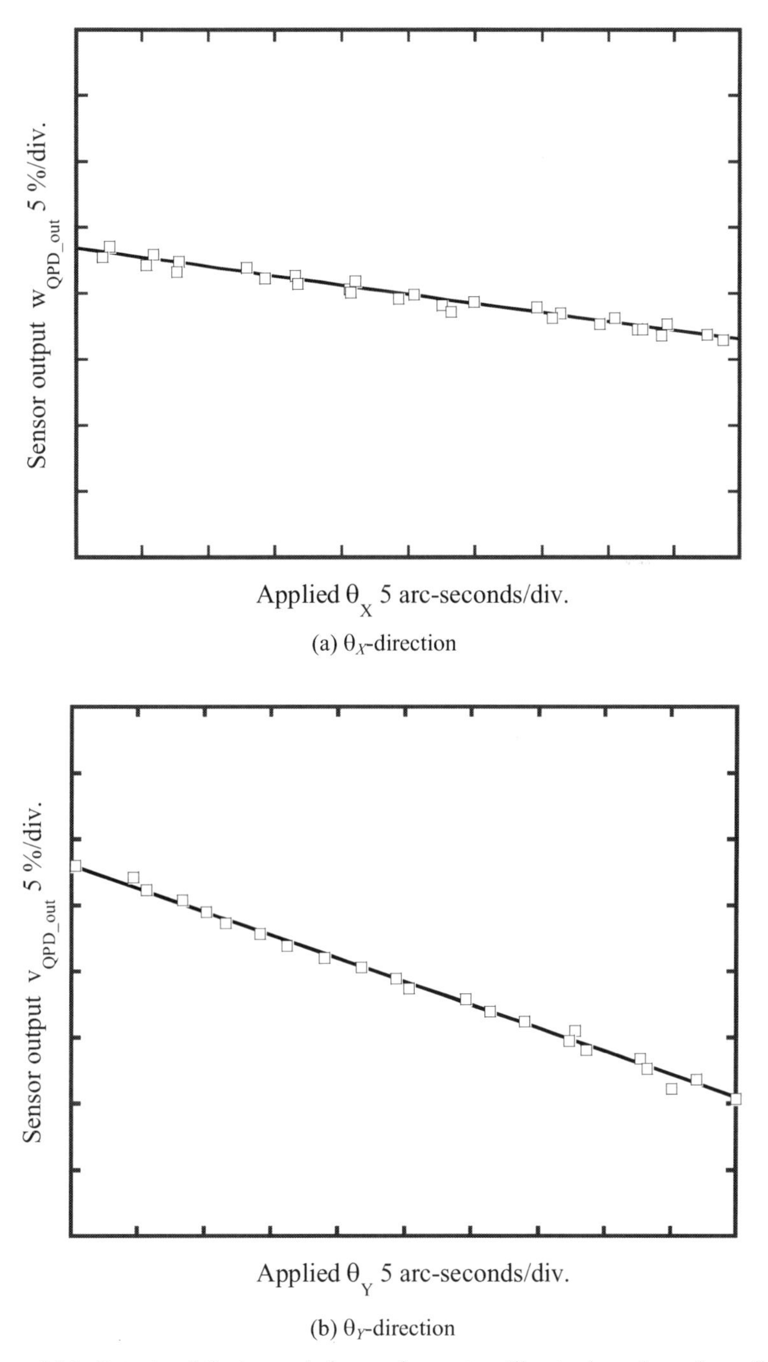

(a) θ_X-direction

(b) θ_Y-direction

Figure 2.15. Outputs of the two-axis laser micro-autocollimator by using a laser diode with an ellipsoidal beam ($D_X = 4$ mm, $D_Y = 1$ mm)

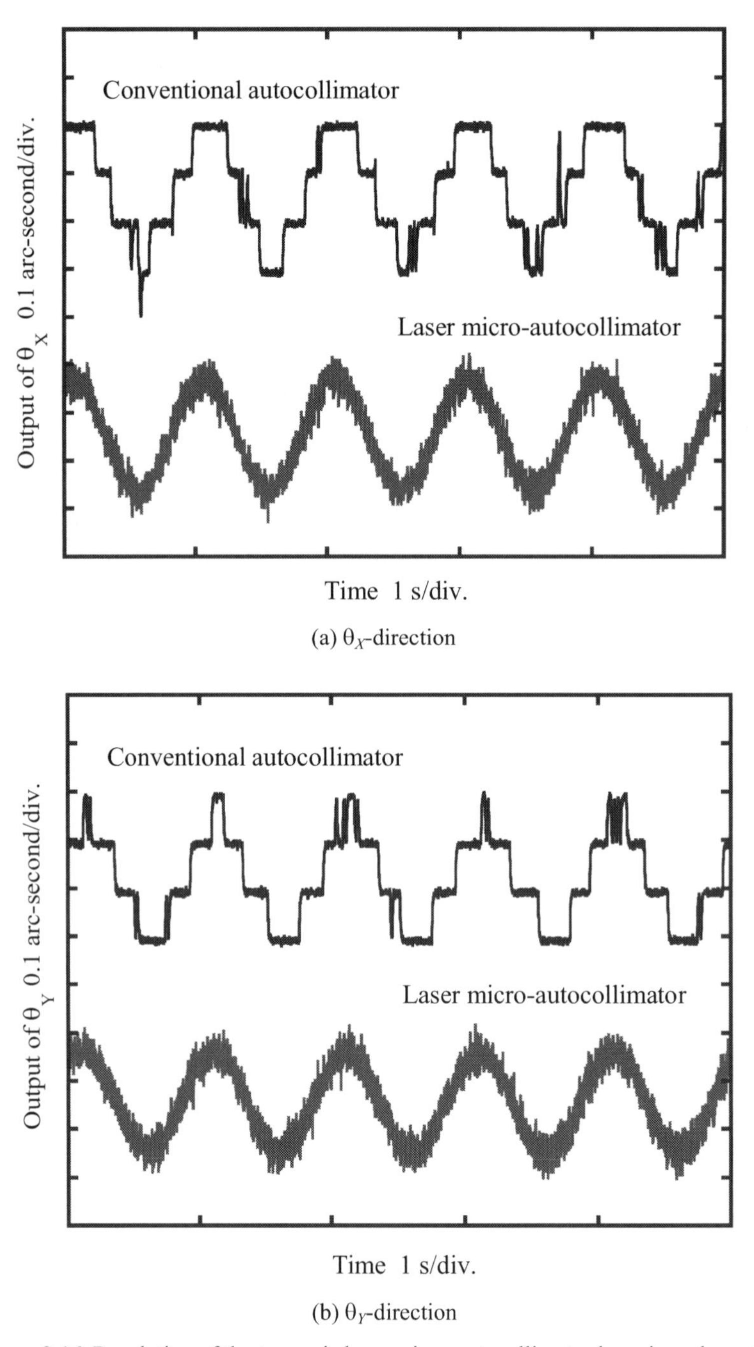

Figure 2.16. Resolution of the two-axis laser micro-autocollimator by using a laser diode with a circular beam ($D = 1$ mm)

2.3 One-lens Laser Micro-autocollimator

Figure 2.17 shows the laser autocollimators with different lens arrangements. In the laser autocollimator with two lenses shown in Figure 2.17 (a), the laser light from a laser diode is first collimated by using lens 1. The reflected beam from the target mirror is bent by the PBS and then received by lens 2. The reflected beam is focused by lens 2 to form a small light spot on the QPD placed at the focal position of lens 2. This optical layout is common for the sensors described in Section 2.2 and Chapter 1.

The optical layout is simplified by combining the two lenses to one lens as shown in Figure 2.17 (b). The first function of the lens in Figure 2.17 (b) is to collimate the laser light from the laser diode. The second function is to focus the reflected beam from the target mirror to the QPD. This configuration minimizes the number of optical components used in the laser autocollimator so that the laser autocollimator can be made more compact. In this case, the optical components can be aligned on a straight line, making the adjustment of the sensor easier [7].

On the other hand, the Gaussian beam from a laser source possesses the property of beam divergence as shown in Figure 2.18. This causes beam refraction at the boundary surfaces of the PBS and the QWP, which can change the focal position of the lens and should be taken into consideration. In Figure 2.18, θ_1 is the divergence angle of the light from the laser diode. The lens is assumed to have a flat surface and a spherical surface as shown in the figure. The term r is the curvature of the spherical surface of the lens. The parameters n_{air}, n_{PBS}, $n_{1/4}$, and n_{lens} are the diffraction indexes of air, the PBS, the QWP, and the lens, respectively. θ_2, θ_3, θ_4, and θ_5 are the refraction angles at the boundary surfaces, which can be calculated from θ_1 and the diffraction indexes based on the law of refraction. The b_2, b_3 and b_3 are the thicknesses of the PBS, the QWP and the lens, respectively.

The actual focal length f_{act} can be expressed as:

$$f_{act} = b_1 + b_2 + b_3 + b_5 , \tag{2.15}$$

where b_5 is the difference between the effective focal length (EFL) and the back focal length (BFL). The term b_1, which is the distance between the emitting point of the laser diode and the PBS, can be calculated as:

$$b_1 = \frac{1}{\tan\theta_1}\left(r\sin\theta_5 - \left(b_2\tan\theta_2 + b_3\tan\theta_3 + b_4\tan\theta_4\right)\right). \tag{2.16}$$

The diameter of the collimated laser beam can be written as:

$$D = 2(b_1\tan\theta_1 + b_2\tan\theta_2 + b_3\tan\theta_3 + b_4\tan\theta_4) . \tag{2.17}$$

Figure 2.19 shows a photograph of the laser autocollimator by using one lens with a nominal focal length of 8 mm. The actual focal length and the beam diameter were calculated to be 9.6 and 2.7 mm, respectively.

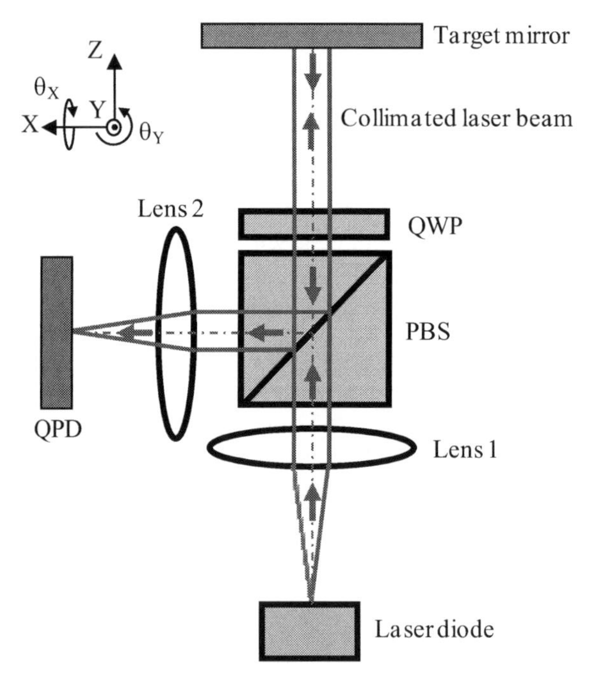

(a) The laser autocollimator with two lenses

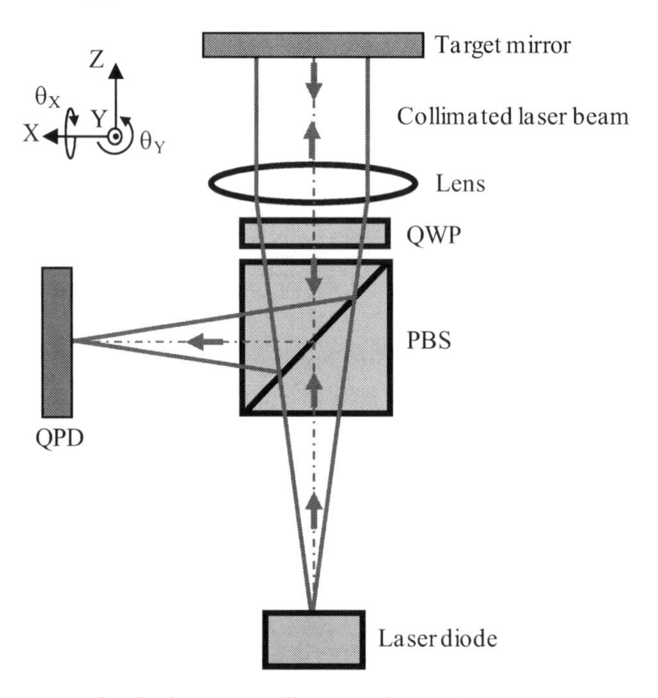

(b) The laser autocollimators with one lens

Figure 2.17. The laser autocollimators with different number of lenses

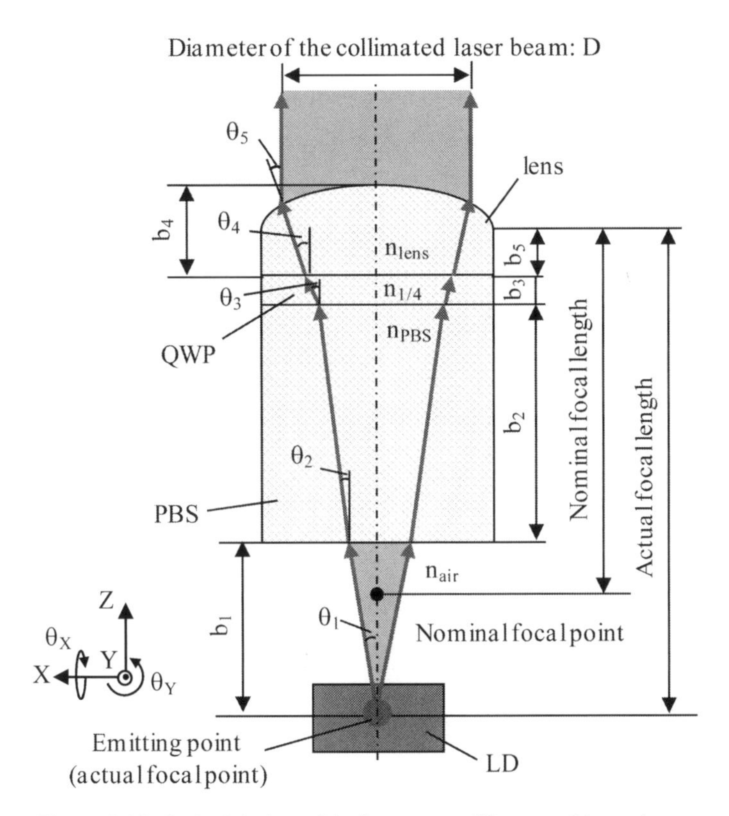

Figure 2.18. Optical design of the laser autocollimator with one lens

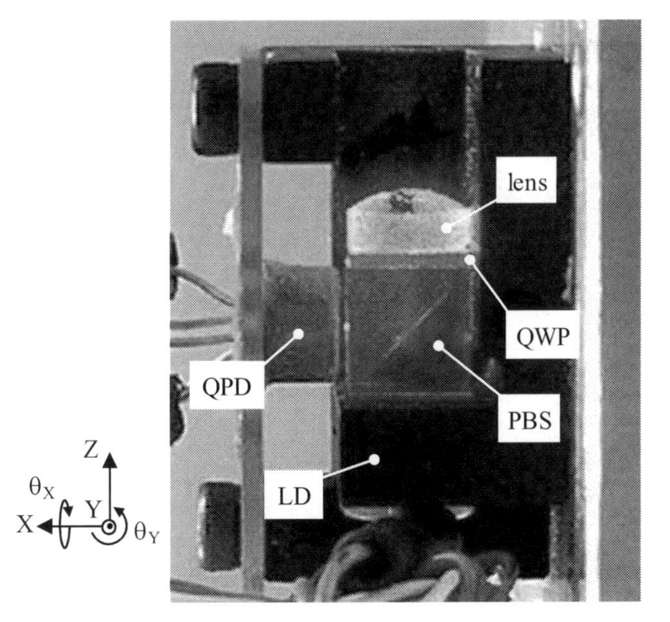

Figure 2.19. The laser micro-autocollimator with one lens (size: 15 mm × 22 mm × 14 mm)

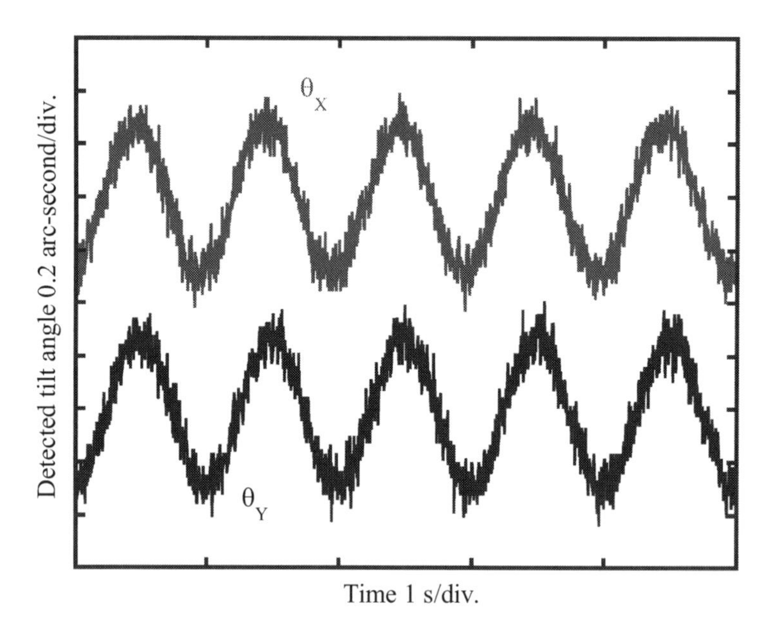

Figure 2.20. Resolution of the laser autocollimator with one lens

Figure 2.20 shows the results of the resolution test. In the test, a sinusoidal signal with a frequency of 0.1 Hz and an amplitude of 0.3 arc-second was applied to the target mirror by using a PZT tilt stage. It can be seen that the sensor has the ability to detect small tilt motions with a resolution on the order of 0.1 arc-second.

2.4 Three-axis Laser Autocollimator

In this section, the laser autocollimation method is improved from two-axis measurement to three-axis measurement. A diffraction grating is employed as the target reflector instead of the plane mirror. Figure 2.21 shows a schematic view of three-axis angle detection [8]. A rectangular grating is chosen as the target reflector to generate the zeroth-order and the positive and negative first-order diffracted beams. The three diffracted beams are received by a lens to form three focused light spots on the focal planes of the lens, respectively. The two-directional displacements of each light spot along the *V*- and *W*-axes are detected by a QPD located at the focal plane of the lens.

The diffraction angle ψ_1 of the positive and negative first-order diffracted beams can be expressed by using the pitch g of the diffraction grating and the wavelength λ of the laser beam as follows:

$$\psi_1 \approx \frac{\lambda}{g}. \tag{2.18}$$

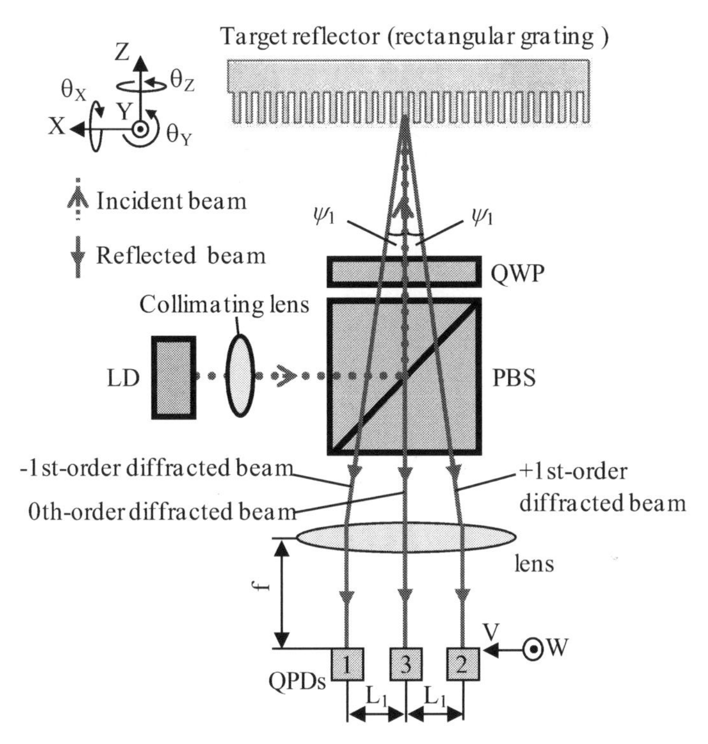

Figure 2.21. Detection of three-axis tilt angles

Assume that distances between the center positions of the focused positive and negative first-order diffracted light spots and that of the zeroth-order diffraction light spot on the QPDs are L_1 when the three-axis angle components of the reflector (θ_X, θ_Y and θ_Z) are zero. L_1 can be calculated as follows:

$$L_1 \approx \frac{f\lambda}{g}. \tag{2.19}$$

Figure 2.22 shows behaviors of the light spots on the QPDs. When the target has a tilt angle θ_Z, the two positive/negative first-order diffracted light spots will rotate about the zeroth-order diffracted light spot while keeping the same distance L_1 as shown in Figure 2.22 (a). Assuming that θ_Z is small, θ_Z can be detected from the displacement $\Delta w_{\theta Z}$ of the light spots on QPD1 and QPD2 along the W-axis as:

$$\theta_Z \approx \frac{W_{-1st\_out}}{L_1} = \frac{\Delta w_{\theta Z}}{L_1}. \tag{2.20}$$

When the target has tilt angles θ_X or θ_Y, the three light spots move simultaneously along the W-direction or V-direction as shown in Figures 2.22 (b) and 2.22 (c).

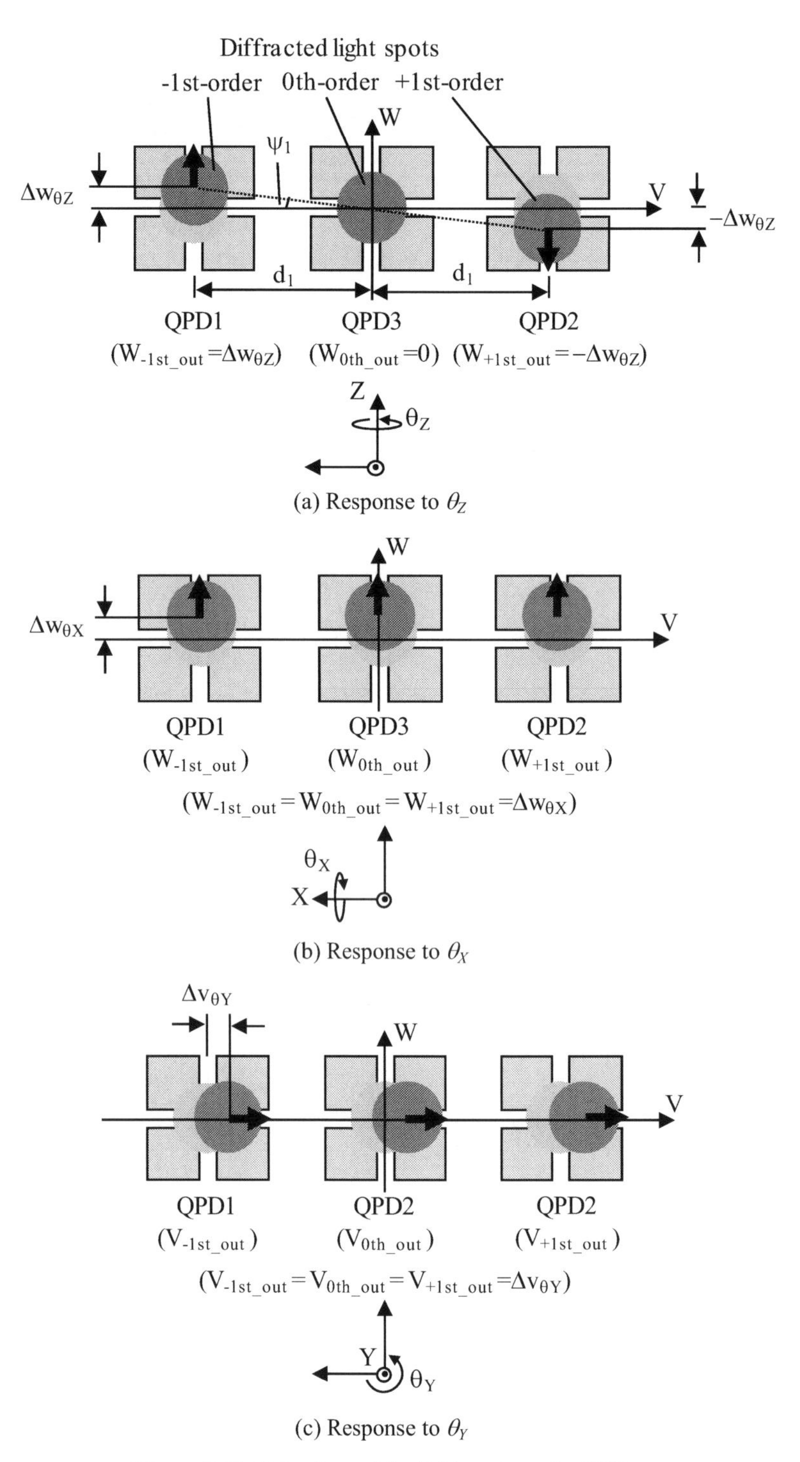

Figure 2.22. Behaviors of the light spots on the QPD

Similar to that in a two-axis laser autocollimator, θ_X and θ_Y can be measured by detecting the W-directional displacement $\Delta w_{\theta X}$ and V-directional displacement $\Delta v_{\theta Y}$ of the zeroth-order diffraction light spot by QPD1 as follows, respectively:

$$\theta_X \approx \frac{w_{0th\_out}}{2f} = \frac{\Delta w_{\theta X}}{2f}, \tag{2.21}$$

$$\theta_Y \approx \frac{v_{0th\_out}}{2f} = \frac{\Delta v_{\theta Y}}{2f}. \tag{2.22}$$

Figure 2.23 shows displacements of the light spots when the target has three-axis tilt angle components θ_X, θ_Y and θ_Z simultaneously. In the measurement of θ_Z, the influence of the displacement $\Delta w_{\theta X}$ caused by θ_X can be removed by using the displacements of the light spots detected by QPD1 and the QPD2 as follows:

$$\theta_Z = \frac{W_{-1st\_out} - W_{+1st\_out}}{2L_1},$$

$$= \frac{(\Delta w_{\theta X} + \Delta w_{\theta Z}) - (\Delta w_{\theta X} - \Delta w_{\theta Z})}{2L_1} = \frac{\Delta w_{\theta Z}}{L_1}. \tag{2.23}$$

Similarly, θ_Z can also be detected by using the displacements of negative first-order diffracted light spot detected by QPD1 (or the positive first-order diffracted light spot detected by QPD2) and the zeroth-order diffracted light spot detected by QPD3 as follows:

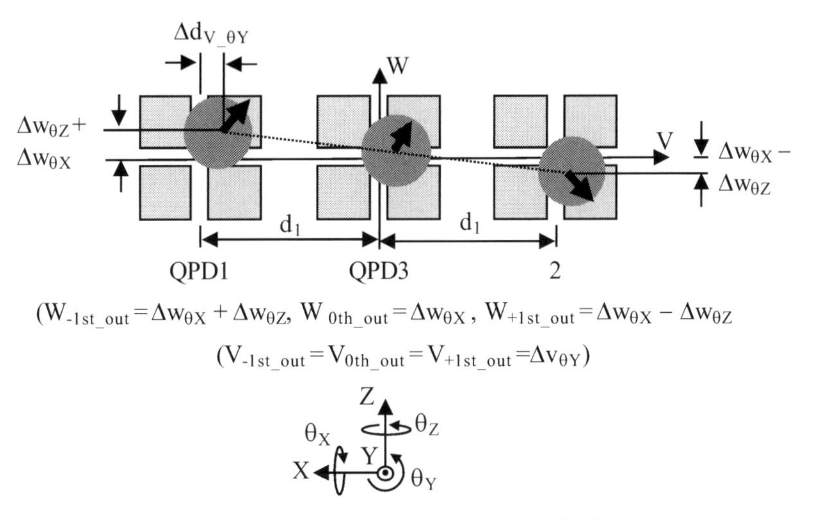

Figure 2.23. Response to the combination of θ_X, θ_Y and θ_Z

$$\theta_Z \approx \frac{W_{\text{-1st\_out}} - W_{0th\_\text{out}}}{L_1},$$

$$= \frac{(\Delta w_{\theta X} + \Delta w_{\theta Z}) - \Delta w_{\theta X}}{L_1} = \frac{\Delta w_{\theta Z}}{L_1}. \tag{2.24}$$

On the other hand, similar to those shown in Equations 2.21 and 2.22, θ_X and θ_Y can be measured from the displacements detected by QPD3. θ_X and θ_Y can also be detected by using the displacements detected by QPD1 and QPD2 as follows:

$$\theta_X \approx \frac{W_{\text{-1st\_out}} + W_{\text{+1st\_out}}}{4f},$$

$$= \frac{(\Delta w_{\theta X} + \Delta w_{\theta Z}) + (\Delta w_{\theta X} - \Delta w_{\theta Z})}{4f} = \frac{\Delta w_{\theta X}}{2f}, \tag{2.25}$$

$$\theta_Y \approx \frac{V_{\text{-1st\_out}} + V_{\text{+1st\_out}}}{4f},$$

$$= \frac{(\Delta v_{\theta Y} + \Delta v_{\theta Z}) + (\Delta v_{\theta Y} - \Delta v_{\theta Z})}{4f} = \frac{\Delta v_{\theta Y}}{2f}. \tag{2.26}$$

Consequently, the three-axis angle components can be detected simultaneously by using QPD1, QPD2 and QPD3. The methods for three-axis angle calculation can be categorized into three types (Table 2.1). Method 1 uses three diffracted light spots, which are the zeroth-order and the positive/negative first-order diffracted light spots, as expressed in Equations 2.21, 2.22 and 2.23. Method 2 uses two diffraction light spots, which are the positive/negative first-order diffraction light spots, from Equations 2.23, 2.25 and 2.26. Method 3 combines the zeroth-order diffraction light spot with the positive first-order diffraction light spot or the negative first-order diffraction light spot, from Equations 2.21, 2.22 and 2.24. Compared with methods 1 and 3, method 2 employs a balanced structure with less diffraction light spots.

The model of QPD shown in Figure 2.1 is employed to evaluate the sensitivity for detection of θ_Z. Similar to Equation 2.1, the diameter of the laser spot on QPD is expressed by:

$$d = \frac{2.44 f\lambda}{D}, \tag{2.27}$$

where D is the diameter of the collimated laser beam projected onto the target reflector.

Table 2.1. Methods for calculation of three-axis tilt angles

	Method 1	Method 2	Method 3
θ_X	$(W_{+1st\_out}-W_{-1st\_out})/2$	$(W_{+1st\_out}-W_{-1st\_out})/2$	$W_{+1st\_out}-W_{0th\_out}$
θ_Y	W_{0th_out}	$(W_{+1st\_out}+W_{-1st\_out})/2$	W_{0th_out}
θ_Z	V_{0th_out}	$(V_{+1st\_out}+V_{-1st\_out})/2$	V_{0th_out}

Assume that the relative output of QPD1 along the W-direction is $w_{-1st\_QPD\_out}$. Referring to Equations 2.3–2.7, $w_{-1st\_QPD\_out}$ can be written as:

$$w_{-1st\_QPD\_out} = \frac{(I_1+I_2)-(I_3+I_4)}{(I_1+I_2+I_3+I_4)} \times 100\%$$

$$= \frac{8\Delta w_{\theta Z}}{\pi d} \times 100\% . \tag{2.28}$$

Denoting the sensitivity of the angle sensor for detection of θ_Z to be $S_{\theta Z}$, $S_{\theta Z}$ can be expressed by:

$$S_{\theta Z} = \frac{w_{-1st\_QPD\_out}}{\theta_Z} . \tag{2.29}$$

Substituting Equations 2.19, 2.20, 2.27 and 2.28 into Equation 2.29 gives:

$$S_{\theta Z} = \frac{4}{1.22\pi}\frac{D}{g} . \tag{2.30}$$

It can be seen that the sensitivity for θ_Z detection is proportional to the diameter D of the laser beam and inversely proportional to the pitch g of the grating. It is not related to the focal length f. It should be noted that the sensitivity is the same for the three different methods of detecting θ_Z shown in Table 2.1.

The sensitivity $S_{\theta X}$ for detection of θ_X and the sensitivity $S_{\theta Y}$ for detection of θ_Y, which are the same as those of a two-axis laser autocollimator shown in Equations 2.10 and 2.12, can be obtained as follows, respectively:

$$S_{\theta X} = \frac{w_{0th\_QPD\_out}}{\theta_X} = \frac{8}{1.22\pi}\frac{D}{\lambda} , \tag{2.31}$$

$$S_{\theta Y} = \frac{v_{0th\_QPD\_out}}{\theta_Y} = \frac{8}{1.22\pi}\frac{D}{\lambda}, \tag{2.32}$$

where $w_{0th\_QPD\_out}$ and $v_{0th\_QPD\_out}$ are the relative outputs of QPD2 along the *W*- and *V*-directions, respectively. It can be seen that the sensitivities $S_{\theta X}$ and $S_{\theta Y}$ are determined by the wavelength λ and the diameter *D* of the collimated laser beam but not related to the focal length *f*. It should be noted that the sensitivities are the same for the three different methods of detecting θ_X and θ_Y shown in Table 2.1.

Figure 2.24 shows a prototype three-axis laser autocollimator designed and fabricated based on Method 2 shown in Table 2.1. The wavelength of the laser diode (LD) is 0.683 μm. The sensor has a similar structure with the two-axis laser autocollimator except for the use of a rectangular grating with 5.5 μm pitch as the target reflector. The beam from the LD is collimated by a collimating lens and shaped by an aperture with a diameter of 2 mm. The collimated laser beam is then projected onto the target grating. The lens shown in Figure 2.21 is replaced by two lenses (lens 1 and lens 2) in order to receive the reflected positive and negative first-order diffracted beams, so that the influence of lens aberration can be reduced. The two lenses have identical focal length and diameter, which are 25.4 and 15.0 mm, respectively. The dimension of the three-axis angle sensor is 59.0 mm × 60.5 mm × 59.0 mm. The cut-off frequency of the sensor electronics is set to be 3 kHz.

Figure 2.25 shows the performance of the three-axis laser autocollimator for detection of θ_Z. Figure 2.25 (a) shows the curve of the applied θ_Z versus the output $w_{\theta Z\_QPD\_out}$ of the laser autocollimator. $w_{\theta Z\_QPD\_out}$ is defined by:

$$w_{\theta Z\_QPD\_out} = \frac{w_{-1st\_QPD\_out} - w_{+1st\_QPD\_out}}{2}, \tag{2.33}$$

where $w_{-1st\_QPD\_out}$ and $w_{+1st\_QPD\_out}$ are the *W*-directional relative outputs of QPD1 and QPD2, respectively. As can be seen in Figure 2.25 (a), the mean sensitivity $S_{\theta Z}$ of θ_Z detection was approximately 0.144%/arc-second.

The resolution test was carried out by applying a periodically changed tilt motion with an amplitude of 0.2 arc-second and a frequency of 1 Hz with a PZT tilt stage. The resolution shown in Figure 2.25 (b) was approximately 0.2 arc-second. As described above, the sensitivity and resolution of θ_Z detection are mainly determined by the diameter of the collimated laser beam and the pitch of the grating and can be improved by shortening the grating pitch.

Figures 2.26 and 2.27 show the performances of the three-axis laser autocollimator for detection of θ_X and θ_Y. The curve of applied θ_X versus the output $w_{\theta X\_QPD\_out}$ of the laser autocollimator is shown in Figure 2.26 (a). The curve of applied θ_Y versus the output $w_{\theta Y\_QPD\_out}$ is shown in Figure 2.26 (b), respectively. $w_{\theta X\_QPD\_out}$ and $w_{\theta Y\_QPD\_out}$ are defined as follows:

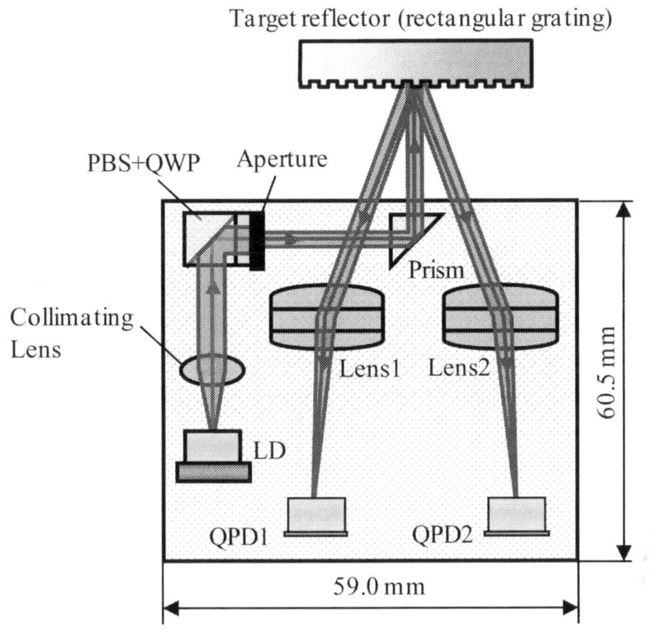

(a) Optical arrangement

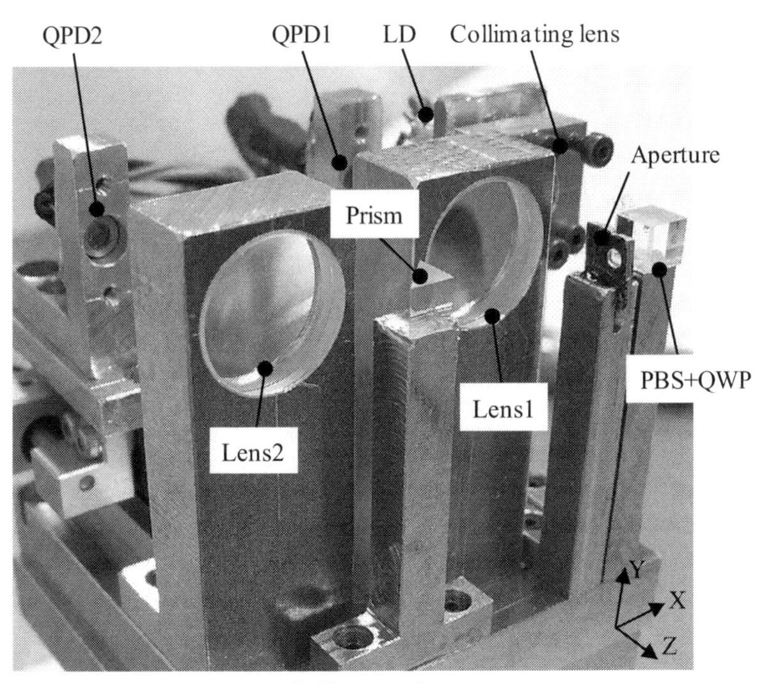

(b) Photograph

Figure 2.24. The three-axis laser autocollimator

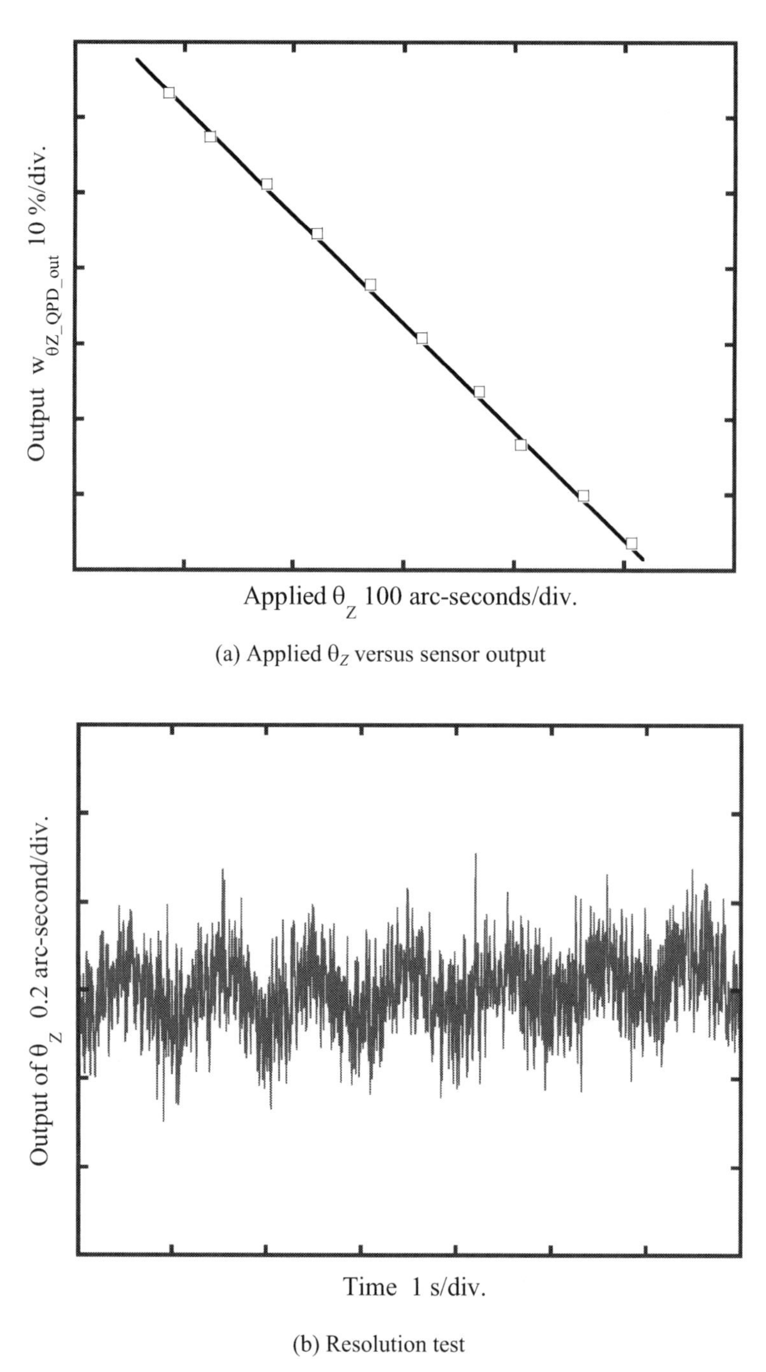

(a) Applied θ_Z versus sensor output

(b) Resolution test

Figure 2.25. Performance of the three-axis laser autocollimator for detection of θ_Z

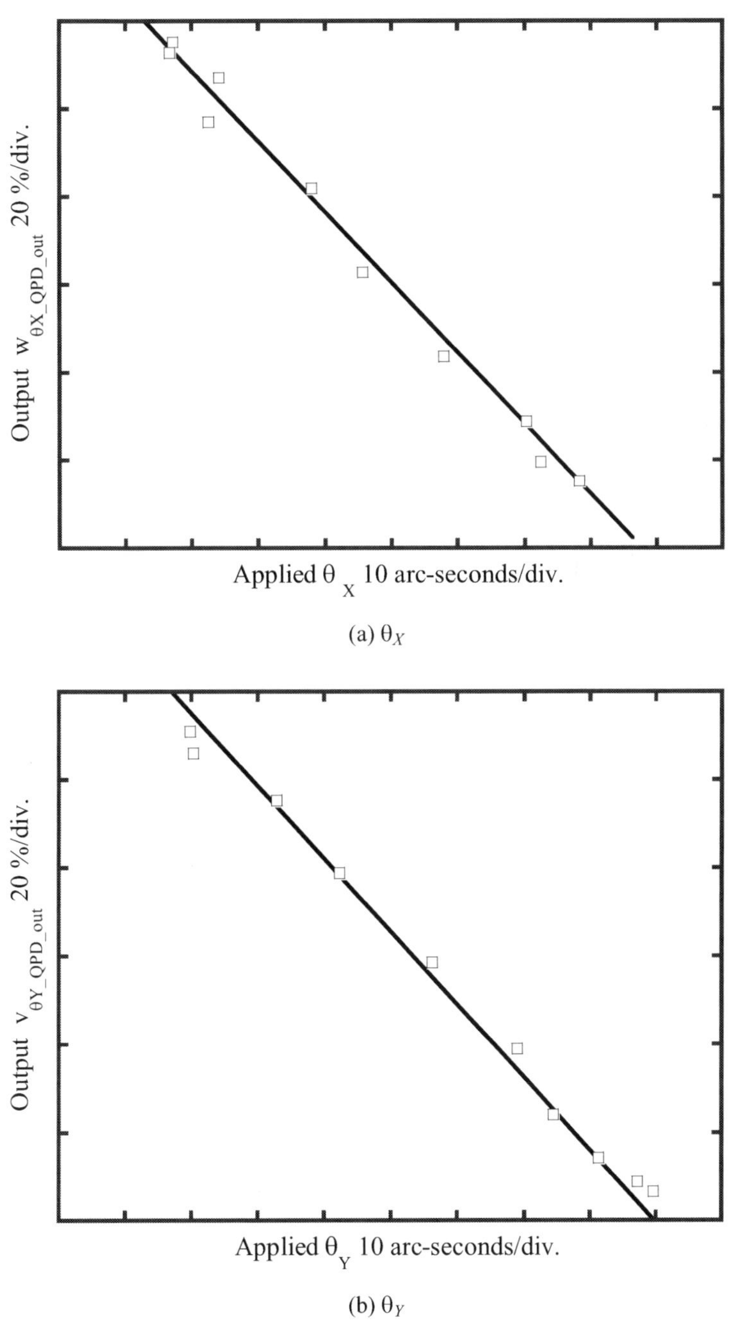

Figure 2.26. Applied θ_X and θ_Y versus sensor outputs

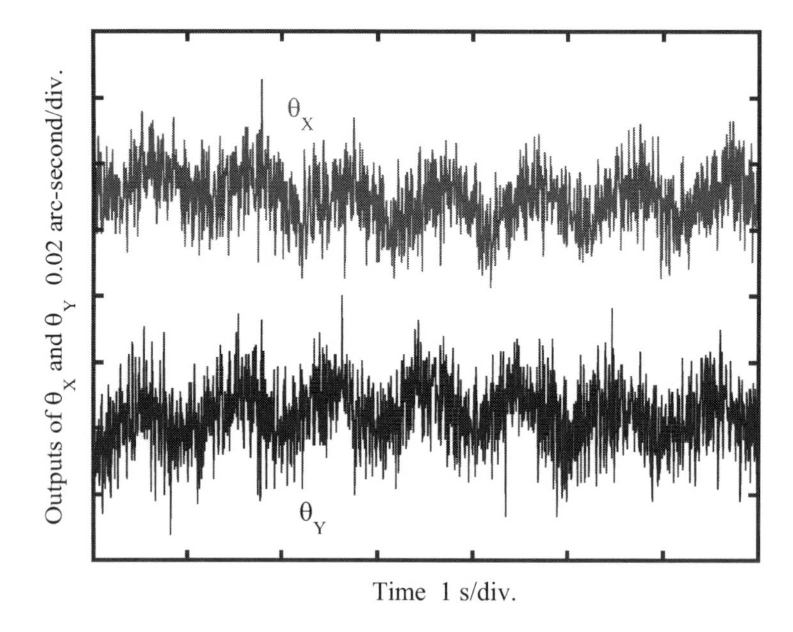

Figure 2.27. Resolution test for detection of θ_X and θ_Y

$$w_{\theta X\_QPD\_out} = \frac{w_{-1st\_QPD\_out} + w_{+1st\_QPD\_out}}{2}, \tag{2.34}$$

$$v_{\theta Y\_QPD\_out} = \frac{v_{-1st\_QPD\_out} + v_{+1st\_QPD\_out}}{2}, \tag{2.35}$$

where $v_{-1st\_QPD\_out}$ and $v_{+1st\_QPD\_out}$ are the V-directional relative outputs of QPD1 and QPD2, respectively. As can be seen from the figure, the mean sensitivity $S_{\theta X}$ of θ_X detection was approximately 1.604%/arc-second, and the mean sensitivity $S_{\theta Y}$ of θ_Y detection was approximately 1.650%/arc-second. $S_{\theta X}$ and $S_{\theta Y}$ are mainly determined by the diameter of the collimated laser beam and the wavelength of the LD. It can be found that sensitivities $S_{\theta X}$ and $S_{\theta Y}$ were ten times higher than $S_{\theta Z}$. This is related to the ratio of the laser wavelength (λ = 0.683 μm) and the grating pitch (g = 5.5 μm). For matching the sensor sensitivities in the three-axes, it is necessary to reduce the grating pitch by a factor of ten.

Figure 2.27 shows the results for testing the resolution of the three-axis laser autocollimator for detection of θ_X and θ_Y. A periodically changed tilt motion with an amplitude of 0.02 arc-second and a frequency of 1 Hz was applied to the sensor as θ_X and θ_Y, respectively. The results of a resolution test shown in the figure indicate that the three-axis laser autocollimator has the ability to detect θ_X and θ_Y with a resolution of 0.02 arc-second, which corresponds to that of a two-axis laser autocollimator shown in Sections 2.2 and 2.3.

2.5 Summary

Laser autocollimators for detection of two-axis and three-axis tilt motions have been described. In the two-axis laser autocollimator, a laser diode (LD) is employed as the light source of the sensor. The laser light is first collimated by a collimating lens before the beam is projected onto a target flat mirror along the Z-axis. A quadrant photodiode placed at the focal plane of a lens is employed for detection of the two-axis displacements of the laser light spot focused on the QPD, which are used for calculation of the tilt angles θ_X and θ_Y about the X- and Y-axes, respectively. The relationship between the input tilt angle (θ_X or θ_Y) and the QPD relative output has been analyzed based on the principle of autocollimation and diffraction of focusing a laser beam. It has been revealed that the sensitivity of the laser autocollimator for detection of θ_X and θ_Y is proportional to the diameter of the collimated laser beam and inversely proportional to the wavelength of the laser. Because the sensitivity is not related to the focal length of the lens, the sensor can be made in a compact size by choosing a lens with a short focal length.

The two-axis laser autocollimator has also been improved for three-axis measurement by replacing the target flat mirror with a grating reflector. In addition to θ_X or θ_Y, the tilt angle θ_Z about the Z-axis can be obtained by detecting the reflected diffracted beams. The methods using different sets of diffracted beams have been discussed. The method using first-order diffracted beams is suggested to be the best method for both sensor structure and performance. It has been determined that the sensitivity for θ_Z detection is proportional to the diameter of the collimated laser beam and inversely proportional to the grating pitch.

References

[1] Bryan JB (1979) The Abbe principle revisited. Precis Eng 1(3):129–132

[2] Koning R, Flugge J, Bosse H (2007) A method for the in situ determination of Abbe errors and their correction. Meas Sci Technol 18(2):476–481

[3] Ennos AE, Virdee MS (1982) High accuracy profile measurement of quasi-conical mirror surfaces by laser autocollimation. Precis Eng 4(1):5–8

[4] Virdee MS (1988) Nanometrology of optical flats by laser autocollimation. Surf Topography 1:415–425

[5] Gao W, Kiyono S, Satoh E (2002) Precision measurement of multi-degree-of-freedom spindle errors using two-dimensional slope sensors. Ann CIRP 51(1):447–450

[6] Jenkins FA, White HE (1976) Fundamentals of optics, chap 10. McGraw-Hill, New York

[7] Saito Y, Gao W, Kiyono S (2007) A single lens micro-angle sensor. Int J Precis Eng Manuf 8(2):14–19

[8] Saito Y, Arai Y, Gao W (2009) Detection of three-axis angles by an optical sensor. Sens Actuators A 150:175–183

3

Surface Encoder for Measurement of In-plane Motion

3.1 Introduction

Precision planar motion (*XY*) stages are widely used in machine tools, photolithography equipment and measuring instruments for nanomanufacturing [1–4]. Measurement of the in-plane motions is essential for evaluation of stage performance and/or for feedback control of the stage. In addition to the position information, tilt motions are also important measurement parameters.

Two-axis *XY*-positions of such a stage are conventionally measured by laser interferometer [5]. Tilt motions can also be measured through increasing the number of interferometers. Laser interferometers have the advantages of high resolution, large measurement range, fast measurement speed, flexible arrangement of optical paths, direct linkage to the length definition, etc. [6, 7]. However, operation in a vacuum is required for the interferometers to avoid influences of air pressure, air temperature and relative humidity on the refractive index. Since most of the two-axis stages are used in air, it is difficult to maintain the measurement accuracy of the interferometer. The high cost of a multi-axis interferometer is another restriction for the application in planar motion stages.

On the other hand, linear encoders, which are more robust to measurement environment and good in cost performance compared with laser interferometers, are suitable for use in industry [8]. The position of a two-axis stage can be measured by using two linear encoders. However, it is necessary to use additional rotary encoders and autocollimators for tilt motion measurement. This makes the multi-degree-of-freedom (MDOF) measuring system complicated. The reading head/scale assembly of a linear encoder also makes it impossible to use linear encoders in a surface motor-driven planar motion stage with a single moving part [9–11].

This chapter describes MDOF surface encoders for measurement of in-plane *XY*-positions and tilt motions. The surface encoder employs a two-axis sinusoidal grid on which periodic sine waves along the *X*- and *Y*-directions are generated. A two-axis slope sensor is used to detect the local slope profiles of the grid surface. The fabrication of the sinusoidal grid and application of the surface encoder in a surface motor-driven planar motion stage are also presented.

3.2 Surface Encoder for MDOF In-plane Motion

3.2.1 Multi-probe-type MDOF Surface Encoder

Figure 3.1 shows a schematic of the basic principle of the surface encoder for *XY* position measurement [12, 13]. The surface encoder consists of a two-axis slope sensor and a two-axis sinusoidal grid, on which two-dimensional sinusoidal waves are generated.

The height profile of the sinusoidal grid, which is a superposition of sinusoidal waves in the *X*- and *Y*-directions, can be expressed as:

$$f(x,y) = H_x \sin(\frac{2\pi}{g_x}x) + H_y \sin(\frac{2\pi}{g_y}y), \tag{3.1}$$

where H_x, H_y are the amplitudes of the sine functions in the *X*-direction and *Y*-direction, respectively. The terms g_x and g_y are the corresponding pitches.

The two-dimensional outputs $\alpha(x)$ and $\beta(y)$ of the slope sensor, which indicate the local slopes of the sinusoidal grid in the *X*-direction and *Y*-direction, can be obtained from the differentiation of $f(x, y)$. The two-axis components x and y of the position can then be determined from the sensor outputs $\alpha(x)$ and $\beta(y)$ as:

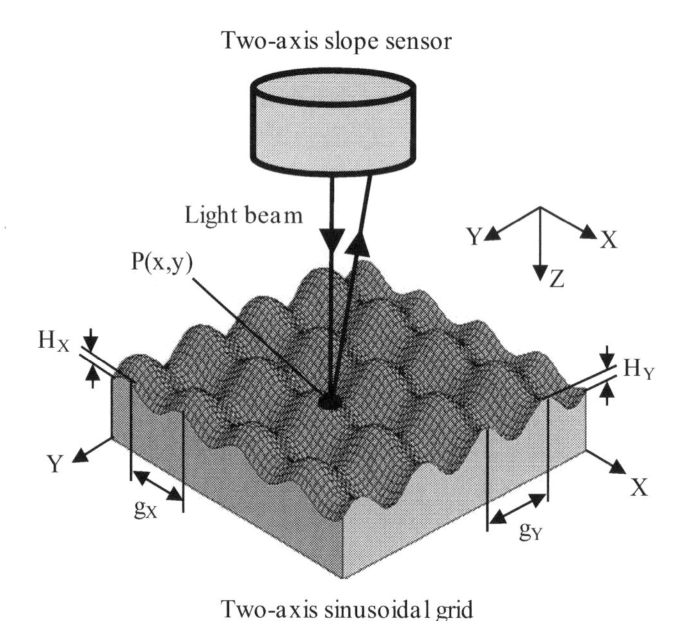

Figure 3.1. The surface encoder for *XY*-position measurement

$$x = \frac{g_x}{2\pi}\cos^{-1}\left(\frac{g_x}{2\pi H_x}\alpha(x)\right), \tag{3.2}$$

$$y = \frac{g_y}{2\pi}\cos^{-1}\left(\frac{g_y}{2\pi H_y}\beta(x)\right). \tag{3.3}$$

Figure 3.2 shows the schematic of the 3-DOF surface encoder for measurement of the *X*-, *Y*-positions and θ_Z-tilt motion. As can be seen in the figure, there are two 2D slope sensors (1 and 2) mounted on a sensor unit. The 2D local slopes of two points *A* and *B* on the sinusoidal grid surface can be simultaneously detected by the two sensors. In addition to *x* and *y*, the rotational displacement θ_Z can also be obtained through the outputs of sensors 1 and 2. Assume that both the translational and rotational displacements are small. When the sensor unit moves from position 1 to position 2, θ_Z can be approximately evaluated as

$$\theta_z \approx \frac{\sqrt{\Delta x_B^2 + \Delta y_B^2} - \sqrt{\Delta x_A^2 + \Delta y_A^2}}{L_{AB}}, \tag{3.4}$$

where $\Delta x_A(= x - x')$ and $\Delta y_A(= y - y')$ are the *X*- and *Y*-displacements of point *A* measured by sensor 1. $\Delta x_B(= x_B - x'_B)$ and $\Delta y_B(= y_B - y'_B)$ are the *X*- and *Y*-displacements of point *B* measured by sensor 2. L_{AB} is the distance between the two sensors. The position of sensor 2 is adjusted in such a way that the output of sensor 2 has a 90° phase difference with respect to sensor 1. The quadrature construction allows the surface encoder algorithm to determine the movement directions.

Figure 3.3 shows the optical configuration and a photograph of the sensor unit of a prototype $XY\theta_Z$ surface encoder. The collimated laser beam with a diameter of 6 mm and the propagation axis along the *Y*-direction is converted into a multi-spot beam after passing through an aperture plate on which a two-dimensional array of micro-square apertures are fabricated by using the lithography process. The multi-spot beam is a bundle of thin beams aligned in the *X*- and *Z*-directions with a pitch spacing of 200 μm, which is twice of the sine function pitch *g* of the sinusoidal grid. The averaging effect of the multi-spot beam is utilized to reduce the influence of the profile errors of the sinusoidal grid surface. A beam splitter (BS) splits the beam into two beams. One beam travels along the *X*-direction and the other beam along the *Y*-direction. The propagation axes of the beams are changed from *X* and *Y* to *Z* by using mirrors with reflection angles of 45° relative to the *XY*-plane so that the beams can be projected onto two different points (points *A* and *B*) of the sinusoidal grid surface aligned in the *XY*-plane. The reflected beams from the sinusoidal grid surface go back to the mirror and are bent by polarization beam splitters (PBSs) before being received by the autocollimation units for angle detection, which are described in Chapters 1 and 2.

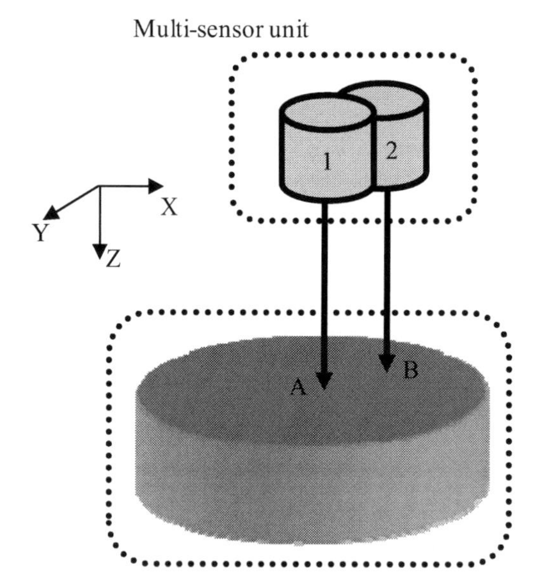

(a) Schematic of the $XY\theta_Z$ surface encoder

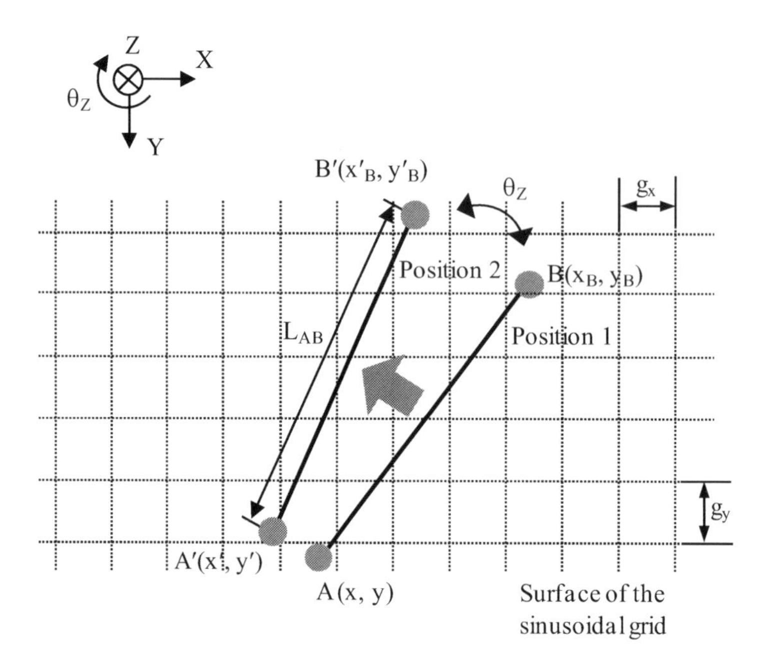

(b) Change of measurement points caused by the θ_Z-motion

Figure 3.2. The multi-probe type 3-DOF surface encoder for measurement of the *X*-, *Y*-displacements and θ_Z-tilt motion

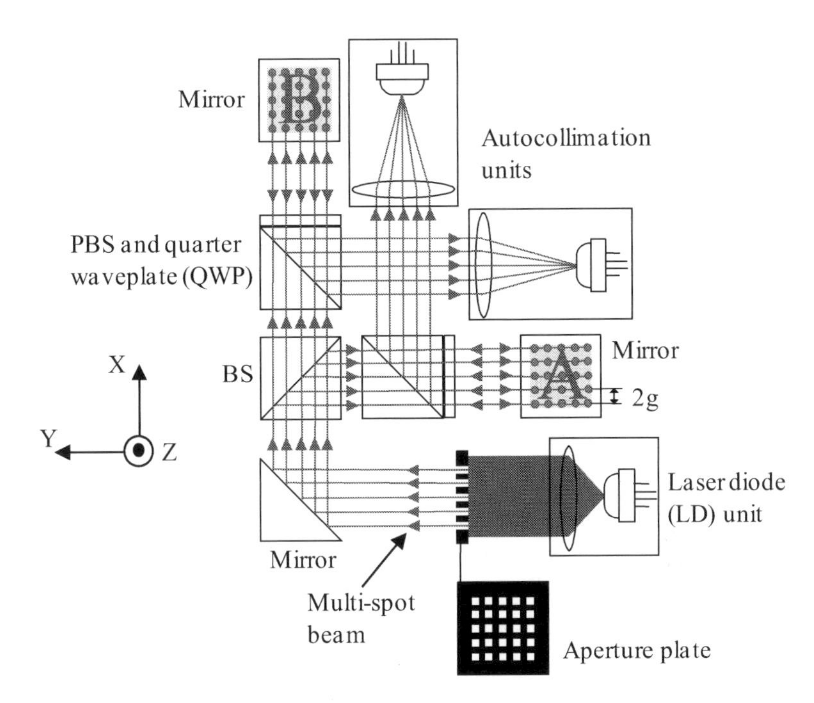

(a) Optical layout

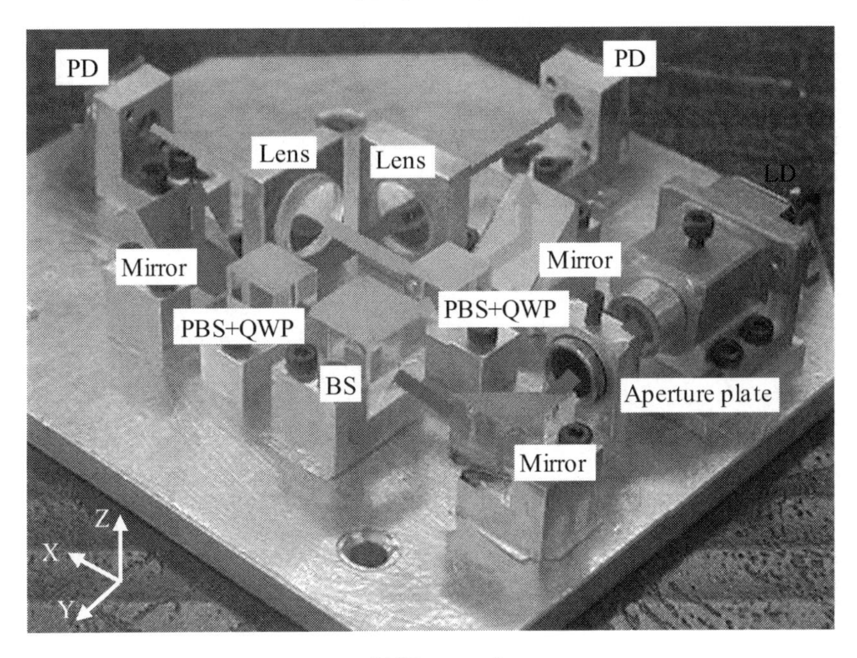

(b) Photograph

Figure 3.3. The sensor unit of a prototype 3-DOF surface encoder

The autocollimation unit consists of a lens and a light position detector (photodetector) placed at the focal plane of the lens. Quadrant photodiodes are used as the photodetectors in the sensor for 2D detection. The autocollimation allows the angle-detection independent from the distance between the sensor unit and the sinusoidal grid surface as described in Chapter 1. The focal length of the lens is 30 mm and the distance between points *A* and *B* is set to 56.6 mm. The sensor unit has a dimension of 90 mm (*L*) × 90 mm (*W*) × 27 mm (*H*). The sensors have a bandwidth of 4.8 kHz.

The basic performance of the surface encoder was investigated in a setup shown in Figure 3.4. The sensor unit and sinusoidal grid were mounted on commercially available air-slide and air-spindle, respectively. The position of the air-slide was used as the reference for *X*- and *Y*-displacements and the air-spindle was used as the reference for θ_Z-displacement. The positioning resolutions of the air-slide and air-spindle were 5 nm and 0.04 arcsec, respectively.

Figure 3.5 shows the result from evaluating the graduation spacing deviation of the surface encoder, which are designed to be 100 μm. The air-slide was moved by a step of 100 μm along the *X*-direction over a range of 40 mm. As can be seen in the figure, the deviation, which is the difference between the surface encoder output and the air-slide position, was approximately 0.2 μm. It should be noted that the positioning error of the air-slide is included in the deviation.

Figure 3.6 shows the sensor outputs when the air-slide moved along the *X*-direction with a step of 1 μm between two graduations of the sinusoidal grid. It can be seen that both sensors output sinusoidal signals with a phase difference of 90°. The two outputs were used to generate the interpolation data shown in Figure 3.7 based on the technique of quadrature logic, which provides the position information between the 100 μm sinusoidal grid graduations. The non-linearity of the interpolation curve, which indicates the interpolation error, was approximately 2.5 μm. The interpolation error can be reduced to the level of the repeatability of the interpolation errors, which was approximately 0.2 μm by a calibration and compensation process. In the case of employing quadrature logic, however, the measurement range of θ_Z was limited to several arcseconds because a larger θ_Z will significantly change the phase difference between the two sensor outputs and reduce the interpolation accuracy.

Figure 3.8 (a) shows the result from testing the resolution of surface encoder in the *X*-direction. The spindle was kept stationary and the air-slide was moved with a step of 20 nm along the *X*-direction. As can be seen in the figure, the surface encoder had a resolution of better than 20 nm. The output of the *Y*-output of the surface encoder is also shown in the figure. The *Y*-output did not vary with the *X*-steps, showing that the surface encoder can detect the *X*- and *Y*-displacement independently. The resolution in the *Y*-direction was tested by rotating the sensor unit by 90° about the *Z*-axis. The result is shown in Figure 3.8 (b). Similar to that in the *X*-direction, the resolution of the surface encoder in the *Y*-direction was also better than 20 nm. Figure 3.9 shows the result from testing the resolution in the θ_Z-direction. The air-slide was kept stationary and the air-spindle was moved with a step of 0.2 arcsec. It can be seen that the 0.2 arcsec steps were successfully detected by the surface encoder.

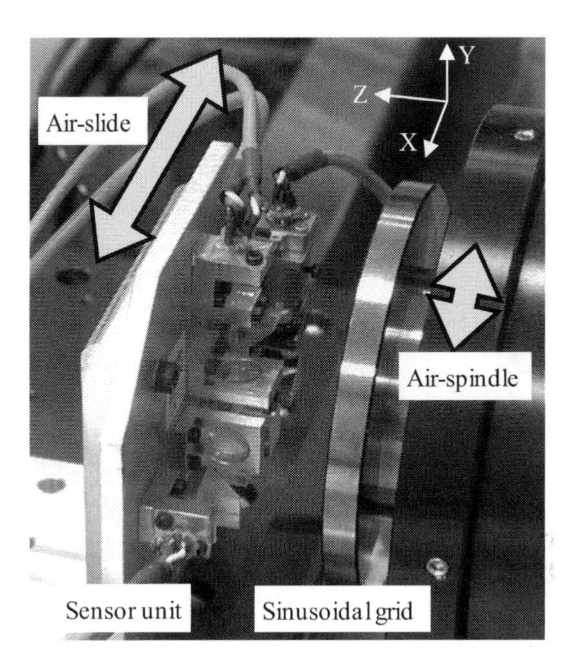

Figure 3.4. Setup for investigating the surface encoder performance

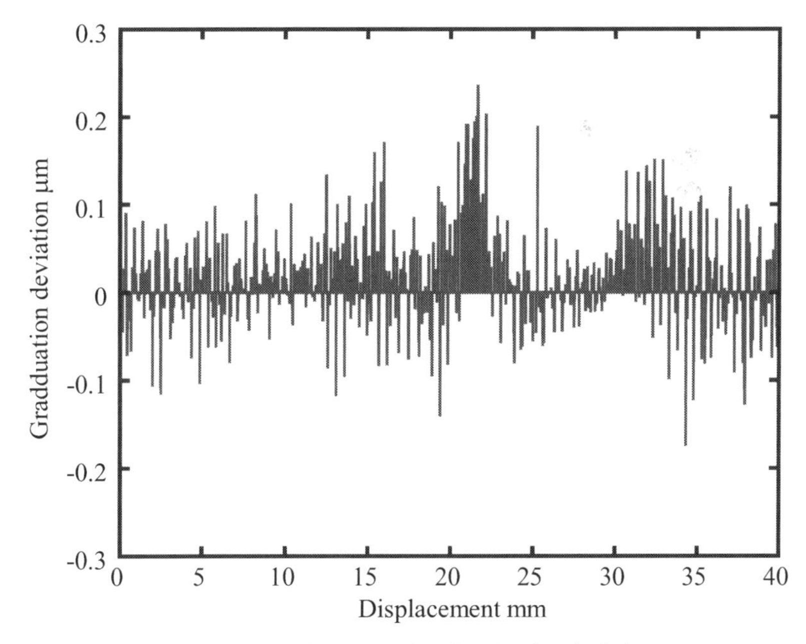

Figure 3.5. Test results of graduation deviation

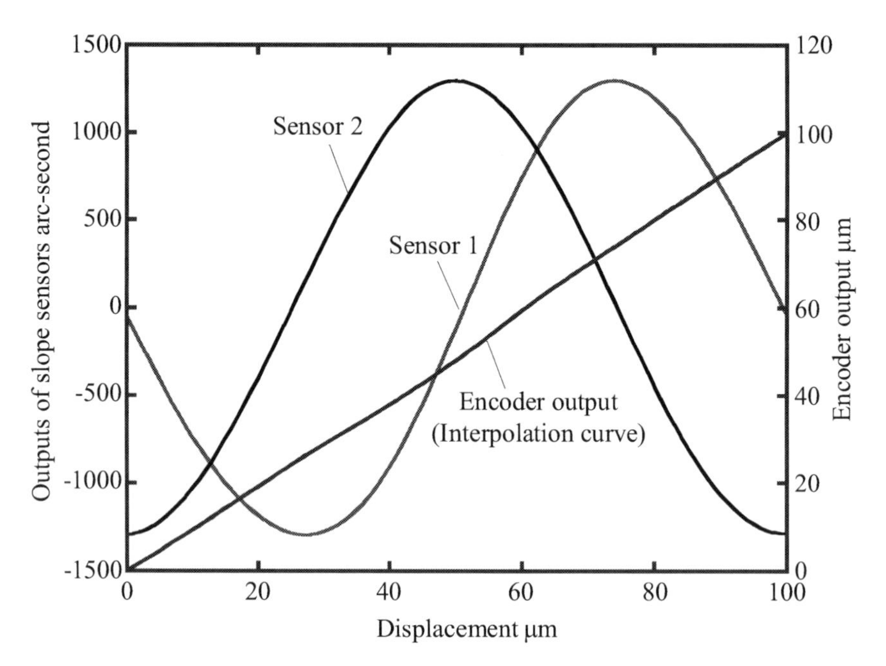

Figure 3.6. Outputs of the slope sensors and the surface encoder

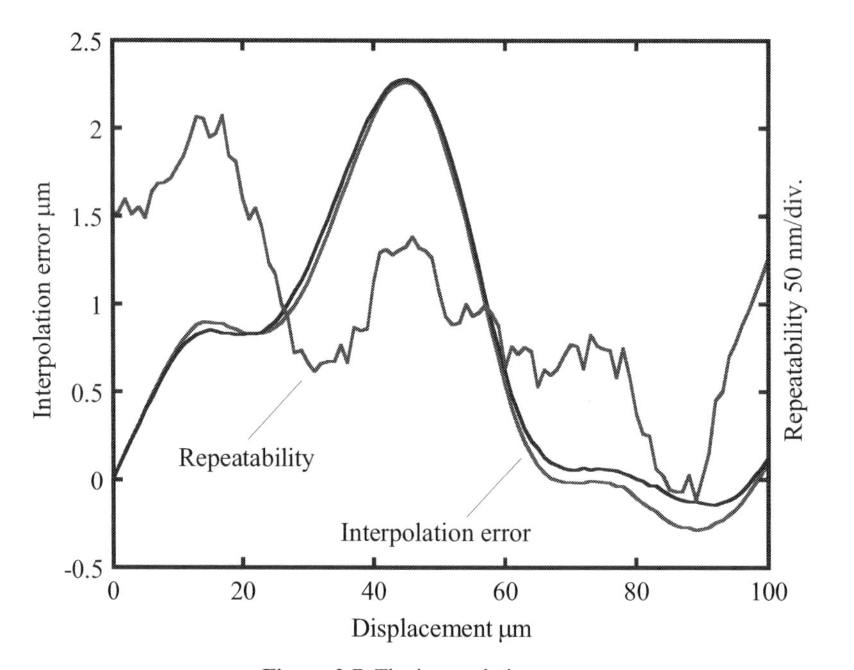

Figure 3.7. The interpolation error

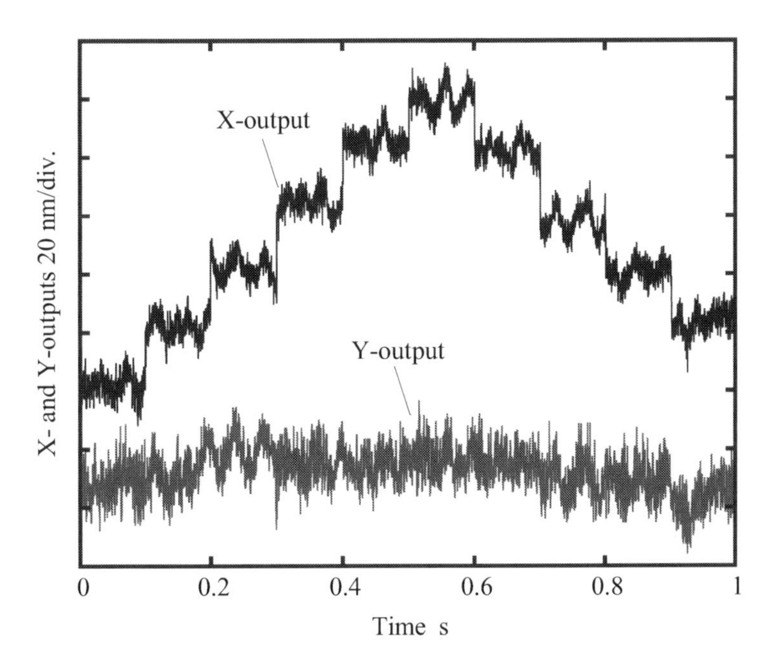

(a) Result of testing the encoder *X*-directional resolution

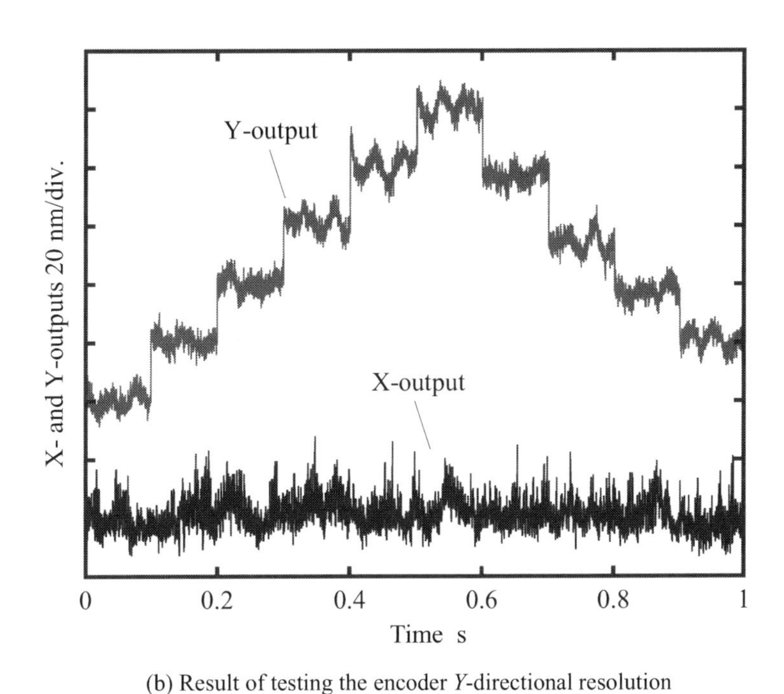

(b) Result of testing the encoder *Y*-directional resolution

Figure 3.8. Testing the encoder resolution in *X*- and *Y*-directions

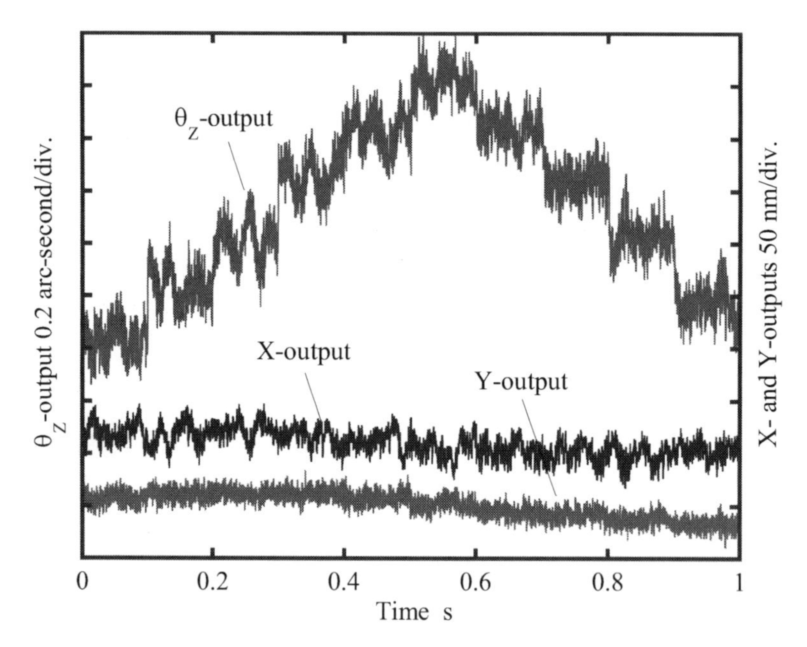

Figure 3.9. Testing the encoder resolution in θ_Z-direction

The multi-probe-type surface encoder can be applied to 5-DOF measurements by adding more sensors, as shown in Figure 3.10 [14]. In the figure, the slope sensors 1 and 2 are the same as those used in the 3-DOF surface encoder. The *X*- and *Y*-displacements and the θ_Z-tilt motion can be obtained from the sensor outputs (m_{1x} and m_{1y}) and (m_{2x} and m_{2y}). The tilt motions θ_X and θ_Y are evaluated by using the output of slope sensor 3, which has a beam diameter larger than the pitch of the sinusoidal grid. Sensor 3 detects θ_X and θ_Y rather than the *XY*-displacement. Therefore, the outputs of sensor 3 are equal to the angular displacement:

$$\theta_Y = m_{3x}, \tag{3.5}$$

$$\theta_X = m_{3y}. \tag{3.6}$$

Thus, the surface encoder can detect the 5-DOF motion of the moving element: two translational motions (x, y) and three rotational motions (θ_X, θ_Y, θ_Z).

Figure 3.11 shows a prototype 5-DOF surface encoder composed of the slope sensor unit and the sinusoidal grid. The grid has a diameter of 150 mm and a thickness of 10 mm. The pitch and amplitude of the pattern are g = 100 μm and H = 0.1 μm. The sensor unit contains the three-slope sensors shown in Figure 3.10 to detect the local slopes of the sinusoidal grid surface. A laser diode is used as the light source, and the emitted laser beam is collimated to a diameter of 7 mm. The laser beam is split into three beams of equal intensity by using a 33% prism (prism 1), a 50% beam splitter (prism 2) and a triangular prism (prism 3). The three laser

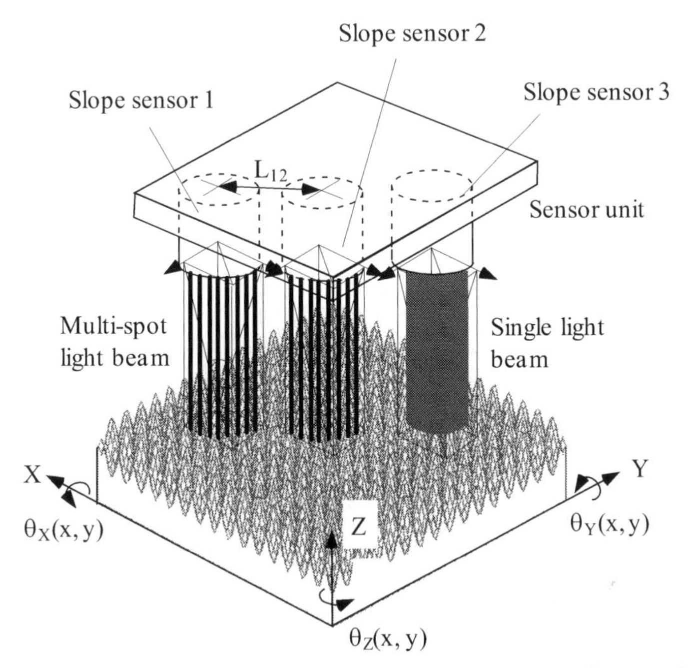

Figure 3.10. Principle of the multi-probe type 5-DOF surface encoder

beams are used for sensors 1, 2 and 3, respectively. For sensors 1 and 2, aperture plates are inserted in the laser path to create bundles of thin laser beams with pitch spacing of 100 μm, equal to the pitch of the sinusoidal grid. The distance L between the detection spots of sensors 1 and 2 on the surface of the sinusoidal grid is set to be 21 mm. The laser beam of sensor 3 is larger in diameter than the pitch of the sinusoidal grid, and is fixed at an arbitrary position to detect the overall inclination of the sinusoidal grid. The three lasers are reflected from the surface of the sinusoidal grid and directed toward three autocollimation units by polarized beam splitters (PBSs). All the optical parts are mounted on a stainless steel (SUS304) sensor base, and covered to avoid disturbance by air movement. The slope sensor unit is designed and assembled into a small size (66 mm in length, 110 mm in width, and 60 mm in height).

Figure 3.12 shows the results of testing the resolution of the surface encoder in the θ_X- and θ_Y-directions. It can be seen that the surface encoder can detect 0.01 arc-second tilt motions in these two directions. Figure 3.13 shows the results of detecting the θ_Y- and θ_Z-motions by the surface encoder. Tilt motions were applied independently about each axis. As shown in Figure 3.13 (a), the output of θ_Y-motion varied in proportion to the θ_Y-input, while the other outputs remained relatively constant. Similar results were obtained for the θ_Z-motion shown in Figure 3.13 (b) and for the X-motion shown in Figure 3.14. The feasibility of measurement of Y- and θ_X-motions was also confirmed through other tests. This demonstration confirms that the surface encoder can detect the 5-DOF motions simultaneously.

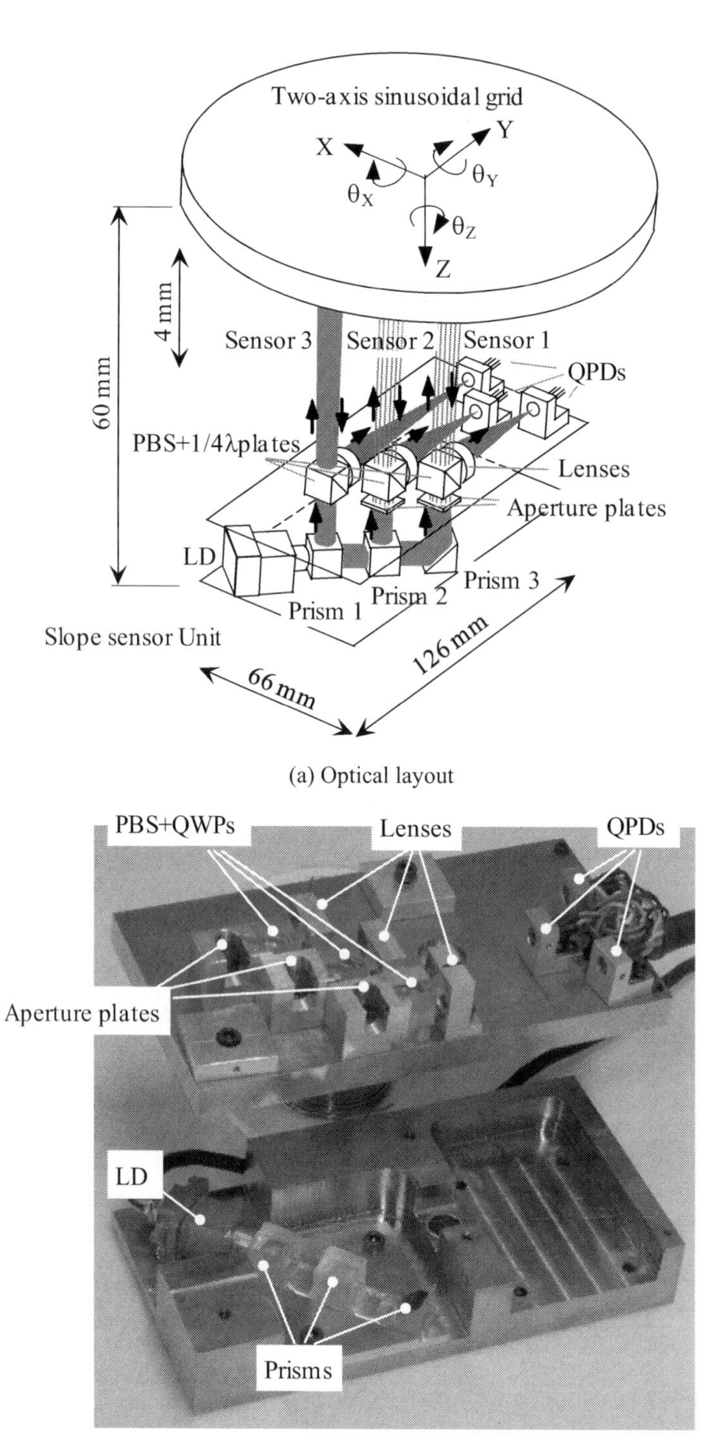

(a) Optical layout

(b) Photograph

Figure 3.11. The sensor unit of a prototype 5-DOF surface encoder

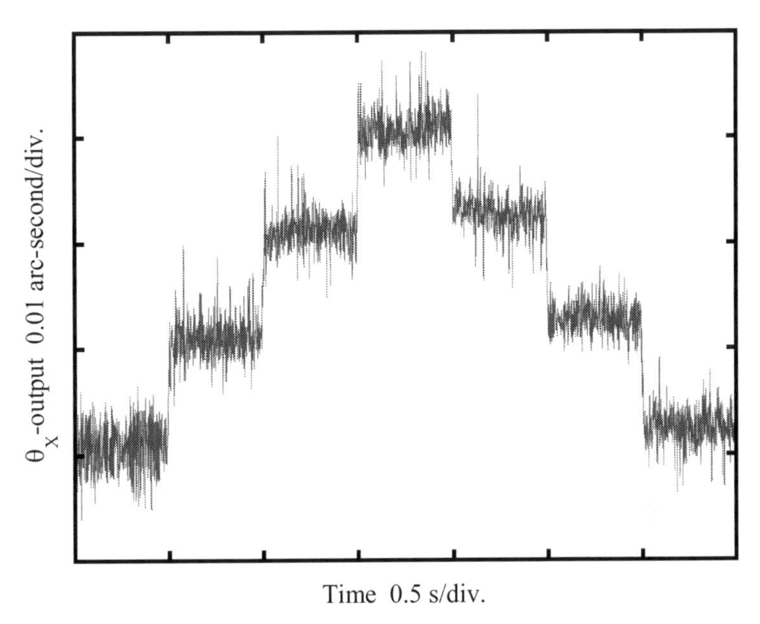

(a) Result of testing the encoder θ_X-directional resolution

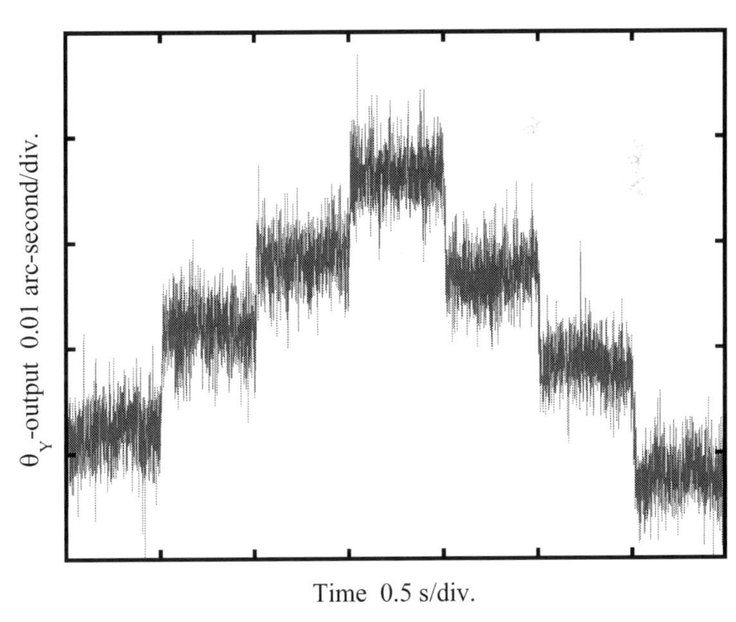

(b) Result of testing the encoder θ_Y-directional resolution

Figure 3.12. Testing the multi-probe type 5-DOF encoder resolution in θ_X- and θ_Y-directions

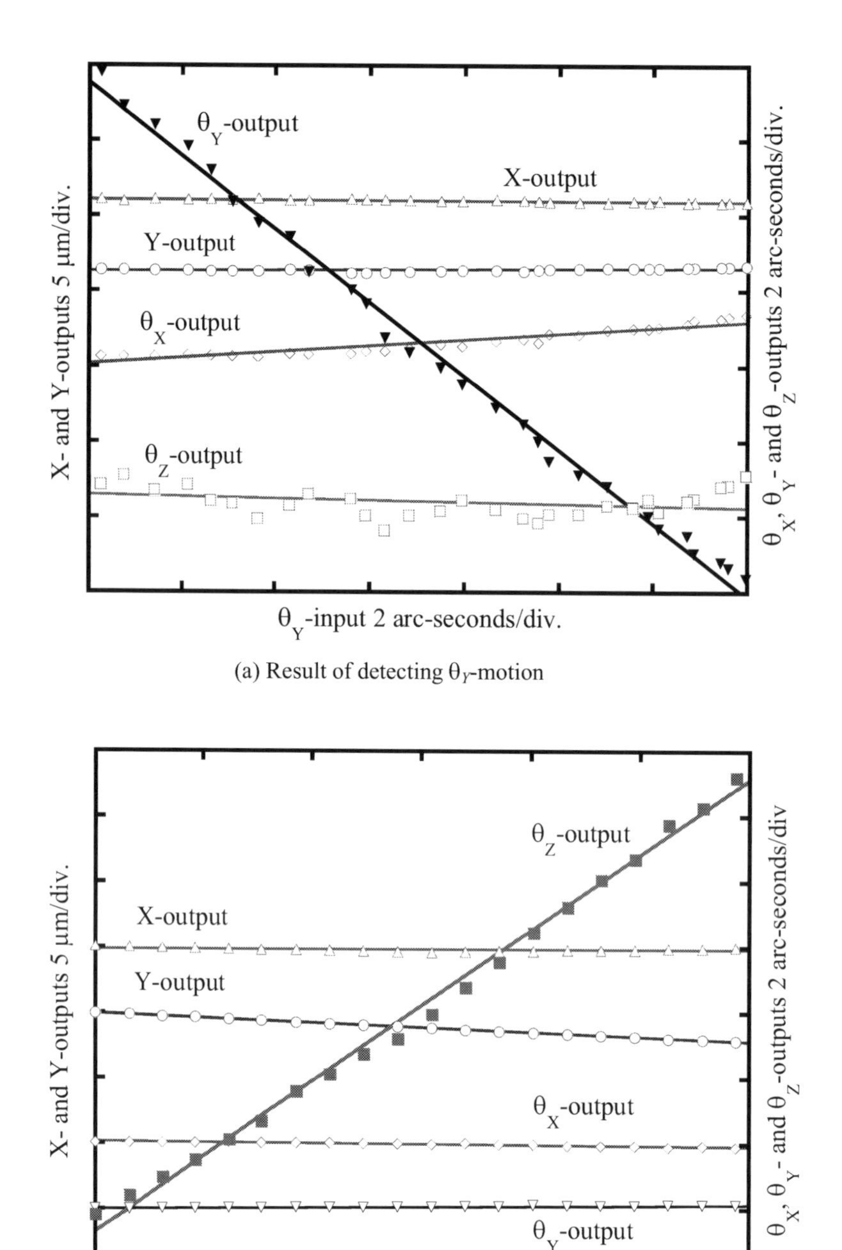

(a) Result of detecting θ_Y-motion

(b) Result of detecting θ_Z-motion

Figure 3.13. Detection of θ_Y- and θ_Z-motions

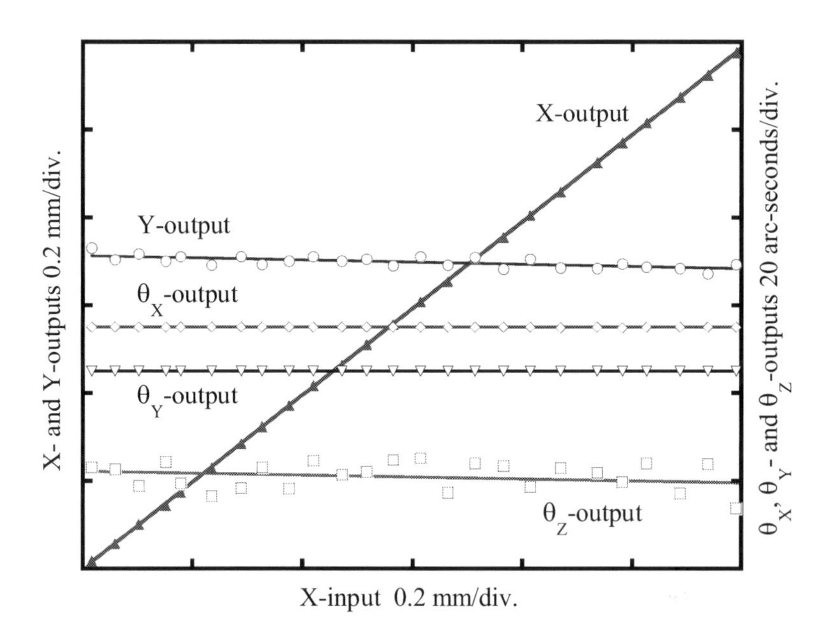

Figure 3.14. Detection of *X*-motion

3.2.2 Scanning Laser Beam-type MDOF Surface Encoder

Figure 3.15 shows the basic principle of the scanning laser beam-type surface encoder. With the beam scanner turned on, the laser beam of the slope sensor periodically scans across the grid surface along a straight line *PP'* at a constant speed *V*. As shown in Figure 3.16, the beam incidence and the scanning length *PP'* are constant for all of the scans. The laser beam is also much faster than the movement speed of the sinusoidal grid so that the sinusoidal grid can be considered to be stationary during each scan. The angle between the line *PP'* and *X*-axis is set to be φ. The *X*- and *Y*-directional zero-lines of the sinusoidal grid, on which $f(x)$ and $g(y)$ are zero, are numbered by i and j, respectively. Since there are two zero-points in each period of the sine function, the distances between two adjacent lines in *X*- and *Y*-directions are $g_x/2$ and $g_y/2$, respectively.

Letting the time for the beam to travel from point *P* to lines i and j be t_X and t_Y, respectively, the position of point *P* can be obtained as follows [15]:

$$x = \left(i - \frac{t_X}{T_{X1} + T_{X2}} \right) \cdot g_x = \left(i - \frac{t_X}{T_X} \right) \cdot g_x , \tag{3.7}$$

$$y = \left(j - \frac{t_Y}{T_{Y1} + T_{Y2}} \right) \cdot g_y = \left(j - \frac{t_Y}{T_Y} \right) \cdot g_y . \tag{3.8}$$

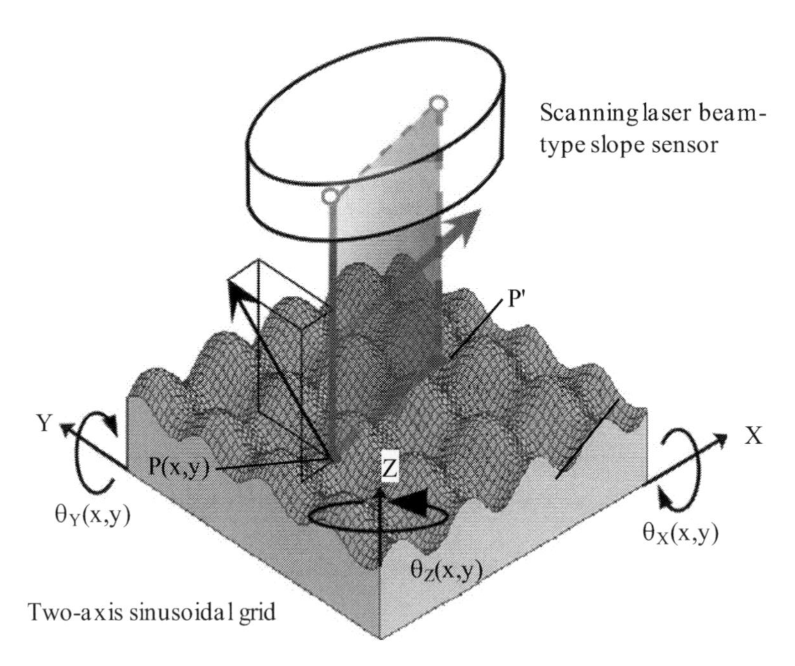

Figure 3.15. Principle of the scanning laser beam-type surface encoder

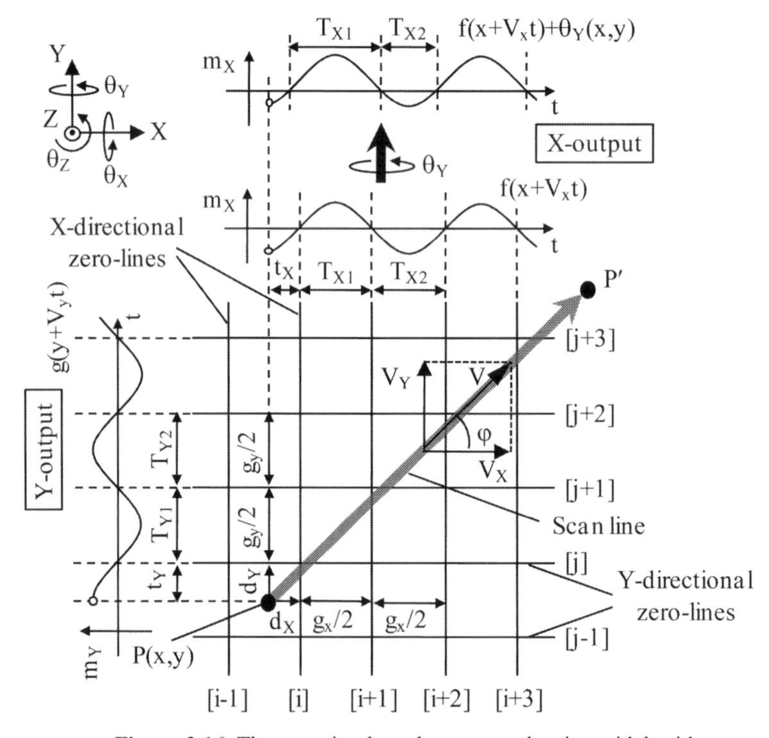

Figure 3.16. The scanning laser beam over the sinusoidal grid

where T_{X1} (T_{X2}) and T_{Y1} (T_{Y2}) are the time for the beam to travel between two adjacent zero-lines in the X- and Y-directions, respectively. V_X and V_Y are the X- and Y-directional components of V.

In addition to the two-axis position measurement, the surface encoder can also be used to detect tilt motions associated with the translational movement. Letting the tilt motion about the Y-axis at point $P(x, y)$ be $\theta_Y(x, y)$, the X-directional output of the slope sensor can be denoted as:

$$m_X(x+V_Xt, y+V_Yt) = f(x+V_Xt)+\theta_Y(x,y)$$
$$= \frac{2\pi H_x}{g_x}\sin\left(2\pi\cdot\frac{x+V_Xt}{g_x}\right)+\theta_Y(x,y). \tag{3.9}$$

T_{X1} and T_{X2} then become:

$$T_{X1}(x,y) = \frac{g_x}{V_X}\left[\frac{1}{2}-\frac{1}{\pi}\arcsin\left(-\frac{g_x\theta_Y}{2\pi H_x}\right)\right], \tag{3.10}$$

$$T_{X2}(x,y) = \frac{g_x}{V_X}\left[\frac{1}{2}+\frac{1}{\pi}\arcsin\left(-\frac{g_x\theta_Y}{2\pi H_x}\right)\right]. \tag{3.11}$$

If $\theta_Y(x, y)$ is small, it can be obtained as:

$$\theta_Y(x,y) = k_Y\cdot\frac{T_{X2}(x,y)-T_{X1}(x,y)}{T_{X2}(x,y)+T_{X1}(x,y)},\ k_Y = \frac{\pi^2 H_x}{g_x}. \tag{3.12}$$

Similarly, the tilt motion $\theta_X(x, y)$ about the X-axis can be evaluated from the corresponding T_{Y1} and T_{Y2}. When a tilt motion $\theta_Z(x, y)$ about the Z-axis exists, the angle between the scanning line PP' and the X-axis will change to $\varphi + \theta_Z(x, y)$, causing a variation in the ratio of V_Y to V_X. $\theta_Z(x, y)$ can thus be obtained as:

$$\theta_Z(x,y) = \arctan\frac{T_{X1}(x,y)+T_{X2}(x,y)}{T_{Y1}(x,y)+T_{Y2}(x,y)}\cdot\frac{g_y}{g_x}-\varphi. \tag{3.13}$$

Since only the zero-lines of the grid surface are utilized for measurement, form errors of the grid surface have less influence on the measurement accuracy.

Figure 3.17 shows a prototype surface encoder designed and built for achieving the 5-DOF measurement principle. The slope sensor was 125 mm (L) × 66 mm (W) × 45 mm (H). The pitches g_x and g_y of the sinusoidal grid surface were both 150 μm. The amplitudes (H_x, H_y) were 100 nm. The size of the grid was 150 mm in diameter, which establishes the position measurement range of the surface encoder.

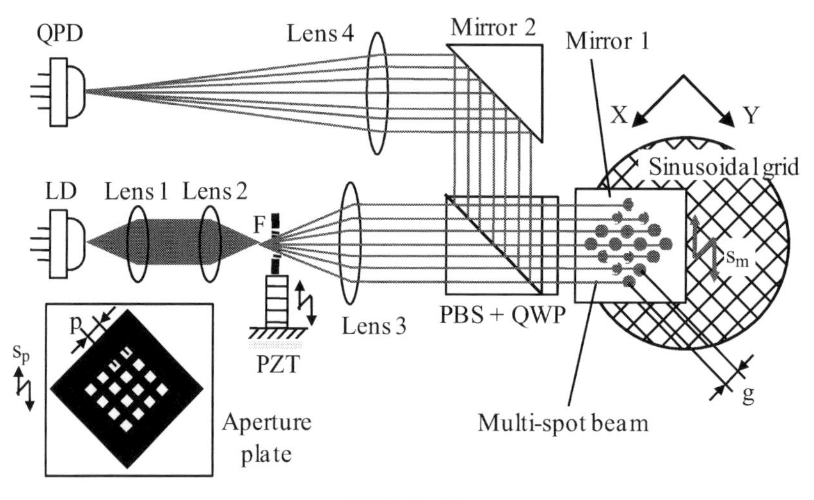

(a) Optical layout

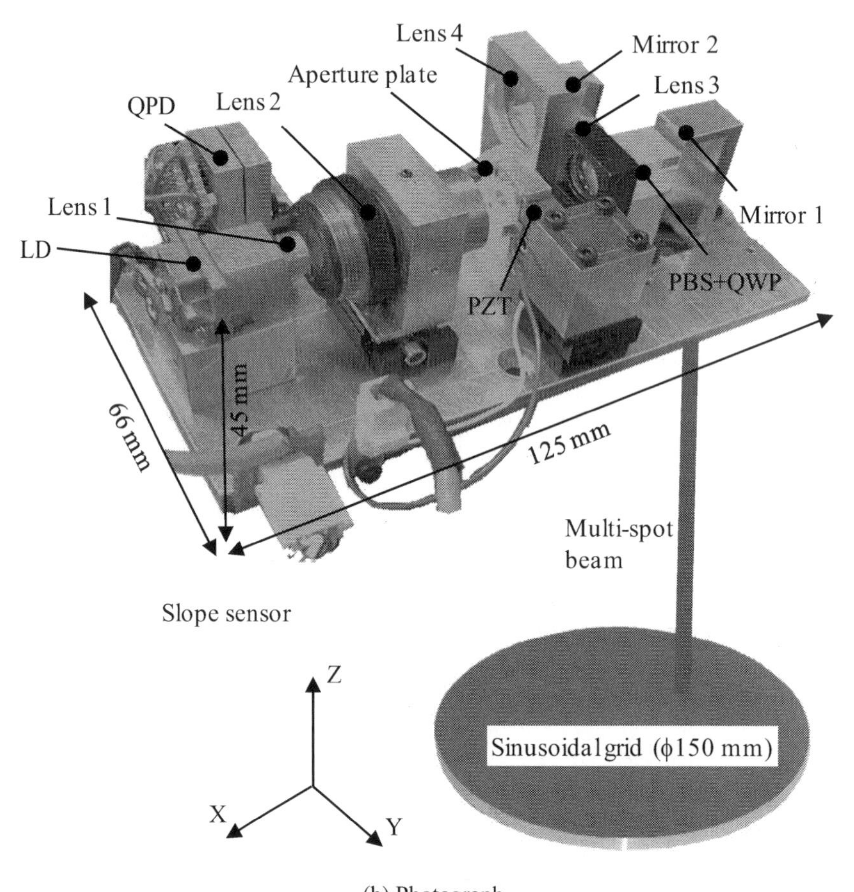

(b) Photograph

Figure 3.17. A prototype of the scanning laser beam type 5-DOF surface encoder

The size of each micro-aperture on the glass aperture plate was 3 μm × 3 μm and the pitch spacing p of the apertures was 6 μm. Point F was the focal point of both lens 2 and lens 3. The NA of lens 2 was 0.1. The aperture plate was located at a position with a distance L from point F. The pitch g of the multiple beams, which is the same as that of the sinusoidal grid, can be determined by

$$g = \frac{f_3}{L} p = k_m p \,, \tag{3.14}$$

where f_3 is the focal distance of lens 3. k_m can be considered as a magnification ratio of the beam pitch g to the aperture pitch p. In the prototype sensor, f_3 was set to be 20 mm and L was adjusted to be 800 μm so that k_m became 25 and g was equal to 150 μm. The number of multiple spots projected on the grid surface was approximately 900.

Scanning of the multiple beams was realized by moving the aperture plate with a piezoelectric actuator (PZT). The movement range of the PZT and the scanning length of the multiple beams across the grid surface are denoted by s_p and s_m, respectively. The ratio of s_m to s_p is also equal to k_m. A PZT with a movement range of 16 μm was employed to produce an s_m of 400 μm. The angle between the scanning line of the multiple beams and the X-axis was set to be 45°.

Figure 3.18 shows a schematic of the technique employed to measure the time parameters t_X, T_{X1} and T_{X2} for calculation of the position and tilt motions. t_Y, T_{Y1} and T_{Y2}, which are not shown in the figure for the sake of clarity, can be measured in the same way. A periodic triangular waveform (A) generated by function generator 1 was applied to the PZT scanner to accomplish the scanning of the multiple beams across the sinusoidal grid surface. A pulse signal (B) was also output from the function generator to determine the starting point of scanning. The corresponding sinusoidal output (C) from the slope sensor was converted to a square wave after passing through a comparator. The output of the comparator (D) was input to a logic AND gate together with a high-frequency clock-pulse signal generated by function generator 2. T_{X1} can then be obtained from the clock-pulse period and the pulse number of the AND gate output counted by a pulse counter. T_{X2} and t_X can also be obtained similarly. The frequencies of function generators 1 and 2 were set to be 100 Hz and 1 MHz, respectively.

Figure 3.19 shows a result of the position measurement. The grid was moved by a two-axis stage along a periodic square root while the slope sensor was kept stationary. The measurement errors were determined to be approximately 3.5 μm by comparison with a two-axis laser interferometer. The error was mainly caused by the movement error of the PZT scanner. Figure 3.20 shows results of the tilt motion measurement. Figure 3.20 (a) shows the result of the θ_X measurement. The encoder output of θ_Y was also detected to evaluate the cross-sensitivity. As can be seen in Figure 3.20, the cross-sensitivity was approximately 3.6%. The error of cross-sensitivity was caused by alignment errors of axes between the sinusoidal grid, the slope sensor and the stages. The out-of-straightness of the PZT movement was another reason for the error of the cross-sensitivity. Similar results were obtained for the θ_Y measurement and θ_Z measurement as shown in Figure 3.20 (b).

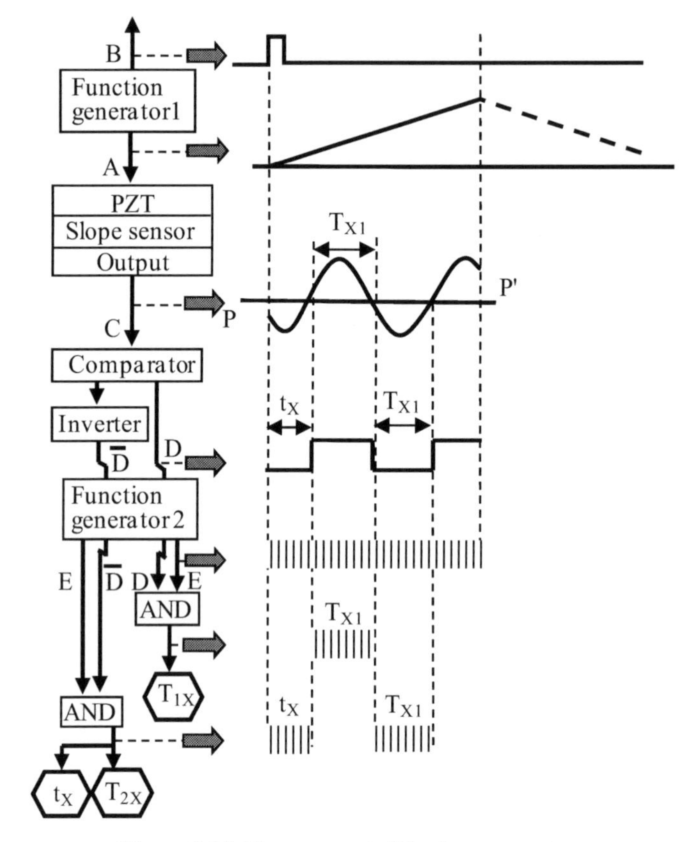

Figure 3.18. Measurement of the time parameters

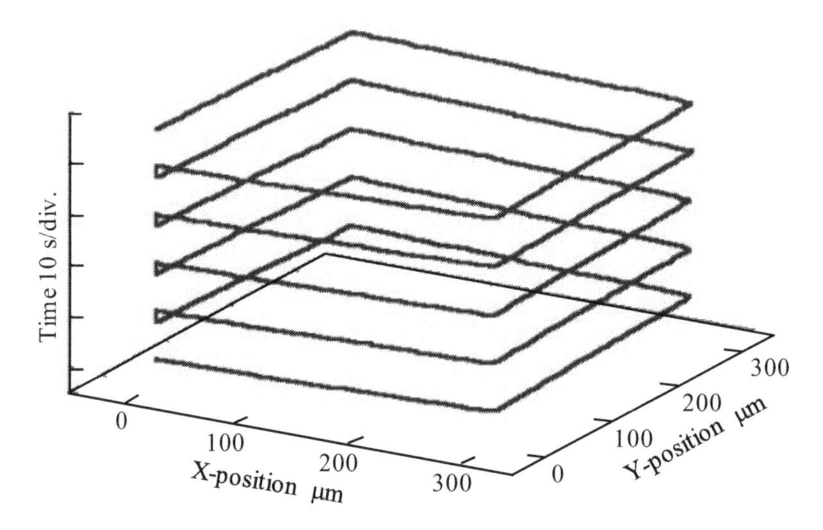

Figure 3.19. Measurement result of *X*- and *Y*-motions by the scanning laser beam type 5-DOF surface encoder

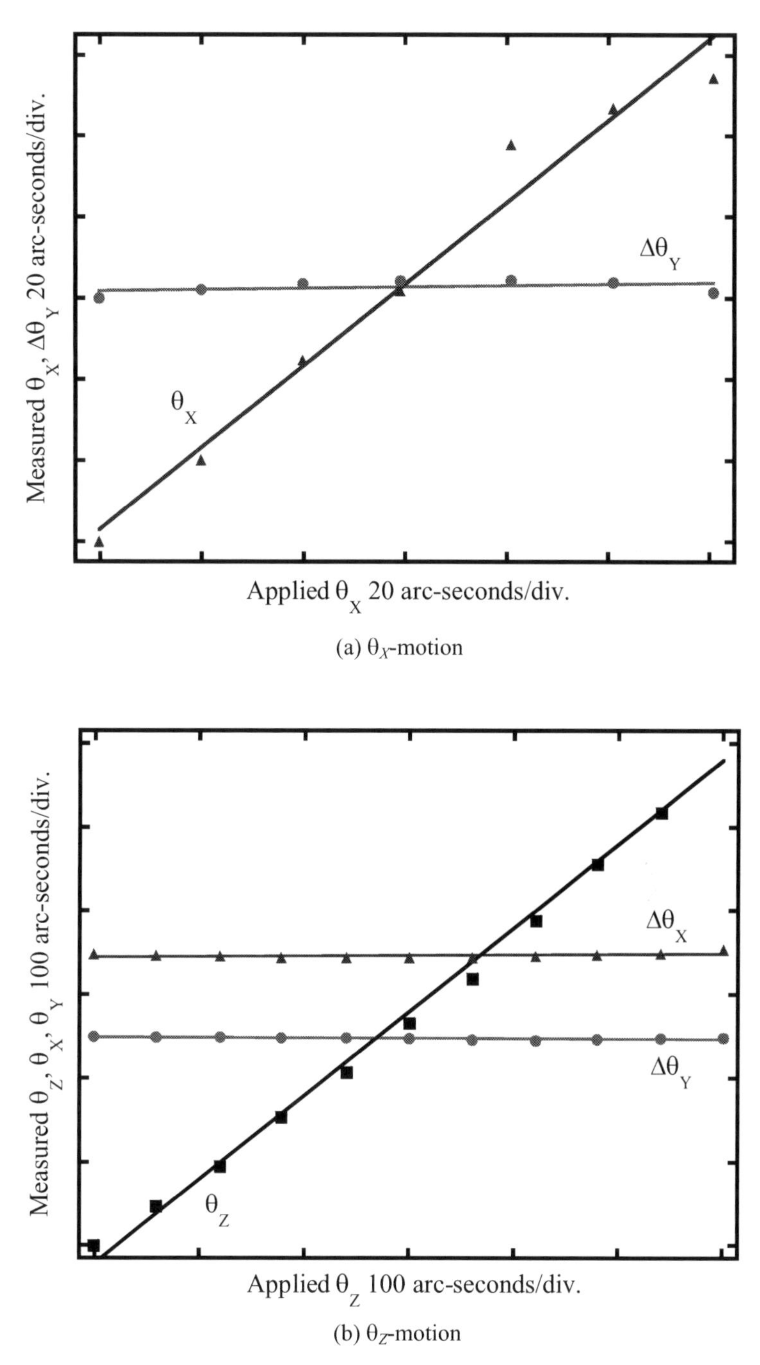

(a) θ_X-motion

(b) θ_Z-motion

Figure 3.20. Measurement results of tilt motions by the scanning laser beam-type 5-DOF surface encoder

3.3 Fabrication of Two-axis Sinusoidal Grid for Surface Encoder

3.3.1 Fabrication System

Figure 3.21 presents the fabrication system for the two-axis sinusoidal grid. The system is composed of a diamond turning machine, a fast-tool-servo (FTS) and a personal computer. The FTS is mounted on the *X*-slide of diamond turning machine, and the workpiece is held on the spindle with its axis along the *Z*-direction by a vacuum chuck. The motion of the *X*-slide and the rotation of the spindle are synchronized. The polar coordinates of the tool tip position in the *XY*-plane [16] is given by:

$$(r_i, \theta_i) = (r_0 - \frac{Fi}{UT}, 2\pi \frac{i}{U}), \; i = 0, 1, ..., N-1, \tag{3.15}$$

where r_0 is the radius of workpiece in millimeters, F is the feed rate of the *X*-slide in mm/min, U is the pulse number of the rotary encoder of the spindle in each revolution in units of pulse/revolution, T is the rotational speed of the spindle in units of revolution/min, i is the ith rotary encoder pulse, and N is the total pulse number of the rotary encoder for the *X*-slide reaching the center. The depth of cut of the diamond tool along the *Z*-direction is controlled by a FTS. The depth of cut at each point is calculated as follows, based on Equations 3.1 and 3.15, and is

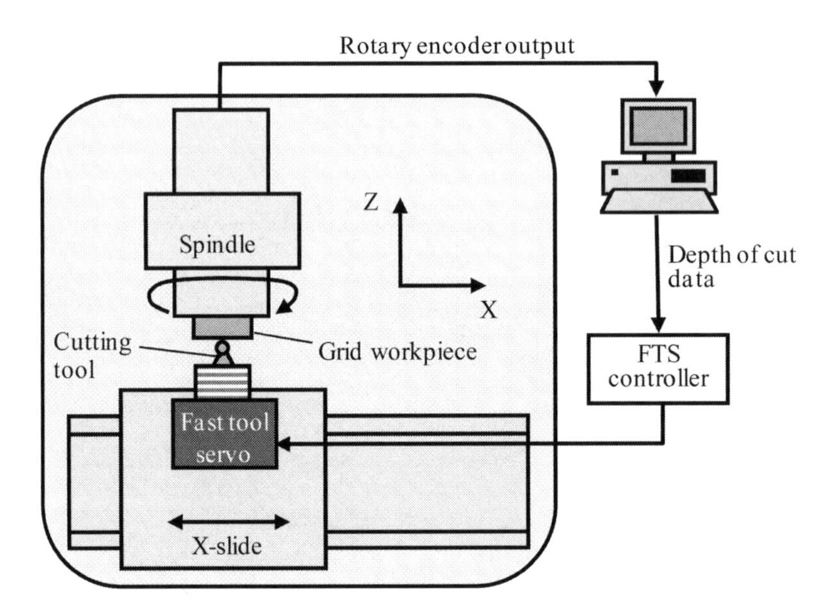

Figure 3.21. The fabrication system for the two-axis sinusoidal grid

stored in a personal computer before fabrication:

$$z(i) = f(r_i \cos\theta_i, r_i \sin\theta_i),$$

$$= f((r_0 - \frac{Fi}{UT})\cos(2\pi \frac{i}{U}), (r_0 - \frac{Fi}{UT})\sin(2\pi \frac{i}{U})). \quad (3.16)$$

When fabrication starts, the cutting data are output to the controller one by one responding to the trigger signal from the rotary encoder of the spindle. In the fabrication experiment, the workpiece material was A5052, the rotary encoder pulse in each revolution was 30,000, the rotational speed of the spindle was 20 rpm, the feed rate of the *X*-slide was 5 μm/rev and the diameter of the workpiece, on which the sinusoidal grid was fabricated, was 150 mm.

3.3.2 Analysis and Compensation of Fabrication Error

An interference microscope was chosen as a profiling instrument for the fabricated grid. Because the microscope had a measurement area of approximately 1 mm × 1 mm, the measurement over the whole area with a diameter of 150 mm was not effective. On the other hand, from the point of view of the turning process shown in Figure 3.22, characteristics of the profile errors of regions A, B, C … along the circumference direction should be similar, and the evaluation result of one of these regions is representative of the others as well.

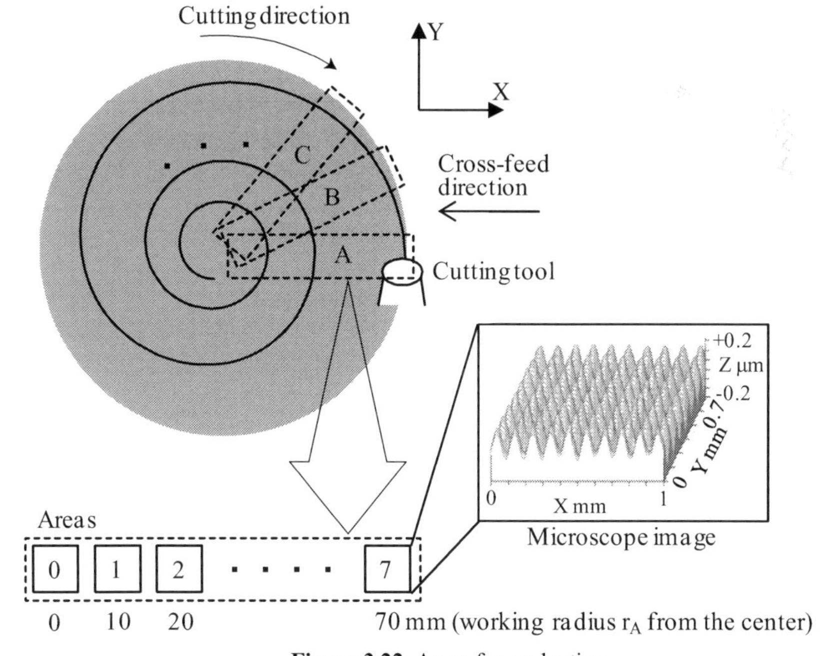

Figure 3.22. Areas for evaluation

Several regions along the radial direction, which were numbered 0, 1, 2, ..., 7 in Figure 3.22, were chosen as representative areas for evaluating profile errors. The area 0 was in the workpiece center and the number 7 was the area at a distance of 70 mm from the center. The interference microscope image of area 7 is shown in Figure 3.22. Because the grid had a sinusoidal profile, the two-dimensional discrete Fourier transform (2D DFT) of the interference microscope image was employed for identification of error components in the surface profile. Figure 3.23 is the spectrum of the interference microscope image by the 2D DFT in area 7. In Figure 3.23, the f_X and f_Y-axes show spatial frequency in the X- and Y-directions, respectively. The $m(f_X, f_Y)$ axis shows the amplitude of the spectrum. The four largest components at the frequency of 0.01 μm^{-1}, which corresponded to a wavelength of 100 μm, showed the desired sinusoidal surface profile components. Other components shown in the spectrum were profile errors. These provided information on the spatial frequency and directions of profile errors. On the basis of the interference microscopy specification, the minimum and maximum spatial wavelengths were 4.6 μm and 1 mm, respectively.

As shown in Figure 3.23, the spectrum had error peaks at 0.02 μm^{-1} in the f_X-axis, which was half of the spatial pitch of the grid. These error peaks were observed in all areas. The amplitudes of the error peak in each area were: (1) 7.8 nm, (2) 9.5 nm, (3) 9.5 nm, (4) 10.4 nm, (5) 10.2 nm, (6) 11.5 nm, and (7) 13.2 nm. Error peaks typically occur only in the tool feed direction and at half the spatial frequency of the profile component. When the cutting data are calculated as intact profile of the grid, the points are different from the programmed cutting position (Figure 3.24). This difference causes the profile error. Because the differential value is zero at the top and bottom of the sinusoidal wave, the profile shows no errors and the wavelength of profile errors becomes equal to a half of the sinusoidal grid pitch. A simulation investigated this error. The results are presented in Figure 3.25. The results showed that the profile error with a wavelength of 50 μm was caused by the difference in cutting points due to the round nose geometry.

The profile error can be compensated based on the model shown in Figure 3.24 on the condition of knowing the local radius R' of the portion of the tool cutting edge with a width of Δx, which actually cut the surface. Δx is calculated by:

$$\Delta x = \frac{2R}{\sqrt{1+\left(\frac{g}{2\pi H}\right)^2}}, \tag{3.17}$$

where R is the nominal tool nose radius, g and H are the pitch and amplitude of the sinusoidal grid, respectively. Assume that R = 1000 μm, g = 100 μm, and H = 0.1 μm. Δx is calculated to be 13 μm. R' in this portion could be quite different from R and also difficult to evaluate accurately. To identify the best value of the local radius for compensation, fabrications were repeatedly carried out on a small area of the workpiece with the compensation of the tool round nose geometry using different radius values. The fabricated sinusoidal surface was imaged by the interference microscope to obtain the DFT spectrum at each time.

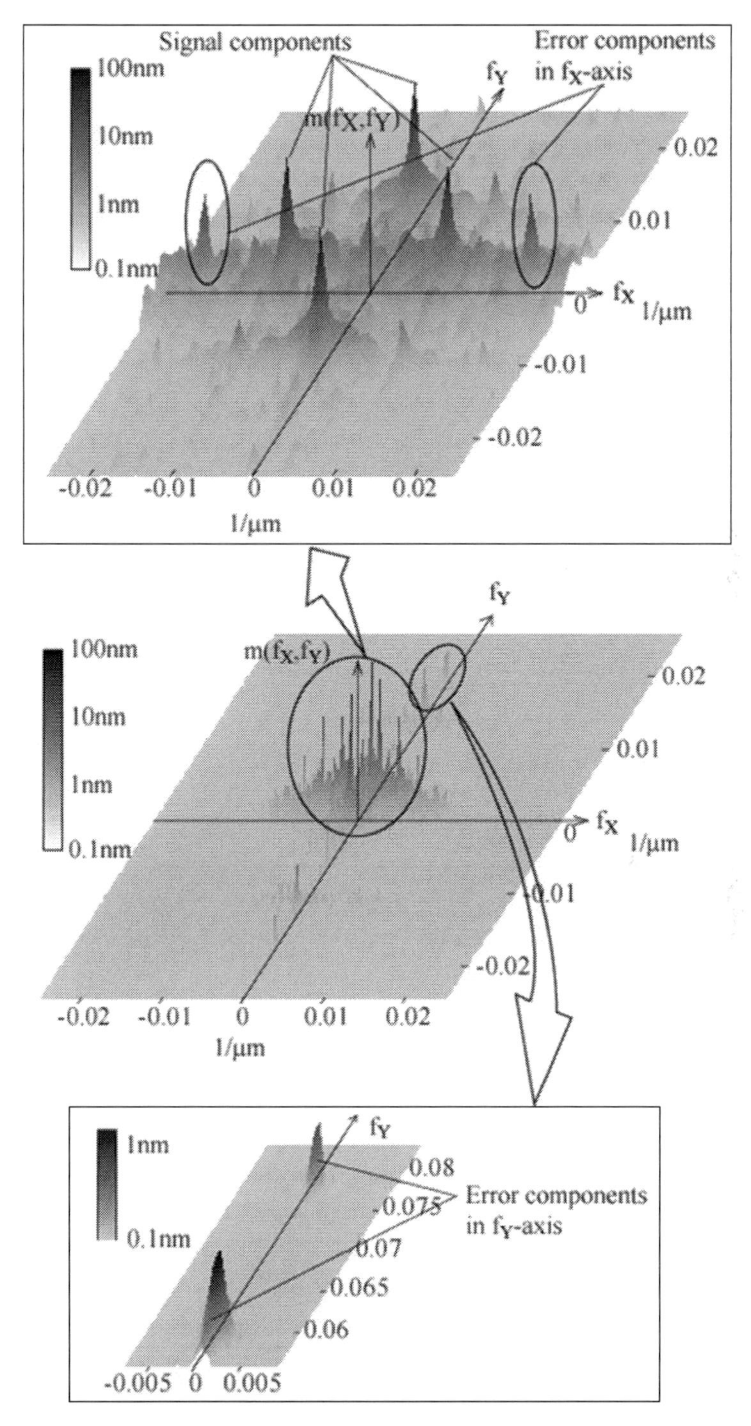

Figure 3.23. Spectrum distribution of the surface profile in area 7

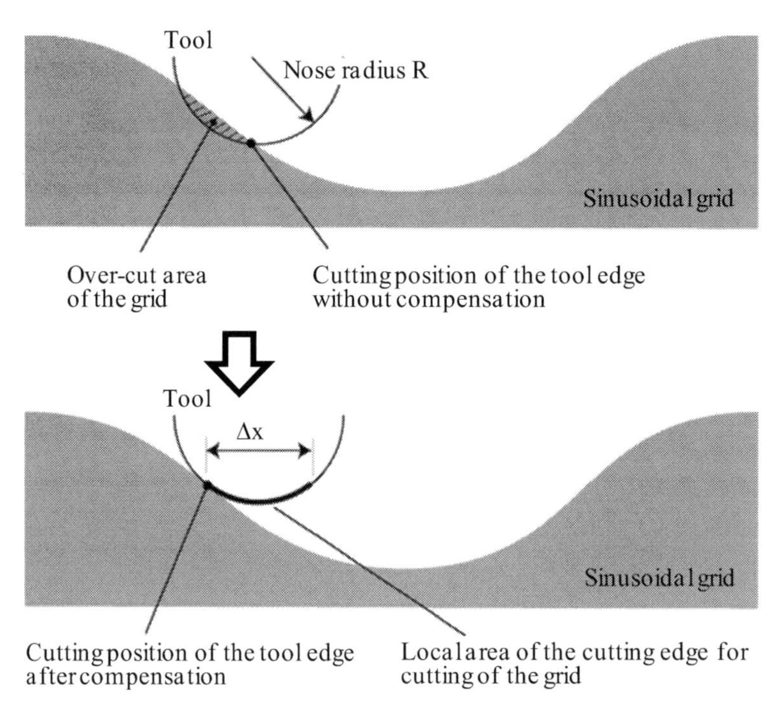

Figure 3.24. Compensation of the cutting position on the cutting edge of the round nose cutting tool

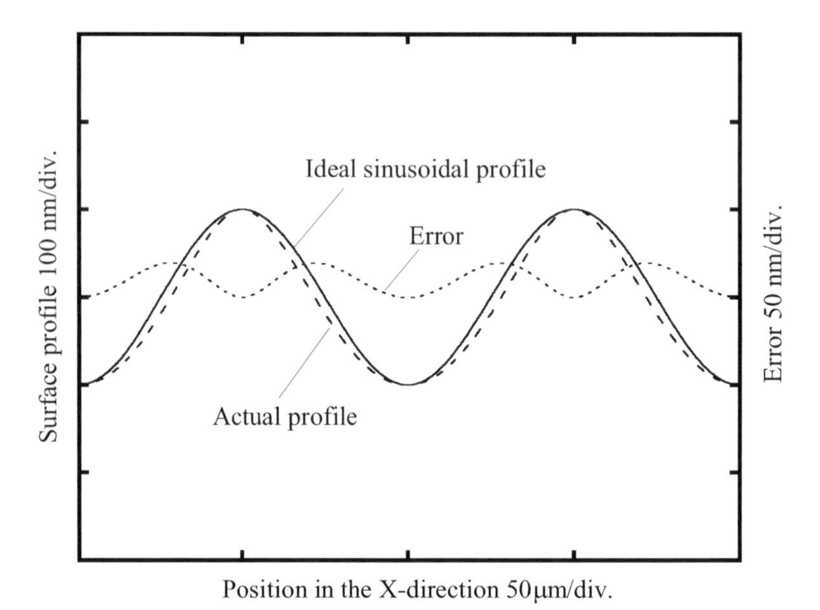

Figure 3.25. Simulation result of the surface profile error caused by the round nose geometry of the cutting tool

Figure 3.26 shows the relationship between the amplitude of the error-component at the wavelength of 50 μm (f_X = 0.02 μm^{-1}) and the local radius used for compensation. As can be seen in the figure, the best value of tool local radius for compensation was evaluated to be approximately 1.05 mm, which had a 50 μm difference from the nominal tool radius of 1 mm. In addition, the local radius after 4.8 km cutting was also obtained for investigation of the influence of the tool wear. It can be seen that the local radius came to be 1.55 mm after the 4.8 km cutting.

Since the tool edge radius corresponding to the tool radius in the *YZ*-plane was on the order of 50 nm, which was much smaller than the smallest curvature of the grid surface, the resultant profile error-component along the cutting direction was small enough to be ignored. That is why no large error-component was observed at f_Y = 0.02 μm^{-1} in Figure 3.23. On the other hand, however, there were some higher-frequency error peaks observed in the f_Y-axis in Figure 3.23. Figure 3.27 shows the amplitude and spatial frequency of error peaks in each of the evaluation areas. It can be seen that the amplitude decreases and the frequency increases with the increase of the working radius r_A. In the circumferential direction f_Y of the workpiece, the interval between adjacent cutting points is proportional to the radius and inversely proportional to the number of rotary encoder pulses of the spindle per revolution. When the rotary encoder pulse is the same for all radii, if the radius is larger, then the interval is larger. Thus, the errors are thought to be caused by the digitization of cutting points, which results in an interval between adjacent cutting points. Figure 3.28 shows the simulation results of the errors in which the fast-tool-servo was assumed to be 2.3 kHz. Although the error peaks differ in size between the measurement and simulation results, the behavior revealed by the simulation is almost the same as that observed from the measurement results.

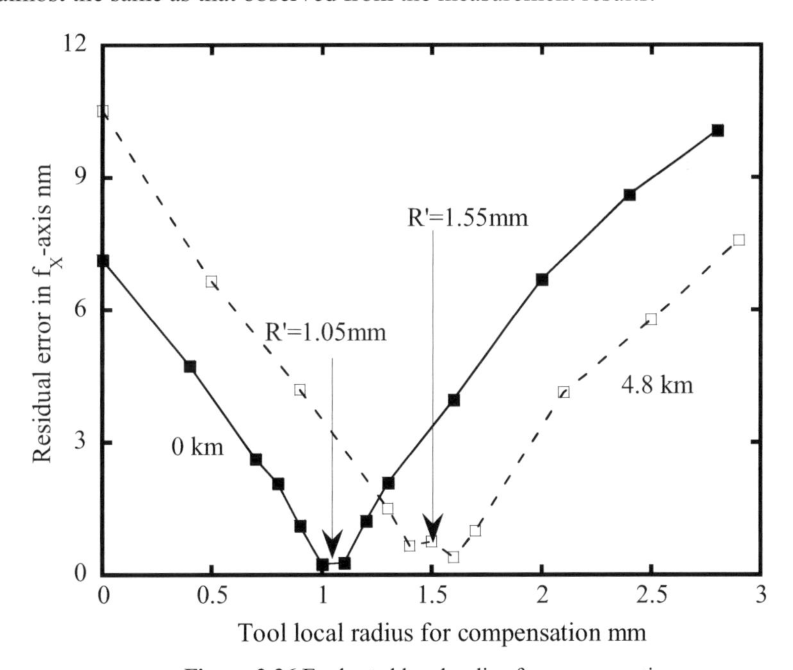

Figure 3.26 Evaluated local radius for compensation

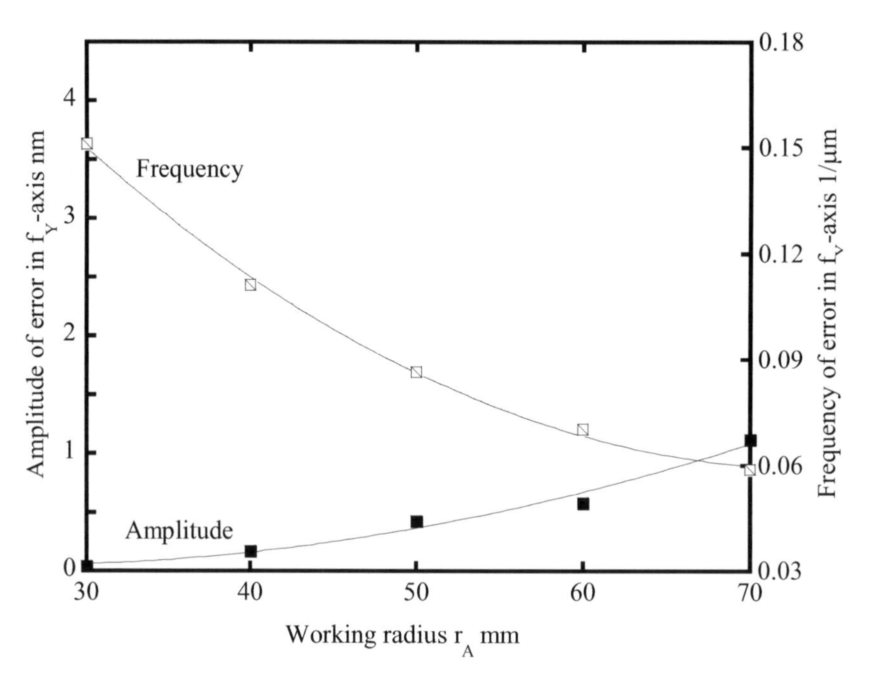

Figure 3.27. Measured result of the error caused by digitization

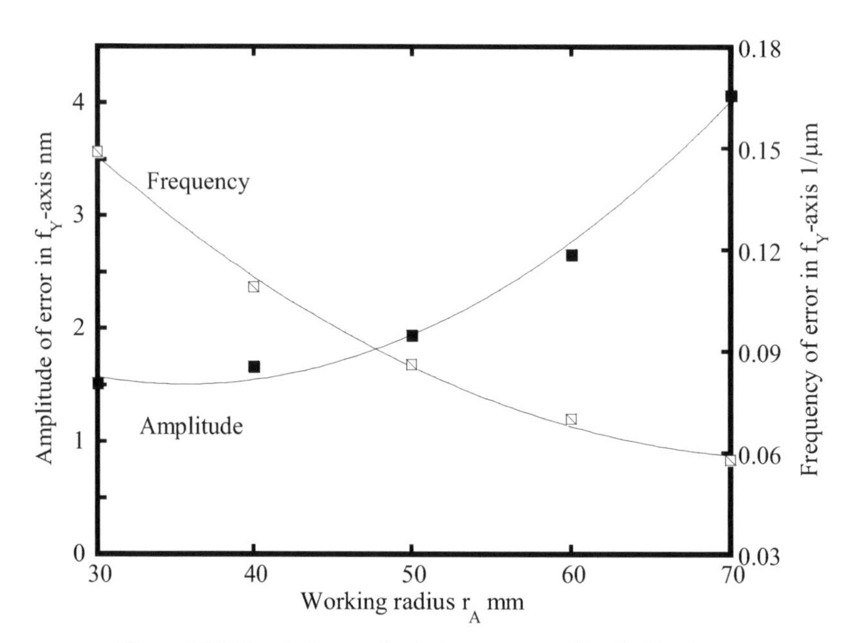

Figure 3.28. Simulation result of the error caused by digitization

The error caused by the digitization of cutting points can be simply compensated by increasing the rotary encoder pulses per revolution and reducing the cutting point interval. However, there are two main restrictions: (1) the maximum number of rotary encoder pulses per revolution, which is determined by the diamond turning machine; (2) the output of the cutting data from the digital-to-analog converter is supposed to cause a delay due to the limitation of the FTS frequency bandwidth. Although this problem can be solved by reducing the rotation speed of the spindle, the fabrication time would increase as a result. In the fabrication experiment described as follows, the fabrication parameters were optimized to reduce the errors caused by digitization of cutting points. The parameters taken into consideration were as follows:

- T, the rotational speed of the spindle with a unit of revolution/min.
- U, the pulse number of rotary encoder of spindle per revolution with a unit of pulse/revolution.
- t_U, the track pitch in X-direction with a unit of μm/revolution.
- r, the radial position in workpiece with a unit of μm.
- $V = 2\pi rT$, cutting speed with a unit of μm/min.
- $F_S = UT/60$, output frequency of cutting data with a unit of Hz.
- $L_S = 2\pi r/U$, interval of the circumferential direction with a unit of μm.
- $F_{FTS} = 2\pi rT/(60g)$, drive frequency of FTS with a unit of Hz, where g is the grid pitch.
- $D_U = Ur_{max}/ t_U$, total number of cutting data points.

The main constraints were:

- $T < 167$, response limit of the rotary encoder of the spindle.
- $U < 180{,}000$, max. number of rotary encoder pulses per revolution.
- $F_{FTS} < 1600$ Hz, bandwidth of the FTS.

Figure 3.29 (a) shows the fabrication conditions before modification. The interval in the circumferential direction L_S was 15.7 μm, which was one-sixth of the pitch of the sinusoidal grid. If the number of rotary encoder pulses per revolution increases three times, L_S will become approximately 5 μm, which is the same as the track pitch t_U. However, the output frequency of the cutting data also increases. The fabrication conditions were modified after investigation of the maximum output frequency in the fabrication system. An output error test of the system was carried out. Output conditions under different output frequencies were investigated by using rectangular data like 0, 1, 0, 1, 0 ... as the cutting data. Exceeding the output frequency limit lead to an absence of data like 0, 1, 0, 0, 1... This was detected by the glitch trigger function of the oscilloscope. No absence of data was observed for 10 min at a frequency less than 120 kHz. Figure 3.29 (b) shows the new fabrication conditions. U was set to 90,000, three times larger than the corresponding value used before modification. The interval of cutting data in the circumferential direction at the outermost area was shortened to 5.2 μm. The rotational speed of the spindle was set to $T = 1500 \times 10^3/\text{r}$ rpm (20–80 rpm) in regions farther than 20 mm away from the center. As a result, the fabrication time was reduced to 6.6 h, which is half the time required before the modification of the fabrication conditions. The maximum rotational speed of the spindle was limited to be 80 rpm by the maximum output frequency of 120 kHz.

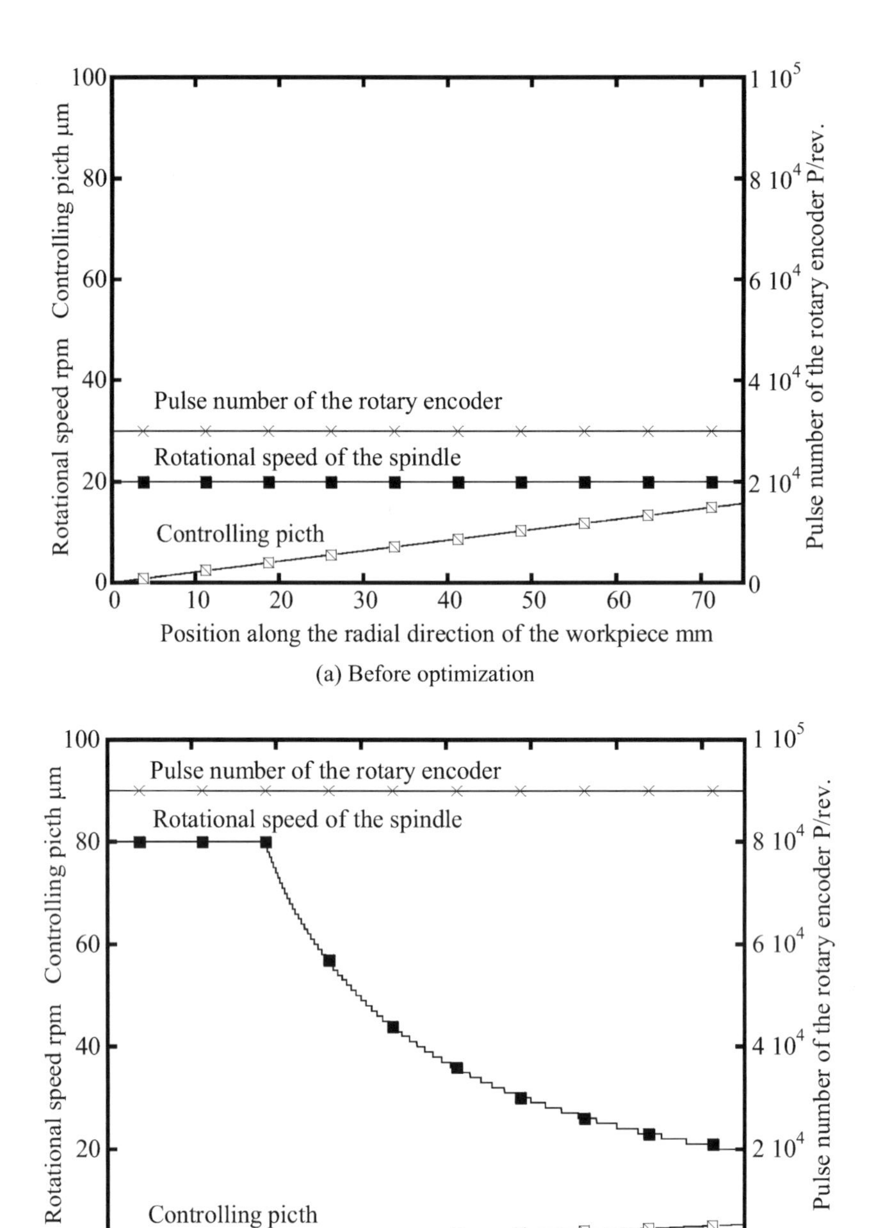

(a) Before optimization

(b) After optimization

Figure 3.29. Optimization of fabrication parameters

Compensated fabrication was performed after the modification of fabrication conditions. A 2D DFT analysis of the sinusoidal grid fabricated was carried out. Figure 3.30 shows the amplitude of the error caused by the round nose geometry of the cutting tool as well as the amplitude of the error caused by digitization of cutting points. As can be seen, the former error was reduced from 13.2 nm to 1.5 nm, while the latter was reduced from 1 nm to 0.1 nm.

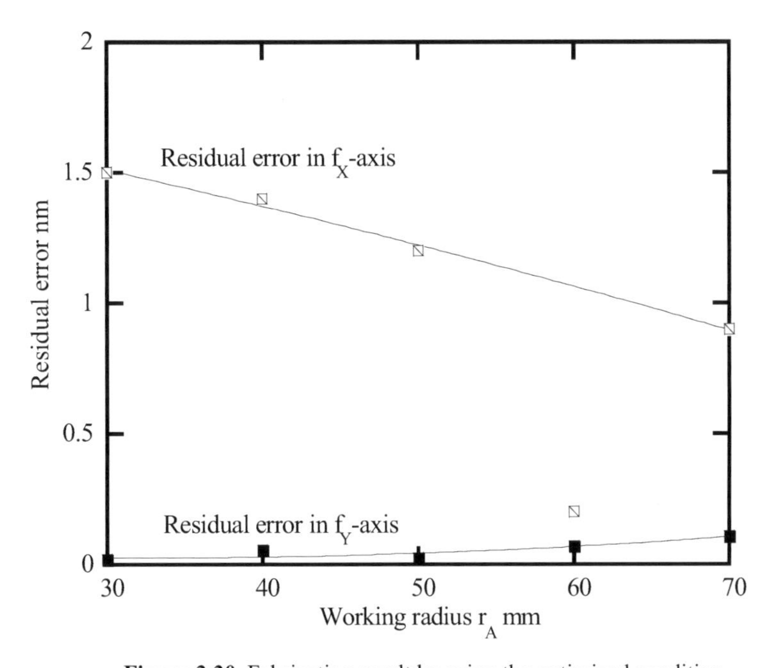

Figure 3.30. Fabrication result by using the optimized condition

3.4 Application of Surface Encoder in Surface Motor-driven Planar Stage

3.4.1 Stage System

Figure 3.31 shows a schematic of the stage system composed of a surface motor planar stage and the three-degree-of-freedom surface encoder shown in Figure 3.3. The planar motion stage is composed of a moving element (platen) and a stage base. The platen is driven by four two-phase DC linear motors of brushless-type. Each motor consists of a paired magnetic array and a stator (coils). The linear motors are symmetrically mounted in the same *XY*-plane, two in the *X*-direction (*X*-motors) and the other two in the *Y*-direction (*Y*-motors). The magnetic arrays and the coils are mounted on the back of the platen and the stage base, respectively. The moving magnet/stationary coil structure avoids the interference of the coil wires with the movement of the platen. The heat generated by the coils is easily removed to the stage base and is not transferred to the platen. Each magnetic array has ten Nd-Fe-B permanent magnets with a pitch spacing of 10 mm. The stator has two coils, which are connected in anti-series with a spacing of 35 mm to construct a two-phase linear motor. Each coil has 126 turns of ϕ0.5 mm wires. The resistance is 2 Ω and the inductance is 610 μH. The coil has a non-magnetic core so that the thrust force ripple and the magnetic attraction force can be minimized. The gap between the magnetic array and the stator is 1 mm. A simple commutation law is applied to the two-phase forces of the linear motor so that a constant force, which is proportional to the applied coil current, can be obtained through out the full range of travel. The full range of travel is 40 mm in both *X*- and *Y*-directions. The thrust force gains of the *X*- and *Y*-linear motor pairs are measured to be 1.6 N/A. The platen is made of an aluminum plate with a dimension of 260 mm × 260 mm × 8 mm. The total weight of the platen assembly, which consists of the platen plate, the magnetic arrays and the sinusoidal grid, is approximately 2.8 kg. The maximum acceleration gain of the platen is calculated to be 0.57 m/s^2/A in the *X*- and *Y*-directions. Simple and low-cost linear motors are constructed because demonstrating the possibility of precision positioning with the surface encoder is the main motivation. The thrust force and acceleration can be improved by employing commercially available high-power linear motors as needed.

The stainless steel plate of the stage base, on which linear motor stators, levitation air-bearings and the sensor unit are mounted, has a dimension of 250 mm

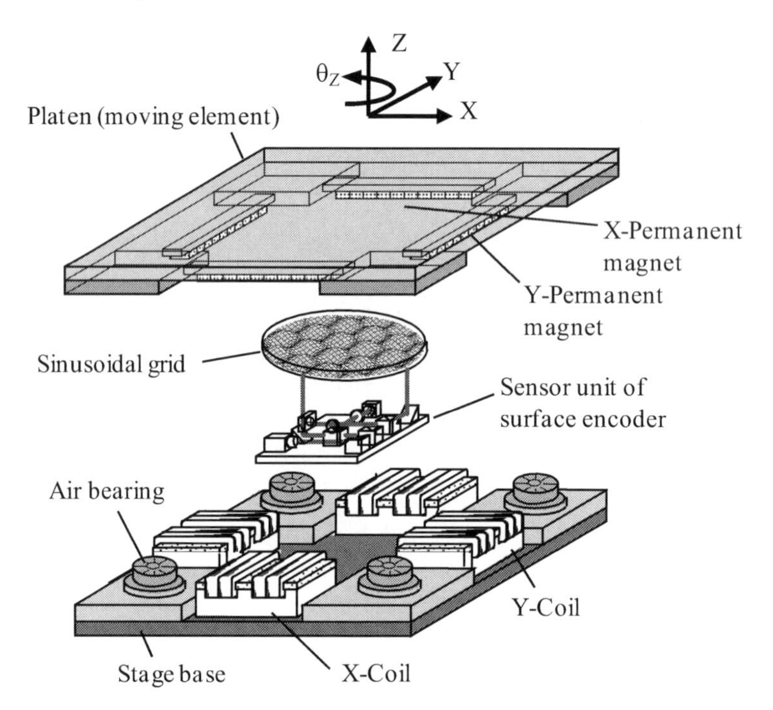

Figure 3.31. The three-degree-of-freedom stage system with a surface motor stage and a surface encoder

× 250 mm × 15 mm and a weight of 7.4 kg. There are four air-bearings used for levitation of the stage platen in the *Z*-direction. The stationary bearing pad structure is employed to avoid interference of the air-hoses on the movement of the stage platen. Grooves are generated on the bearing pad surface so that a uniform levitation force can be obtained over the pad surface. The force on each of the air-bearings, for supporting the weight of the platen, is 6.9 N. A permanent magnet is also used in the air-bearing so that the load for each of the air-bearings is increased to 20.6 N. This results in a higher stiffness of 1.4 μm/N in the *Z*-direction. Taking the movement range of the platen and the size of the bearing pad into consideration, the dimension of the opposing stainless steel pad on the platen is designed to be 70 mm × 70 mm × 8 mm.

Figure 3.32 shows a model for actuating the stage platen. Assume that the force vector $\mathbf{F}(t)$ and the stator coil current vector $\mathbf{I}(t)$ of the linear motors at time t are defined as:

$$\mathbf{F}(t) = \begin{bmatrix} f_{X1}(t) \\ f_{X2}(t) \\ f_{Y1}(t) \\ f_{Y2}(t) \end{bmatrix}, \tag{3.18}$$

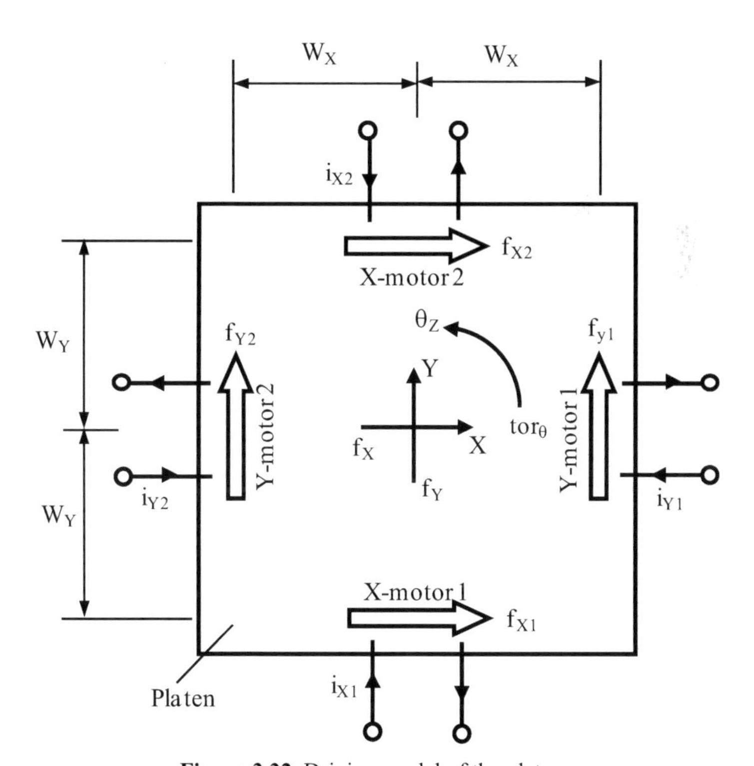

Figure 3.32. Driving model of the platen

$$\mathbf{I}(t) = \begin{bmatrix} i_{X1}(t) \\ i_{X2}(t) \\ i_{Y1}(t) \\ i_{Y2}(t) \end{bmatrix}. \tag{3.19}$$

The relationship between the thrust force vector and the coil current vector can be expressed by

$$\mathbf{F}(t) = \mathbf{K}_{\text{force}}\mathbf{I}(t), \tag{3.20}$$

where $\mathbf{K}_{\text{force}}$ is the force constant matrix defined as

$$\mathbf{K}_{\text{force}} = \begin{bmatrix} k_{X1} & 0 & 0 & 0 \\ 0 & k_{X2} & 0 & 0 \\ 0 & 0 & k_{Y1} & 0 \\ 0 & 0 & 0 & k_{Y2} \end{bmatrix}. \tag{3.21}$$

The motion equations of the platen can be written as:

$$\begin{bmatrix} f_x(t) \\ f_y(t) \\ tor_\theta(t) \end{bmatrix} = \mathbf{M}_{iner} \begin{bmatrix} \dfrac{d^2x(t)}{dt^2} \\ \dfrac{d^2y(t)}{dt^2} \\ \dfrac{d^2\theta_z(t)}{dt^2} \end{bmatrix}, \tag{3.22}$$

where

$$\mathbf{M}_{iner} = \begin{bmatrix} m & 0 & 0 \\ 0 & m & 0 \\ 0 & 0 & J \end{bmatrix}, \tag{3.23}$$

is called the mass and inertial matrix. The m(= 2.8 kg) and J(= 0.04 kgm^2) are the mass and inertia of the platen; $f_X(t)$, $f_Y(t)$ and $tor_\theta(t)$ are the resultant forces and torque for driving the platen in the X-, Y- and θ_Z-directions, respectively. The following combinations of linear motor forces are used to generate $f_X(t)$, $f_Y(t)$ and $tor_\theta(t)$:

$$\begin{bmatrix} f_X(t) \\ f_Y(t) \\ tor_\theta(t) \end{bmatrix} = \begin{bmatrix} 1 & 1 & 0 & 0 \\ 0 & 0 & 1 & 1 \\ W_X & -W_X & 0 & 0 \end{bmatrix} \begin{bmatrix} f_{X1}(t) \\ f_{X2}(t) \\ f_{Y1}(t) \\ f_{Y2}(t) \end{bmatrix}. \tag{3.24}$$

To relate the resultant forces and torque to the coil currents, Equation 3.24 is rewritten as:

$$\begin{bmatrix} f_X(t) \\ f_Y(t) \\ tor_\theta(t) \\ 0 \end{bmatrix} = \begin{bmatrix} 1 & 1 & 0 & 0 \\ 0 & 0 & 1 & 1 \\ W_X & -W_X & 0 & 0 \\ 0 & 0 & W_Y & -W_Y \end{bmatrix} \begin{bmatrix} f_{X1}(t) \\ f_{X2}(t) \\ f_{Y1}(t) \\ f_{Y2}(t) \end{bmatrix}. \tag{3.25}$$

Substituting Equation 3.20 into Equation 3.25 yields:

$$\begin{bmatrix} f_X(t) \\ f_Y(t) \\ tor_\theta(t) \\ 0 \end{bmatrix} = \begin{bmatrix} 1 & 1 & 0 & 0 \\ 0 & 0 & 1 & 1 \\ W_x & -W_x & 0 & 0 \\ 0 & 0 & W_Y & -W_Y \end{bmatrix} \mathbf{K}_{force} \begin{bmatrix} i_{X1}(t) \\ i_{X2}(t) \\ i_{Y1}(t) \\ i_{Y2}(t) \end{bmatrix}. \tag{3.26}$$

The coil currents for generating the necessary resultant forces and torque can then be obtained as:

$$\begin{bmatrix} i_{X1}(t) \\ i_{X2}(t) \\ i_{Y1}(t) \\ i_{Y2}(t) \end{bmatrix} = \frac{\mathbf{K}_{force}^{-1}}{2} \begin{bmatrix} f_X(t) + \dfrac{tor_\theta(t)}{W_X} \\ f_X(t) - \dfrac{tor_\theta(t)}{W_X} \\ f_Y(t) \\ f_Y(t) \end{bmatrix}, \tag{3.27}$$

where $\mathbf{K}_{\text{force}}^{-1}$ is the matrix inverse of $\mathbf{K}_{\text{force}}$.

The positions of the platen in the *X*-, *Y*- and θ_Z-directions, which are measured by the surface encoder, are sent to a personal computer (PC) via an AD converter as feedback signals. The measured positions are compared with the command positions in the PC to obtain the error signals. After passing through a software PID controller in the PC [17], the error signals are used to determine the necessary stator coil currents for generating the actuating forces and torque. The coil currents, which are controlled by the PC through a DA converter and current amplifiers, are then applied to the linear motors to drive the platen to the command positions. The values of the parameters in the PID controller are first determined by computer simulation based on the stage dynamics model. These values are then adjusted in the experiments by a trial-and-error process.

3.4.2 Stage Performance

Positioning experiments of the planar motion stage were carried out in a setup shown in Figure 3.33. The surface encoder outputs were used as the position feedback signals. A three-axis interferometer system was also employed to monitor the actual movement of the stage.

Figure 3.34 shows the results from testing the positioning resolutions of the stage in the *X*- and *Y*-directions. The stage was moved first in the *X*-direction with a step of 200 nm while the *Y*- and θ_Z-positions were kept stationary. Then the stage was kept stationary in the *X*- and θ_Z-positions while being moved in the *Y*-direction with the same step. The outputs of the surface encoder, which were used as the feedback signals, are shown in Figure 3.34 (a). Figure 3.34 (b) shows the interferometer outputs, which illustrates the actual movements of the stage. It can be seen that positioning resolutions in the *X*- and *Y*-directions were approximately 200 nm. Figure 3.35 shows the result from testing the stage resolution in the θ_Z-position. The resolution in the θ_Z-direction, which was determined by those in the *X*- and *Y*-directions, was approximately 1 arc-second. The variations in the *X*- and *Y*-outputs of the interferometer are due to the offset of the measurement point from the rotational center. The variation in the θ_Z-output of the interferometer was caused by the imperfection of the stage positioning. The results shown in Figures 3.34 and 3.35 also demonstrate that the stage can be controlled in the *X*-, *Y*- and θ_Z-directions independently.

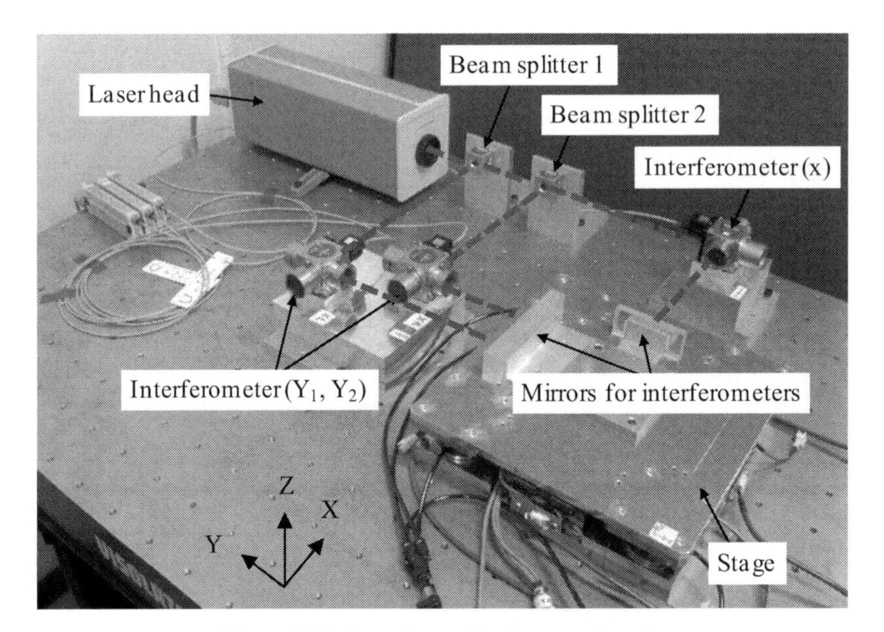

Figure 3.33. Setup for positioning experiments

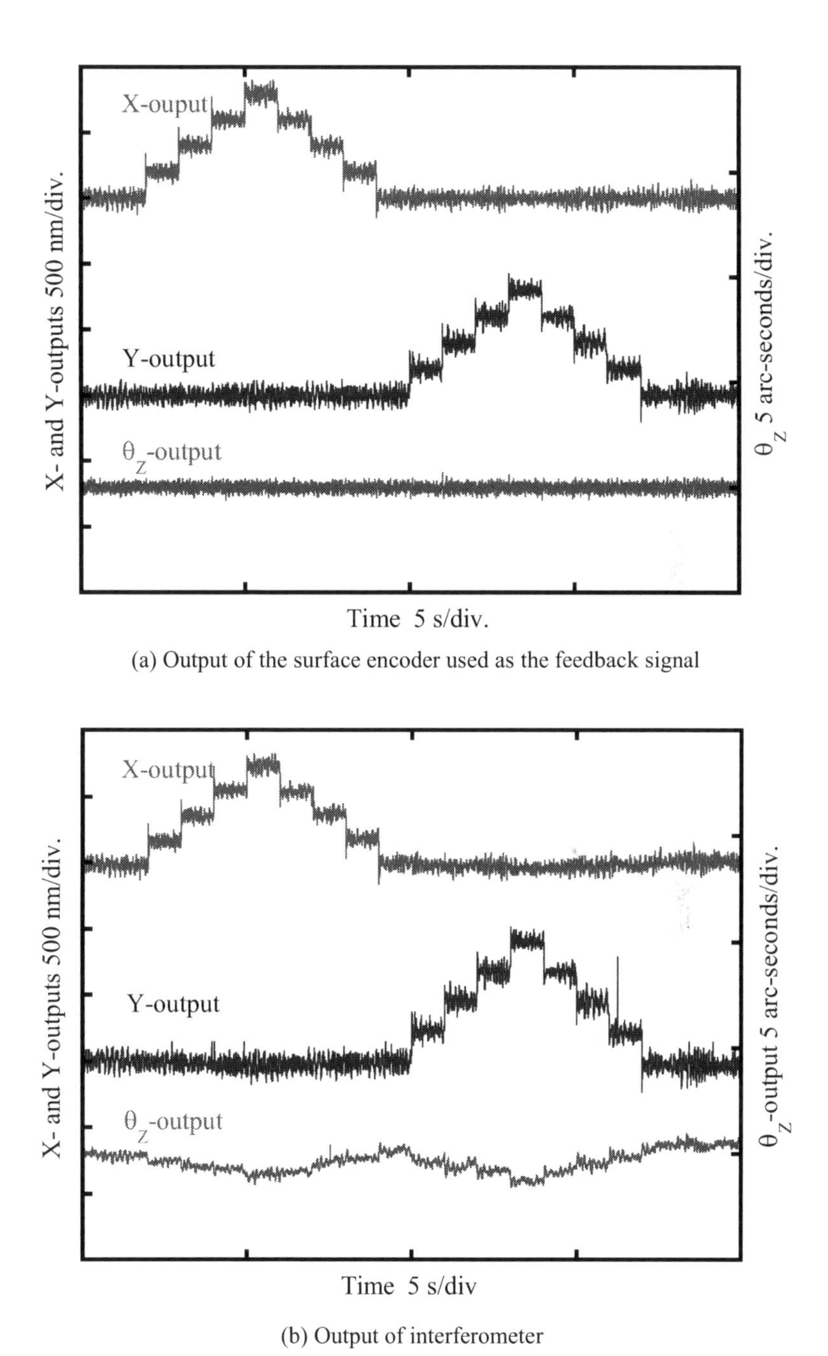

(a) Output of the surface encoder used as the feedback signal

(b) Output of interferometer

Figure 3.34. Testing the positioning resolution in *X*- and *Y*-directions

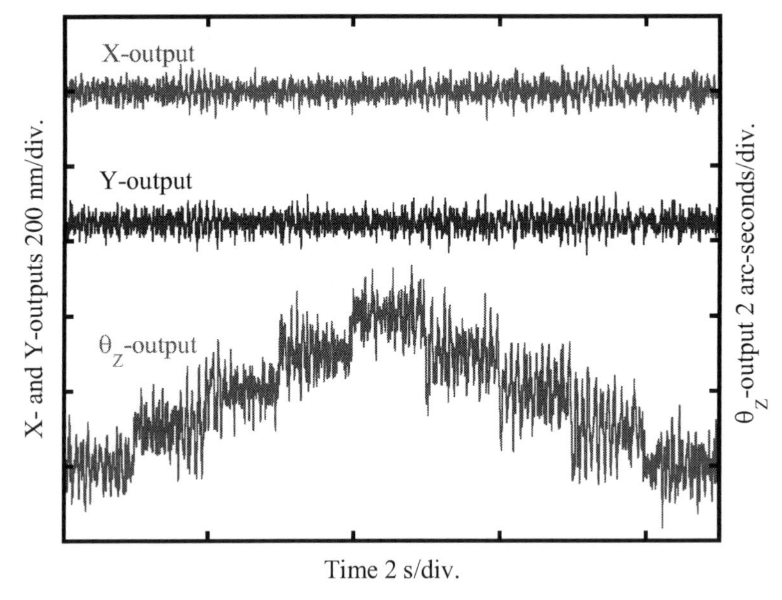

(a) Output of the surface encoder used as the feedback signal

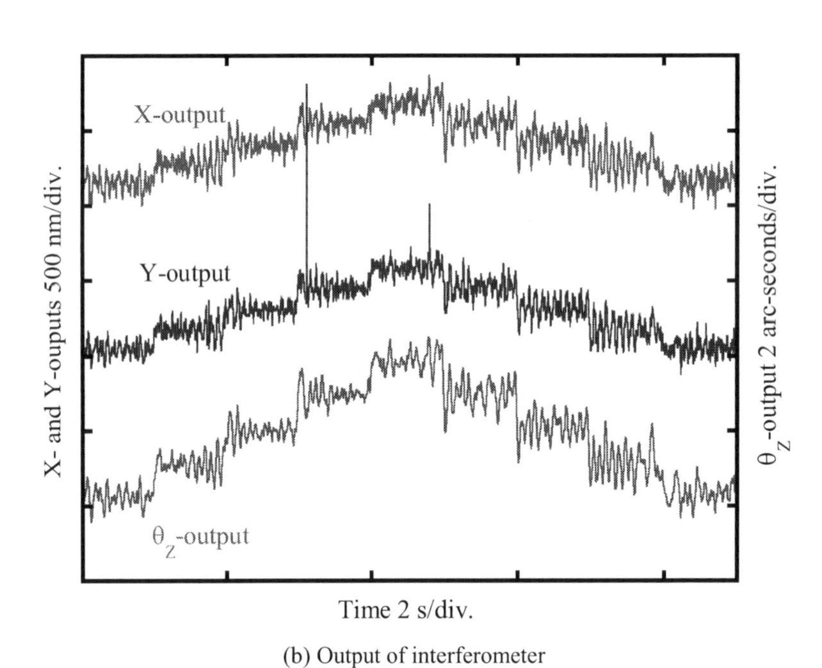

(b) Output of interferometer

Figure 3.35. Testing the positioning resolution in θ_Z-direction

3.5 Summary

Multi-degree-of-freedom (MDOF) surface encoders have been described for measurement of in-plane motions. The surface encoder consists of a sinusoidal grid and a slope sensor unit. Two types of slope sensor units are presented for MDOF in-plane motion measurement. They are the multi-probe-type and the scanning laser beam-type. Two two-axis slope sensors have been employed in the slope sensor unit of the multi-probe-type 3-DOF surface encoder to measure the *XY*-positions of two points simultaneously. The θ_Z-tilt motion is calculated from the *XY*-positions of the two points. A two-axis slope sensor with a large light beam, which can directly detect the $\theta_X\theta_Y$-tilt motions, has been added in the sensor unit for the measurement of 5-DOF in-plane motions. In the scanning laser beam-type surface encoder, a laser beam periodically scans across the grid surface along a straight line at a constant speed. The *XY*-positions and $\theta_X\theta_Y\theta_Z$-tilt motions can be obtained from the time parameters of crossing the lines with zero-slopes. Compared with the multi-probe-type surface encoder, the scanning laser beam-type has simpler structure but lower measurement speed.

A fabrication system based on diamond turning with fast-tool-servo has been presented for the sinusoidal grid surface. An efficient evaluation technique based on two-dimensional DFT of the interference microscope image of the sinusoidal grid surface has been employed for the fabricated grid surface. The error-component caused by the round nose geometry of the diamond tool and cutting data digitization has been identified by the evaluation technique. The amplitudes of the error-components have been reduced to several nanometers through a compensation procedure based on the evaluation result.

The surface encoder has been integrated into a planar motion stage. The platen of the stage, which is levitated by air-bearings in the *Z*-direction, is driven by four two-phase linear motors, i.e., two pairs in the *X*- and *Y*-directions. The stage has a stroke of 40 mm in the *X*- and *Y*-directions. Positioning experiments to demonstrate the performance of the stage have been described.

References

[1] Fan KC, Ke ZY, Li BK (2009) Design of a novel low-cost and long-stroke co-planar stage for nanopositioning. Technisches Messen 76(5):248–252

[2] Hocken RJ, Trumper DL, Wang C (2001) Dynamics and control of the UNCC/MIT sub-atomic measuring machine. Ann CIRP 50(1):373–376

[3] Kim WJ, Trumper DL (1998) High-precision magnetic levitation stage for photolithography. Precis Eng 22(2):66–77

[4] Weckenmann A, Estler WT, Peggs G, McMurtry D (2004) Probing systems in dimensional metrology. Ann CIRP (53)2:657–684

[5] Kunzmann H, Pfeifer T, Flugge J (1993) Scales versus laser interferometers, performance and comparison of two measuring systems. Ann CIRP 42(2):753–767

[6] Büchner HJ, Stiebig H, Mandryka V, Bunte E, Jäger G (2003) An optical standing-wave interferometer for displacement measurements. Meas Sci Technol 14:311–316

[7] Steinmetz CR (1990) Sub-micron position measurement and control on precision machine tools with laser interferometry. Precis Eng 12(1):12–24

[8] Teimel A (1992) Technology and applications of grating interferometers in high-precision measurement. Precis Eng 14(3):147–154
[9] Tomita Y, Koyanagawa Y, Satoh F (1994), Surface motor-driven precise positioning system. Precis Eng 16(3):184–191
[10] Gao W, Dejima S, Yanai H, Katakura K, Kiyono S, Tomita Y (2004) A surface motor-driven planar motion stage integrated with an $XY\theta_Z$ surface encoder for precision positioning. Precis Eng 28(3):329–337
[11] Dejima S, Gao W, Shimizu H, Kiyono S, Tomita Y (2005) Precision positioning of a five degree-of-freedom planar motion stage. Mechatronics 15(8):969–987
[12] Kiyono S, Cai P, Gao W (1999) An angle-based position detection method for precision machines. Int J Jpn Soc Mech Eng 42(1):44–48
[13] Kiyono S (2001) Scale for sensing moving object, and apparatus for sensing moving object. US Patent 6,262,802, 17 July 2001
[14] Gao W, Dejima S, Kiyono S (2005) A dual-mode surface encoder for position measurement. Sens Actuators A 117(1):95–102
[15] Gao W, Dejima S, Shimizu Y, Kiyono S (2003) Precision measurement of two-axis positions and tilt motions using a surface encoder. Ann CIRP 52(1):435–438
[16] Gao W, Araki T, Kiyono S, Okazaki Y, Yamanaka M (2003) Precision nano-fabrication and evaluation of a large area sinusoidal grid surface for a surface encoder. Precis Eng 27(3):289–298
[17] Dejima S, Gao W, Katakura K, Kiyono S, Tomita Y (2005) Dynamic modeling, controller design and experimental validation of a planar motion stage for precision positioning. Precis Eng 29(3):263–271

4

Grating Encoder for Measurement of In-plane and Out-of-plane Motion

4.1 Introduction

In nanomanufacturing, it is necessary to generate not only *XY* in-plane motions but also *Z*-directional out-of-plane motions [1]. A lot of stages employed for generation of both in-plane and out-of-plane motions require large strokes in the *XY*-axes and a relatively small stroke in the *Z*-axis. Displacement measurement along the three-axes with nanometric resolution is essential for precision positioning of such a stage. Three-axis displacement measurement can be realized by combining a surface encoder described in Chapter 3 for measurement of the translational motion in the *XY*-plane and a short-range displacement sensor for measurement in the *Z*-axis. Compared with laser interferometers, the combination of the surface encoder and the displacement sensor are more thermally stable and less expensive, which are important for practical use. However, it is difficult for the two sensors to measure the same point, resulting in large Abbe errors. The difference in the sensor type also causes difficulties in the stage controlling system. It is thus desired to improve the surface encoder from *XY* in-plane measurement to three-axis in-plane (*XY*) and out-of-plane (*Z*) measurement.

On the other hand, the requirement for measurement of multi-axis in-plane and out-of-plane translational motions can also be found in precision linear stages [2, 3]. Conventionally, a linear encoder is employed for measurement of the position of a precision linear stage along the moving axis (*X*-axis) [4, 5]. The output signal of the linear encoder is usually employed for feedback control of the stage position to reduce the positioning error, which is one of the most important factors influencing the stage performance. The out-of-straightness of the movement of the precision linear stage, which is defined as the error motion along the axis vertical to the moving axis (*Z*-axis), is another important factor influencing the stage performance [6, 7]. The straightness error motion, which is a small amount of displacement on the order of several hundred nanometers, is typically measured with nanometric resolution by employing the straightness measuring kit of a laser interferometer [8–10] or combining a short range displacement sensor with a straightedge [11, 12]. However, the multiple sensor systems for position and straightness make the measurement of a precision stage complicated and inefficient. It is thus desired to develop a two-degree-of-freedom (2-DOF) linear encoder that can measure not only the X-directional position but also the Z-directional straightness of a precision stage.

In this chapter, a new type of grating encoder is described for measurement of stage in-plane and out-of-plane motions. Instead of plane mirrors in a conventional Michelson interferometer [13] for Z-directional displacement measurement, the grating encoder employs two grating mirrors with identical pitches as the stationary reference mirror and the moving target mirror, respectively. The plus and minus first-order diffraction lights from the two grating mirrors interference with each other to generate interference signals, from which the in-plane and out-of-plane displacements of the target grating mirror can be simultaneously obtained.

4.2 Two-degree-of-freedom Linear Grating Encoder

Figure 4.1 shows the schematic of the 2-DOF linear grating encoder for measuring two-axis displacement. This encoder is composed of a reflective-type scale grating with a pitch of g and an optical sensor head [14]. The optical sensor head consists of a laser diode (LD), a collimating lens, an aperture, four prisms, a non-polarizing beam splitter (BS), a reference grating and a detector unit. The arrangement of the scale grating and the reference grating, which are identical except in scale length, is similar to that of the moving plane mirror and reference plane mirror of a Michelson interferometer, respectively. The detector unit is composed of two photodetectors (PDs), which are PD X+1 and PD X−1.

The laser beam with a wavelength of λ emitted from the LD is collimated by the collimating lens. The collimated laser is divided into two beams by the BS. One beam is projected onto the scale grating and the other onto the reference grating. Positive and negative first-order diffracted beams from the two gratings with a diffraction angle of θ are bent by the prisms and superimposed to generate interference signals. The prisms and the PDs are arranged in such a way that only the first-order diffracted beams can be received by the detector unit, in which the interference signal generated by the positive first-order diffracted beams is detected by PD_{X+1} and that generated by the negative first-order diffracted beams is detected by PD_{X-1}, respectively.

Assume that the wavefront functions of the positive and negative first-order diffracted beams from the scale grating and the reference grating are represented by $Es_{X\pm1}$ and $Er_{X\pm1}$, respectively. Consider the case when the scale grating moves a displacement of Δx along the X-axis as shown in Figure 4.2. In this figure, the scale grating and the reference grating are placed parallel with each other for simplicity. Similarly to a conventional grating interferometer [15], Δx changes the phases of $Es_{X\pm1}$ and the optical path difference as follows:

$$Es_{X+1} = E_0 \exp(i\varphi), \tag{4.1}$$

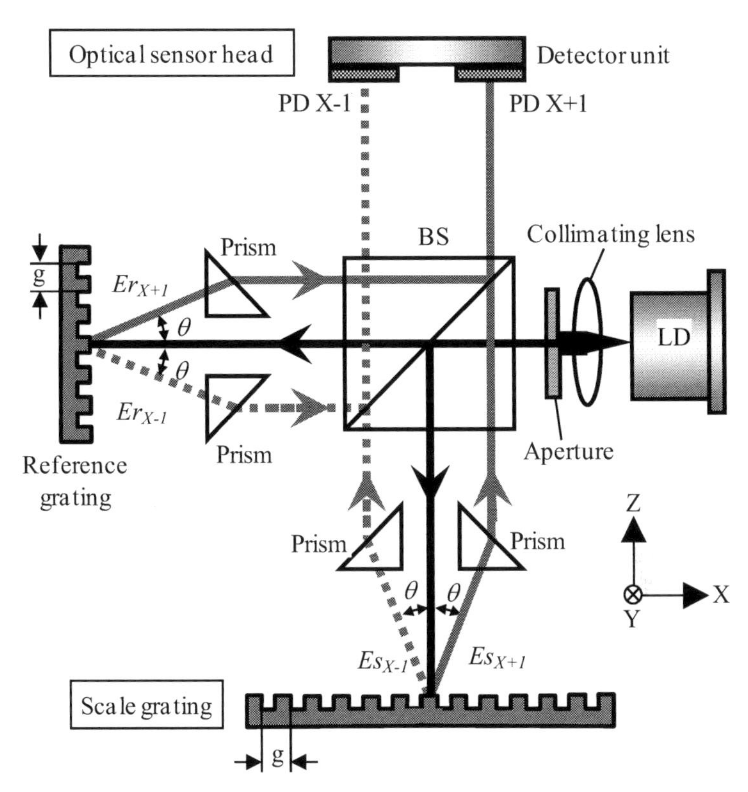

Figure 4.1. Schematic of the 2-DOF linear grating encoder

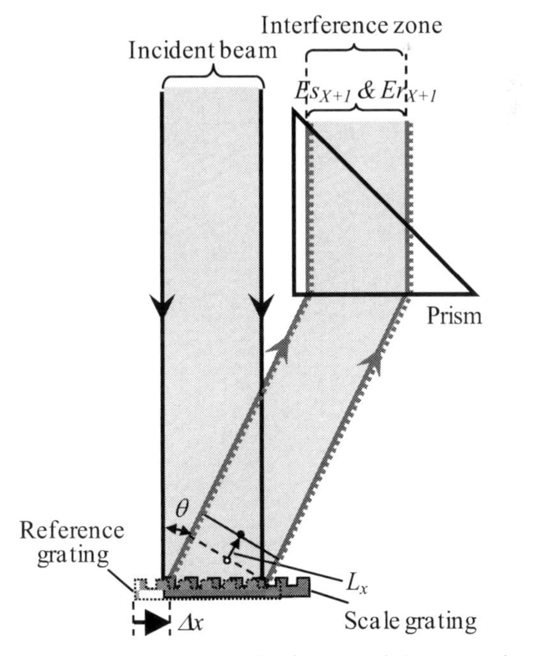

Figure 4.2. Schematic of the optical path of Es_{X+1} and Er_{X+1} associated with Δx

$$Es_{X-1} = E_0 \exp(-i\varphi), \tag{4.2}$$

$$\varphi = \frac{2\pi}{g}\Delta x, \tag{4.3}$$

$$L_X = \frac{\lambda}{g}\Delta x, \tag{4.4}$$

where E_0 is the amplitude of the positive and negative first-order diffracted beams which is proportional to the intensity of the laser from the LD. φ is the phase difference of $Es_{X\pm1}$ caused by Δx. L_x is the optical path difference of the beams.

The Z-directional displacement Δz also produces a phase change in the diffracted beam $Es_{X\pm1}$. Figure 4.3 shows a schematic of the optical path of Es_{X+1} and Er_{X+1} associated with Δz. When Δz is applied, the diffracted beam Es_{X+1} will have a lateral shift with respect to the beam Er_{X+1}. The light ray b_s in the beam Es_{X+1} and the light ray b_r in the beam Er_{X+1} superimpose with each other to generate the interference signal. The phase difference ω and the optical path difference L_Z corresponding to Δz can be obtained by:

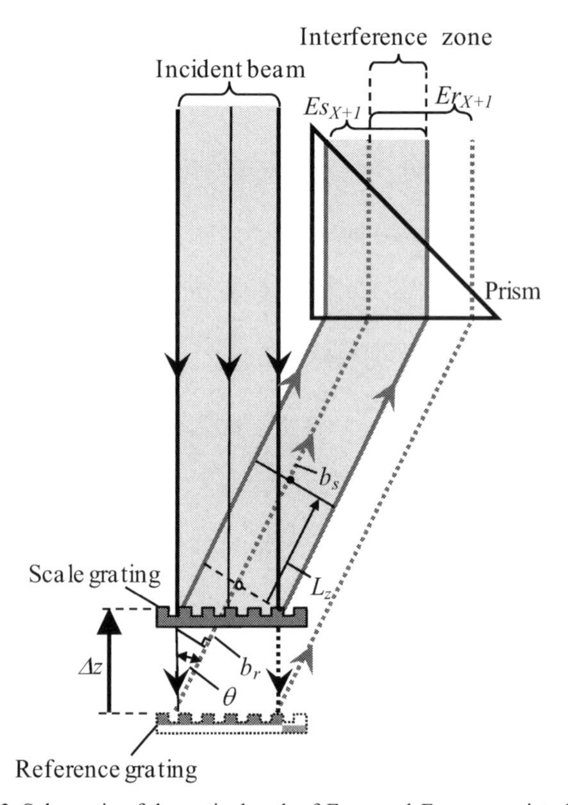

Figure 4.3. Schematic of the optical path of Es_{X+1} and Er_{X+1} associated with Δz

$$\omega = (1+\cos\theta)\cdot\frac{2\pi}{\lambda}\cdot\Delta z\,, \tag{4.5}$$

$$L_z = (1+\cos\theta)\cdot\Delta z\,. \tag{4.6}$$

Consequently, $Es_{X\pm1}$ and $Er_{X\pm1}$ can be obtained as follows when both Δx and Δz are applied to the scale grating:

$$Es_{X+1} = E_0\exp(i\varphi)\cdot\exp(i\omega), \tag{4.7}$$

$$Er_{X+1} = E_0\,, \tag{4.8}$$

$$Es_{X-1} = E_0\exp(-i\varphi)\cdot\exp(i\omega)\,, \tag{4.9}$$

$$Er_{X-1} = E_0\,. \tag{4.10}$$

Let the wavefront function of the interference beam superimposing the positive first-order diffracted beams from the scale grating and the reference grating be E_{X+1}, and that superimposing the negative first-order diffracted beams be E_{X-1}. E_{X+1} and E_{X-1} can be expressed as:

$$E_{X+1} = Es_{X+1} + Er_{X+1}\,, \tag{4.11}$$

$$E_{X-1} = Es_{X-1} + Er_{X-1}\,. \tag{4.12}$$

The X- and Z-directional displacements Δx and Δz can thus be calculated from the intensities $I_{X\pm1}$ of the interference beams shown in the following equations:

$$I_{X+1} = E_{X+1}\cdot\overline{E_{X+1}} = 2{E_0}^2\{1+\cos(\varphi+\omega)\}\,, \tag{4.13}$$

$$I_{X-1} = E_{X-1}\cdot\overline{E_{X-1}} = 2{E_0}^2\{1+\cos(-\varphi+\omega)\}\,. \tag{4.14}$$

It can be seen from Equations 4.3, 4.5, 4.13 and 4.14 that the interference signals $I_{X\pm1}$ have signal periods [15] of g and $\lambda/(1+\cos\theta)$ with respect to Δx and Δz, respectively.

Figure 4.4 shows the configuration of the optical sensor head improved for removing the influence of the variation of E_0 and recognizing the directions of the scale grating displacements. In addition to the optical components shown in Figure 4.1, three polarizing beam splitters (PBSs), five quarter-wave plates (QWPs), one BS, and four detector units are added. The fast axes of the QWPs are indicated in the figure. The interference waves $E_{X\pm1}$ are divided into four sub-waves by the BS and PBSs. The phases and polarization directions of the sub-waves are controlled by the PBSs and QWPs in such a way that the corresponding intensities detected

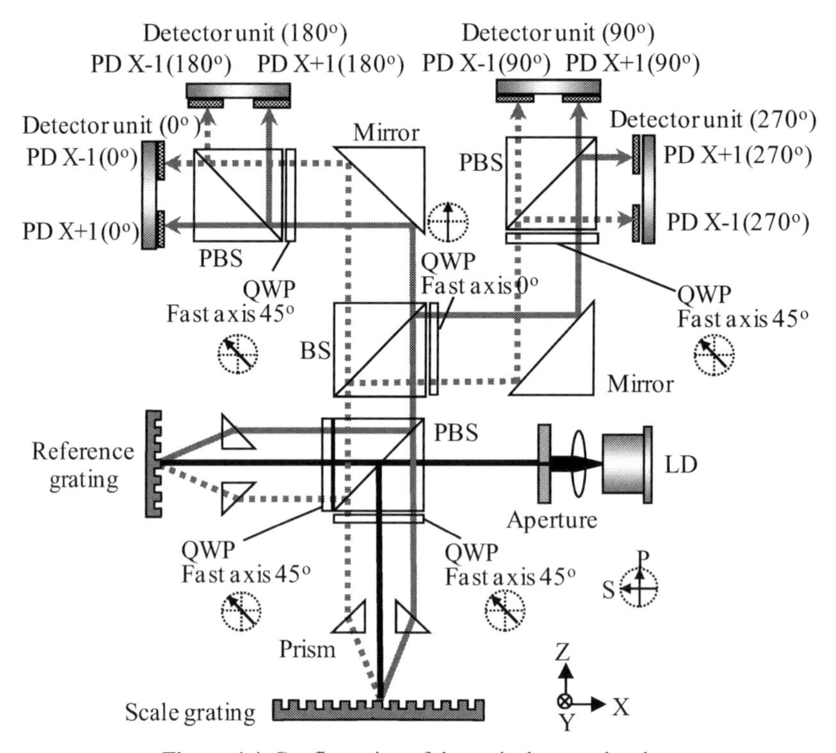

Figure 4.4. Configuration of the optical sensor head

by the detector unit (0°), unit (90°), unit (180°) and unit (270°) have phase differences of 0°, 90°, 180° and 270°, respectively. The quadrature interference signals Q_α and Q'_α ($\alpha = X + 1, X - 1$) with a phase difference of 90° are obtained from the intensities detected by the PDs as:

$$Q_{X+1} = \frac{I_{X+1}(0^o) - I_{X+1}(180^o)}{I_{X+1}(0^o) + I_{X+1}(180^o)} = \sin(\varphi + \omega), \tag{4.15}$$

$$Q'_{X+1} = \frac{I_{X+1}(90^o) - I_{X+1}(270^o)}{I_{X+1}(90^o) + I_{X+1}(270^o)} = \cos(\varphi + \omega), \tag{4.16}$$

$$Q_{X-1} = \frac{I_{X-1}(0^o) - I_{X-1}(180^o)}{I_{X-1}(0^o) + I_{X-1}(180^o)} = \sin(-\varphi + \omega), \tag{4.17}$$

$$Q'_{X-1} = \frac{I_{X-1}(90^o) - I_{X-1}(270^o)}{I_{X-1}(90^o) + I_{X-1}(270^o)} = \cos(-\varphi + \omega). \tag{4.18}$$

It can be seen that the influence of E_0 is removed from Q_α and Q'_α. In addition, the directions of the scale grating displacements can be recognized from the quadrature interference signals Q_α and Q'_α with a 90° phase difference [14]. The phase differences φ and ω can also be obtained from Q_α and Q'_α as:

$$\varphi + \omega = \arctan\left(\frac{Q_{X+1}}{Q'_{X+1}}\right), \tag{4.19}$$

$$-\varphi + \omega = \arctan\left(\frac{Q_{X-1}}{Q'_{X-1}}\right). \tag{4.20}$$

Finally, X- and Z-directional sensor outputs ($S_{\Delta x}$, $S_{\Delta z}$) of the displacements can be calculated as follows:

$$S_{\Delta x} = \frac{1}{2}\cdot\frac{g}{2\pi}\cdot\left\{\arctan\left(\frac{Q_{X+1}}{Q'_{X+1}}\right) - \arctan\left(\frac{Q_{X-1}}{Q'_{X-1}}\right)\right\} = \Delta x, \tag{4.21}$$

$$S_{\Delta z} = \frac{1}{2}\cdot\frac{1}{1+\cos\theta}\cdot\frac{\lambda}{2\pi}\cdot\left\{\arctan\left(\frac{Q_{X+1}}{Q'_{X+1}}\right) + \arctan\left(\frac{Q_{X-1}}{Q'_{X-1}}\right)\right\} = \Delta z\,. \tag{4.22}$$

Similar to a conventional linear encoder based on grating diffraction [15], the resolution and range of the proposed 2-DOF linear encoder along the X-axis, which are basically determined by the pitch and length of the scale grating, are expected to be better than 1 nm and up to several hundred millimeters, respectively. Similar to a conventional Michelson interferometer employing an LD as the light source, the resolution and range of the proposed 2-DOF linear encoder along the Z-axis are expected to be better than 1 nm and up to 100 μm [16]. These specifications, which are the design goals for the prototype 2-DOF linear encoder described in the following, are qualified for the position and straightness measurement of a precision linear stage.

Figure 4.5 shows a photograph of the prototype 2-DOF linear encoder designed and constructed based on the optical configuration shown in Figure 4.4. The size of the optical sensor head was about 50 mm (X) × 50 mm (Y) × 30 mm (Z). An LD with a wavelength of 685 nm was employed as the light source. The diameter of the collimated laser beam through the aperture was approximately 1 mm. Commercial gratings with a pitch of 1.6 μm and a length of 12.7 mm were employed as the scale grating and the reference grating. It should be noted that a longer grating can be employed as the scale grating for larger range measurement of displacement along the X-axis. The scale grating was mounted on a two-axis PZT stage. The working distance between the scale grating and the optical sensor head was about 5 mm. Current signals from the PDs, which correspond to the intensities of the interference beams, were first converted to voltage signals by a current-to-voltage converter circuit. The amplitudes of the voltage signals were

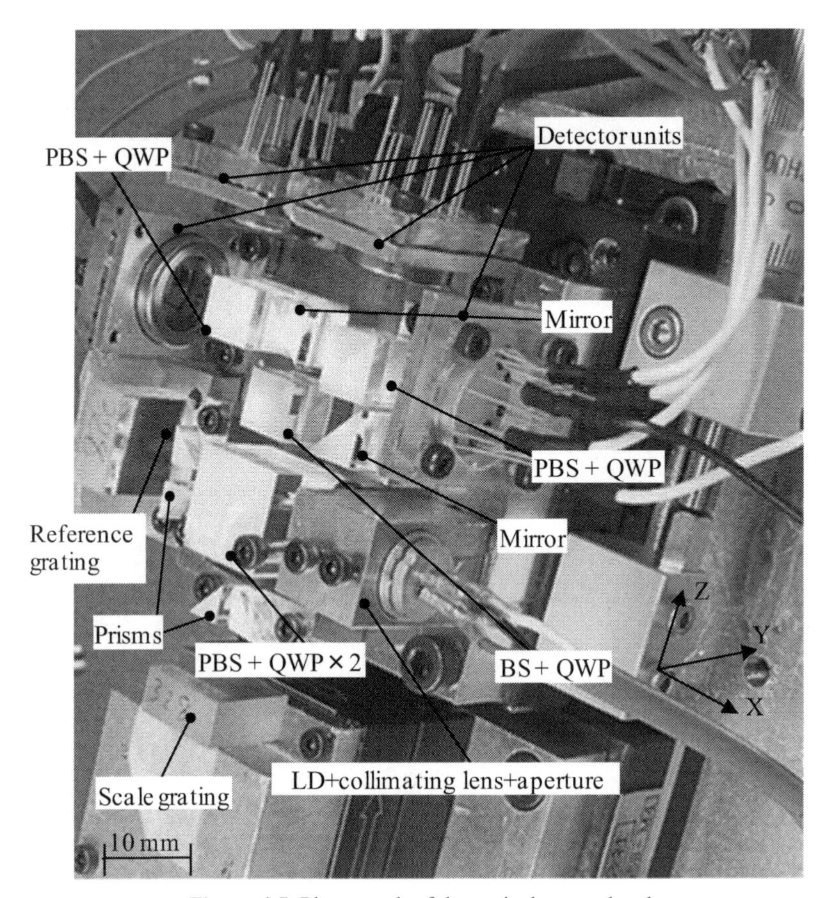

Figure 4.5. Photograph of the optical sensor head

adjusted to be identical by using an adjustment circuit. Q_α and Q'_α were taken into a personal computer via a 16-bit analog-to-digital converter. The *X*- and *Z*-directional sensor outputs ($S_{\Delta x}$, $S_{\Delta z}$) were calculated in the personal computer based on Equations 4.21 and 4.22. The bandwidth of the analog devices in the sensor electronics was about 100 kHz at –3 dB.

Resolutions of the prototype 2-DOF linear encoder were tested through detecting the vibration of the two-axis PZT stage, which was kept stationary. The vibration was also detected by a commercial displacement sensor with a resolution of 0.08 nm [17] at the same time. The sampling rate of data acquisition was 100 kHz. Because the commercial displacement sensor could only detect one-axis displacement, the position of the displacement sensor was changed for the measurement along different axes as shown in Figures 4.6 and 4.7. The results are plotted in Figure 4.8. The vibration components with a frequency of approximately 250 Hz and an amplitude of approximately 0.5 nm were detected by both the sensors, indicating that the prototype 2-DOF linear encoder had resolutions better than 0.5 nm in both the *X*- and *Z*-axes.

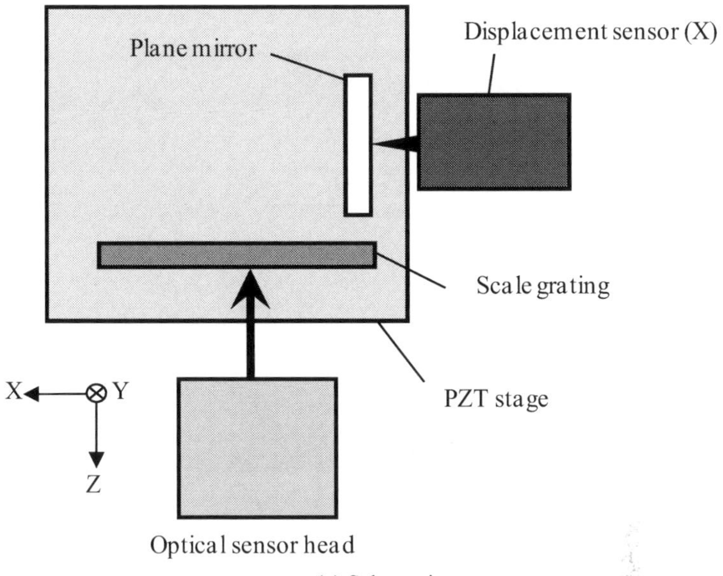

(a) Schematic

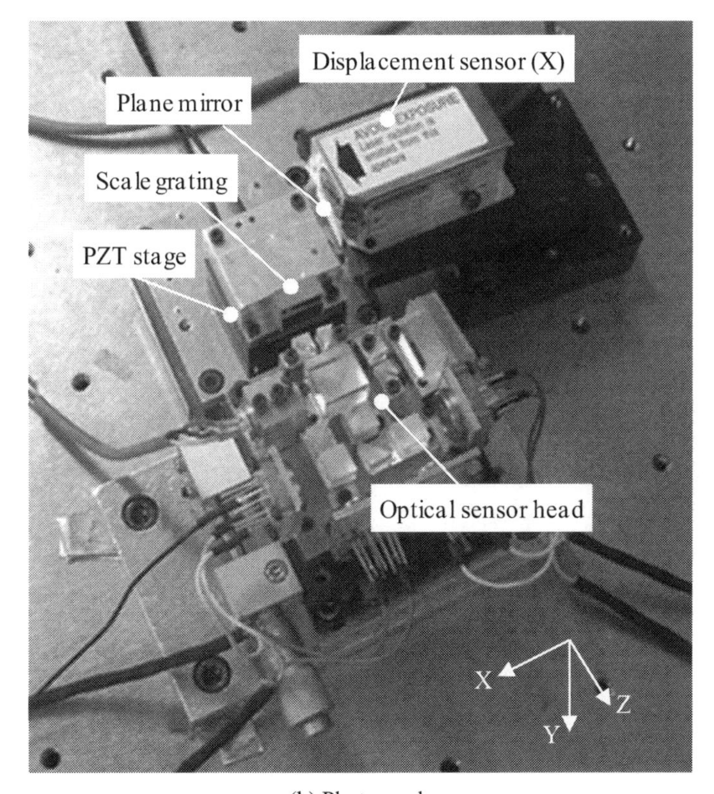

(b) Photograph

Figure 4.6. Setup for monitoring X-directional vibration of a PZT stage

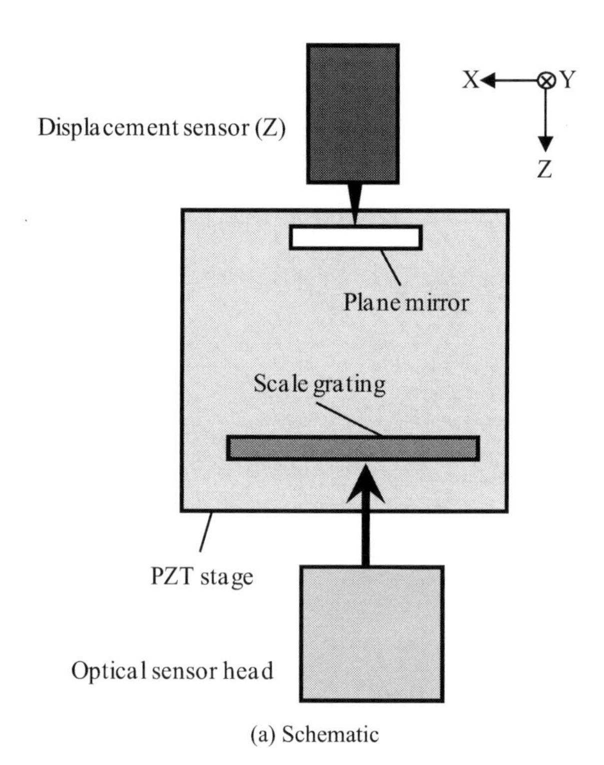

(a) Schematic

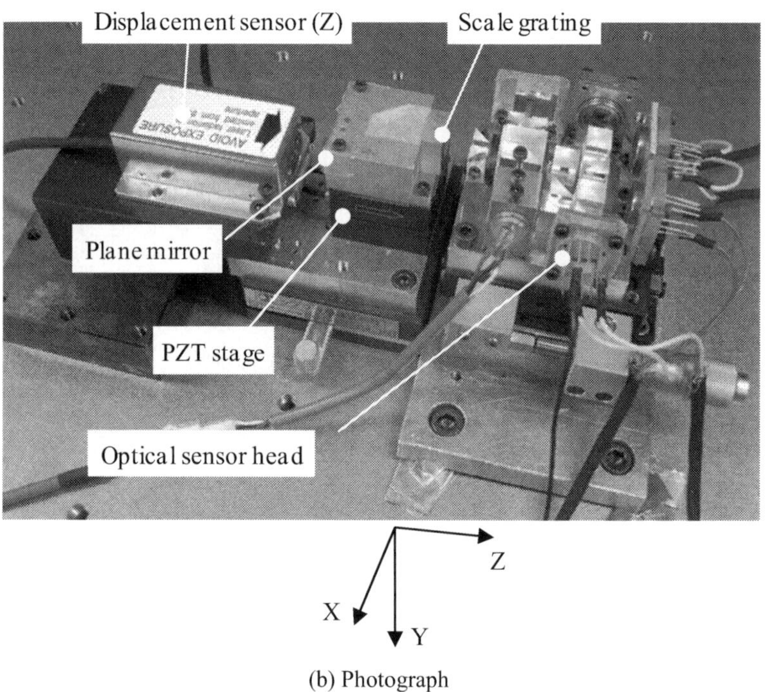

(b) Photograph

Figure 4.7. Setup for monitoring Z-directional vibration of a PZT stage

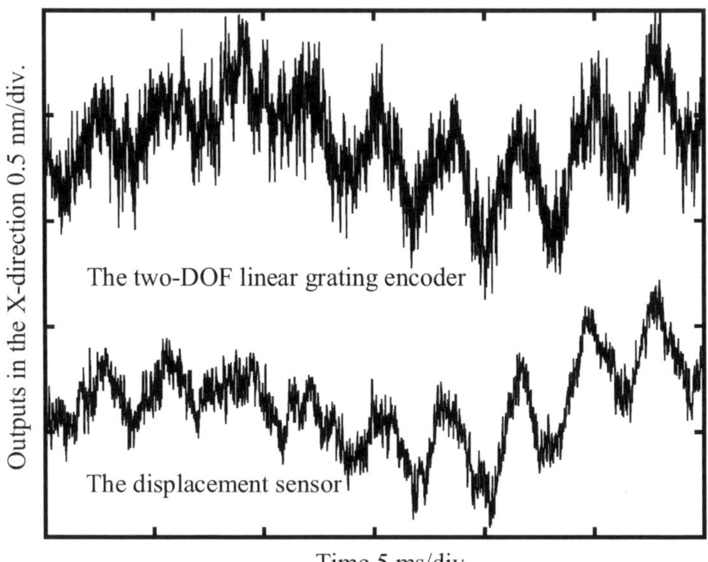

(a) Outputs in the *X*-direction

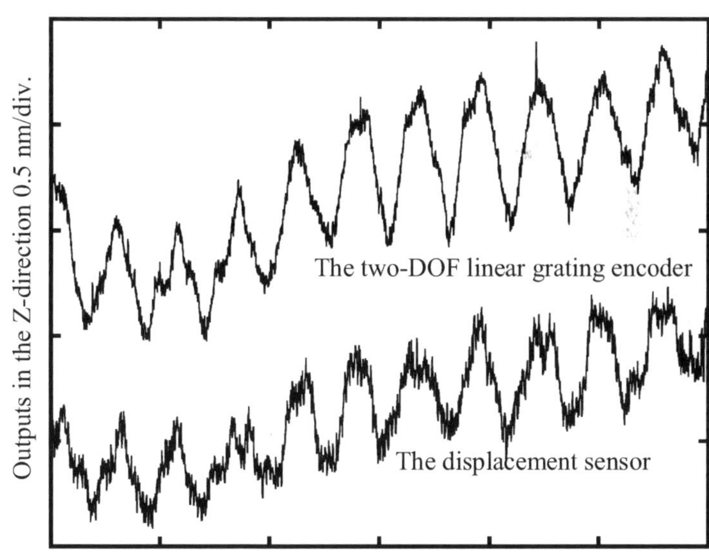

(b) Outputs in the *Z*-direction

Figure 4.8. Results of vibration detection

Figure 4.9 shows a setup for evaluating interpolation errors of the prototype 2-DOF linear encoder, which are associated with the subdivision of the quadrature interference signals within a signal period. Two commercial interferometers [16] were employed as references to measure the *X*- and *Z*-directional motions of the two-axis PZT stage simultaneously. Figures 4.10 and 4.11 show the quadrature interference signals Q_α and Q'_α and corresponding interpolation errors when the scale grating was moved over five signal periods by the PZT stage along the *X*- and *Z*-axes, respectively. The periodic interpolation errors were within ±1% of the signal period of the quadrature interference signals. In Figures 4.12 and 4.13, the measurement errors of displacements in the *X*- and *Z*-directional sensor outputs, which were caused by the interpolation errors, are plotted. It can be seen that the measurement errors of displacements caused by the interpolation errors were 20 nm and 5 nm in the *X*- and *Z*-axes, respectively.

Figure 4.14 shows the amplitudes of Q_α and Q'_α as well as corresponding measurement errors of displacements when the scale grating had different offsets Δw_d from the working position of the optical sensor head, which was designed to be about 5 mm along the *Z*-axis. The measurement errors remained almost constant, although the amplitudes of the quadrature interference signals varied significantly when Δw_d changed over a range of 1 mm. This indicates that the optical sensor head had a tolerance of approximately ±0.5 mm for the working position along the *Z*-axis. Figures 4.15 to 4.17 show the results for the change of the angular deviations $\Delta\theta_X$, $\Delta\theta_Y$ and $\Delta\theta_Z$ between the scale grating and the optical sensor head about the *X*-, *Y*- and *Z*-axes, respectively. The angular displacements were applied to the scale grating by manual tilt stages. The angular tolerances between the optical sensor head and the scale grating are approximately ±80 arcsec ($\Delta\theta_X$), ±80 arcsec ($\Delta\theta_Y$) and ±320 arcsec ($\Delta\theta_Z$), respectively.

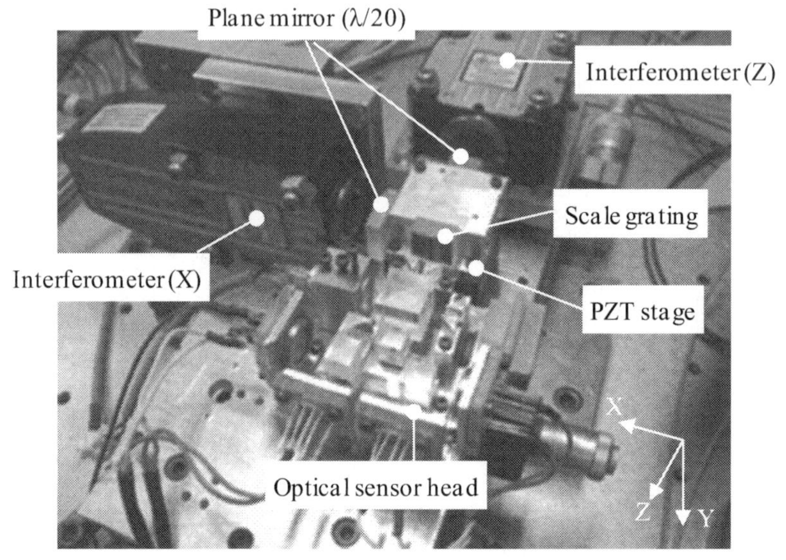

Figure 4.9. Setup for evaluation of interpolation errors

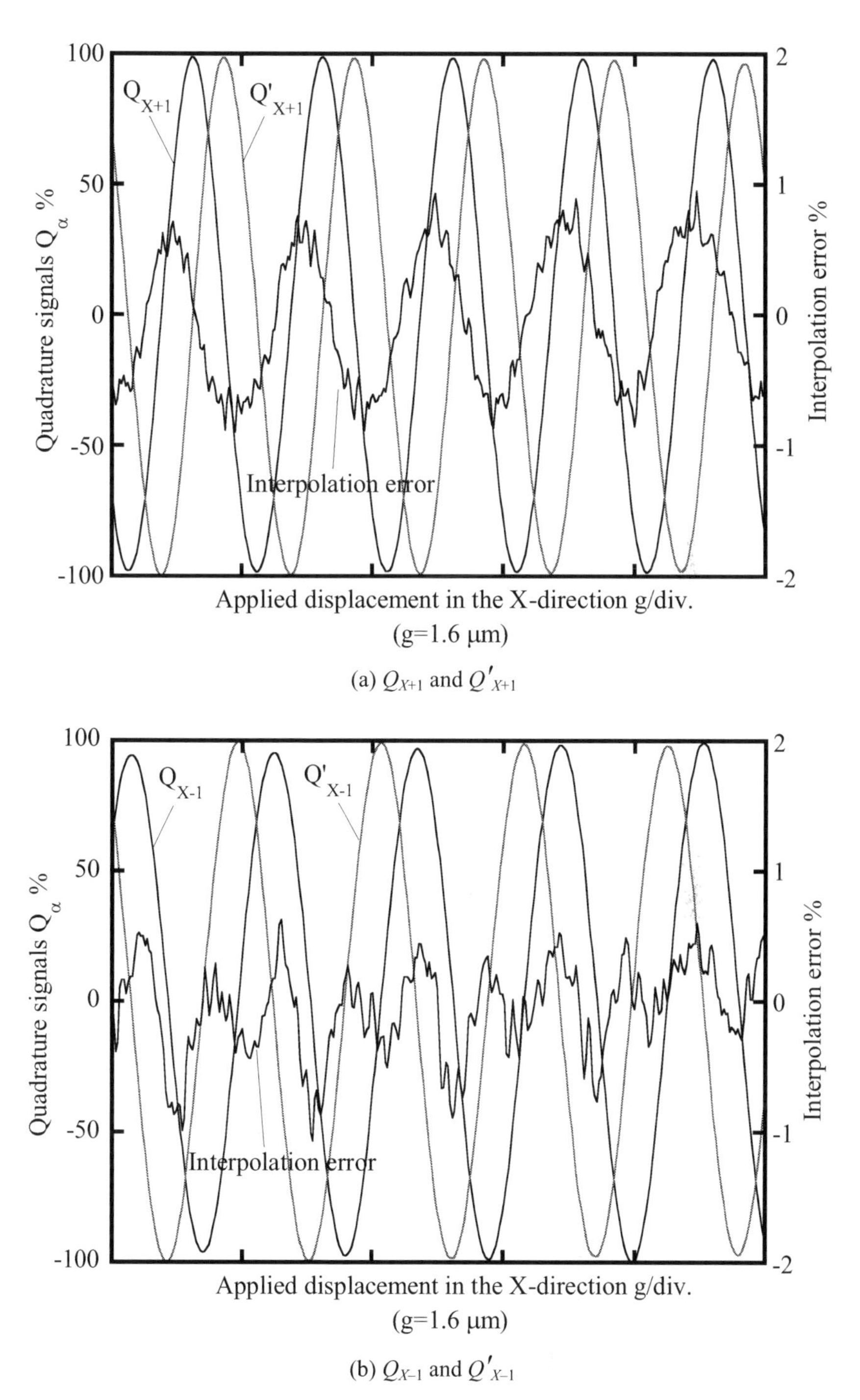

(a) Q_{X+1} and Q'_{X+1}

(b) Q_{X-1} and Q'_{X-1}

Figure 4.10. Interpolation errors of quadrature interference signals for *X*-directional measurement

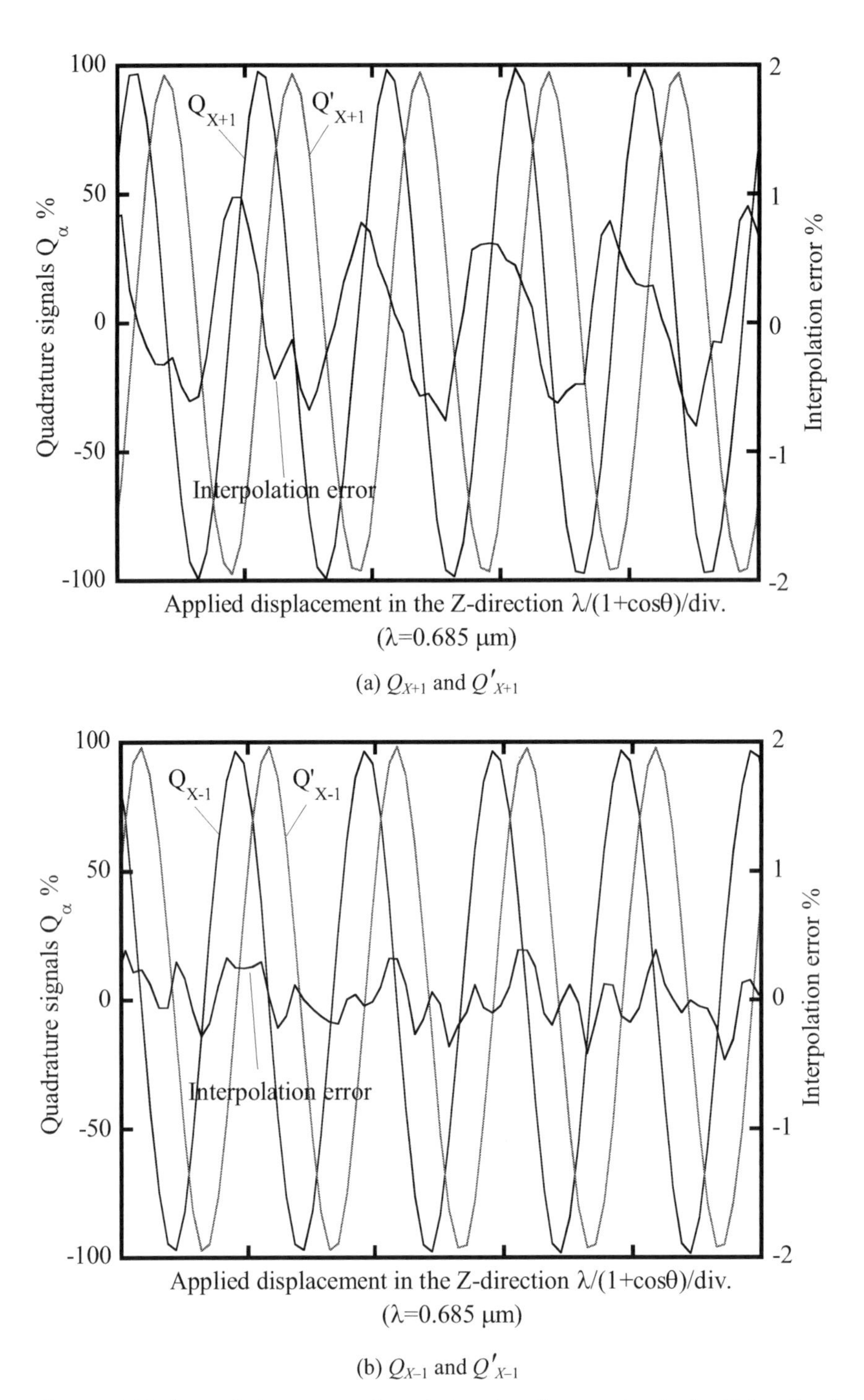

Figure 4.11. Interpolation errors of quadrature interference signals for *Z*-directional measurement

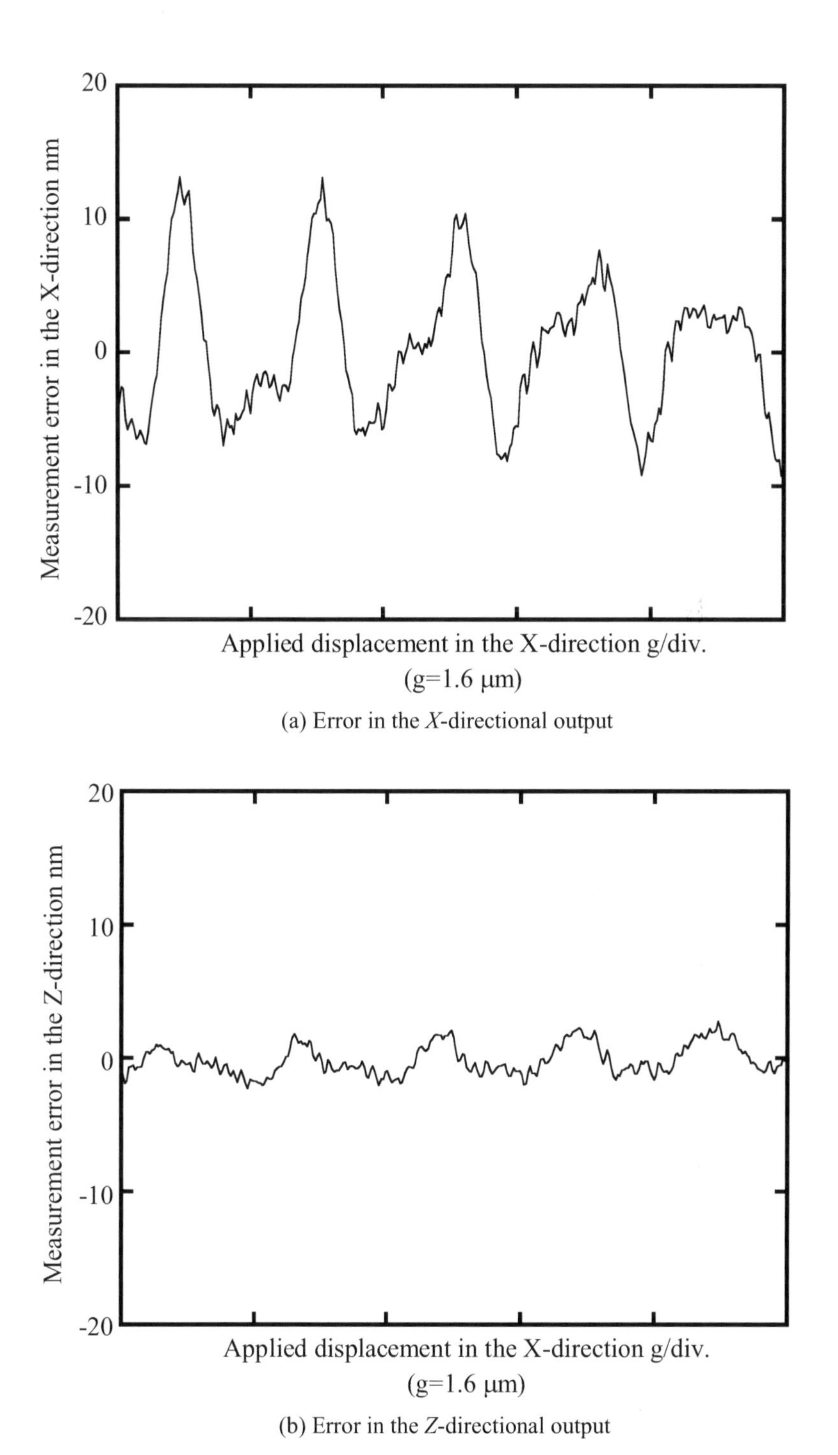

(a) Error in the *X*-directional output

(b) Error in the *Z*-directional output

Figure 4.12. Measurement errors of the 2-DOF linear encoder when a displacement is applied in the *X*-direction

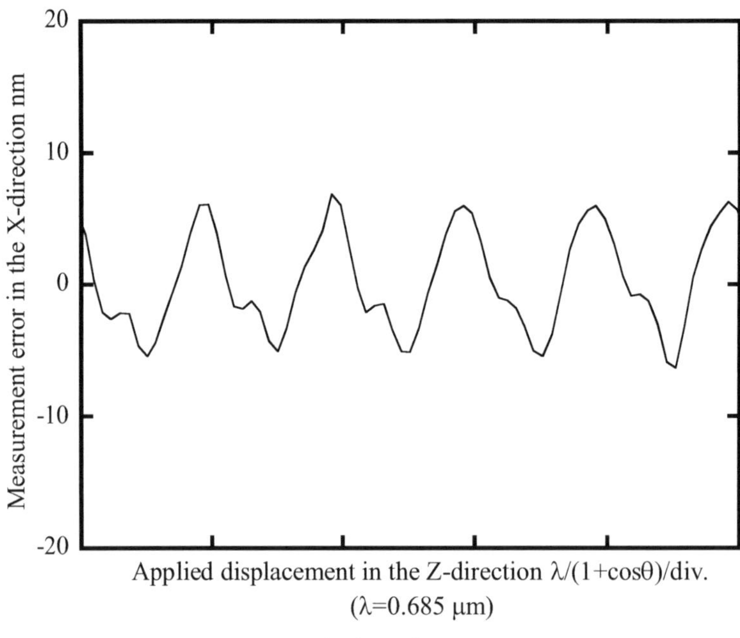

(a) Error in the *X*-directional output

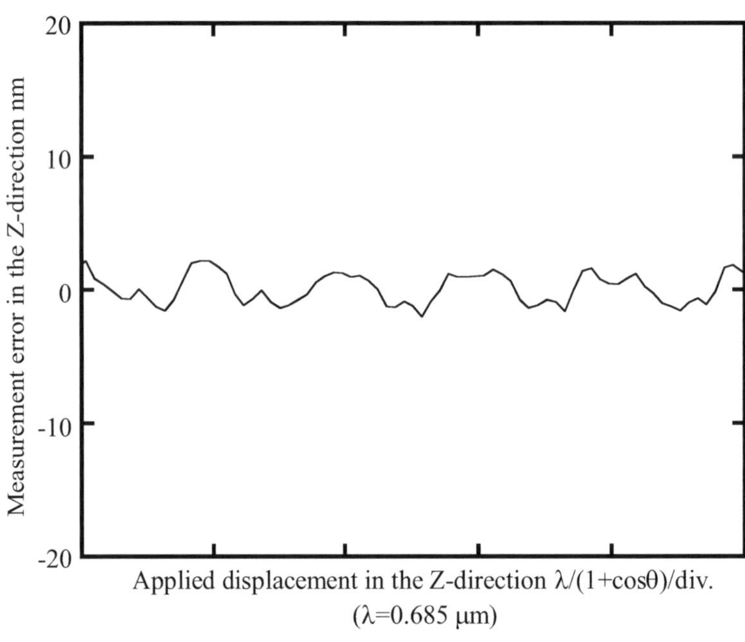

(b) Error in the *Z*-directional output

Figure 4.13. Measurement errors of the 2-DOF linear encoder when a displacement is applied in the *Z*-direction

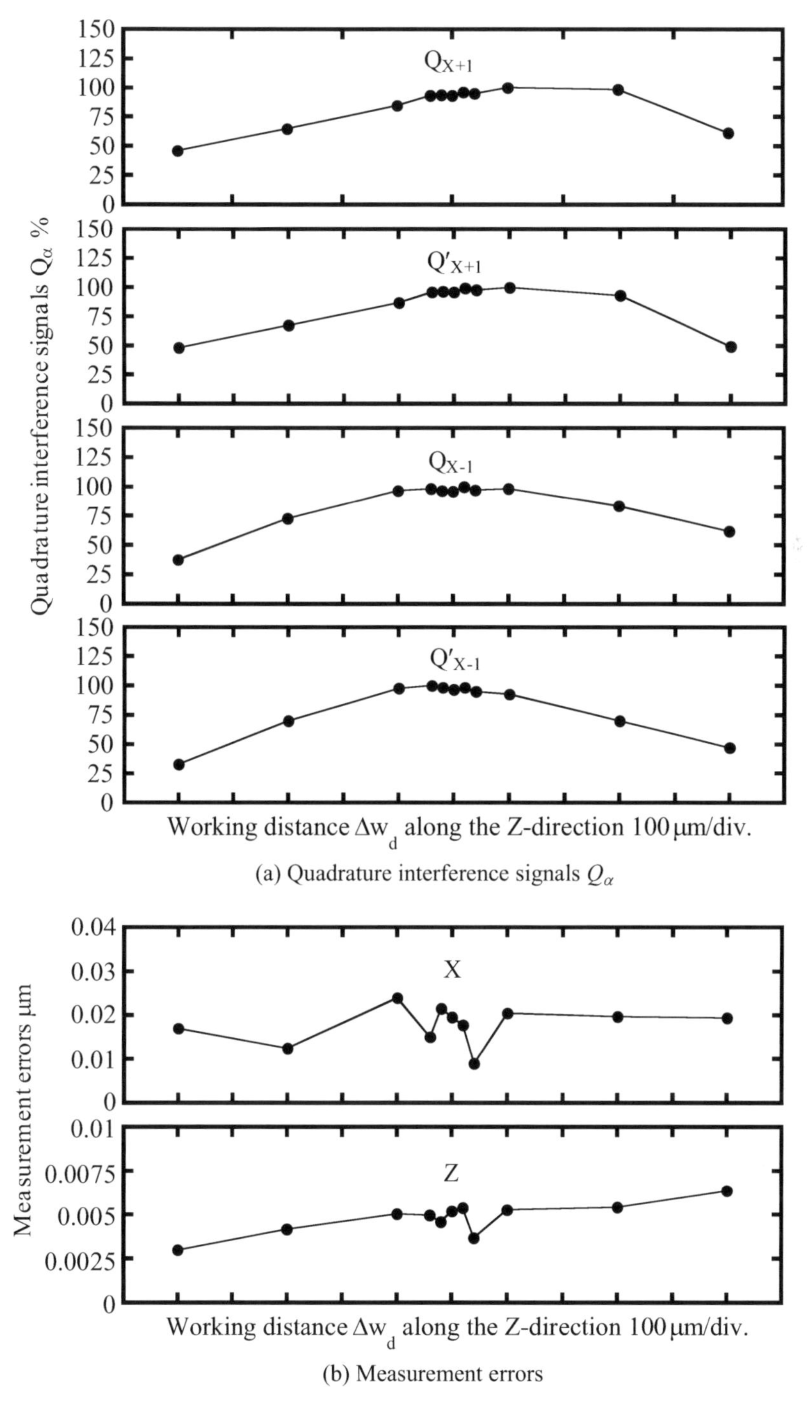

(a) Quadrature interference signals Q_α

(b) Measurement errors

Figure 4.14. Amplitudes of the quadrature interference signals and measurement errors with respect to the change of the working distance Δw_d along the Z-axis

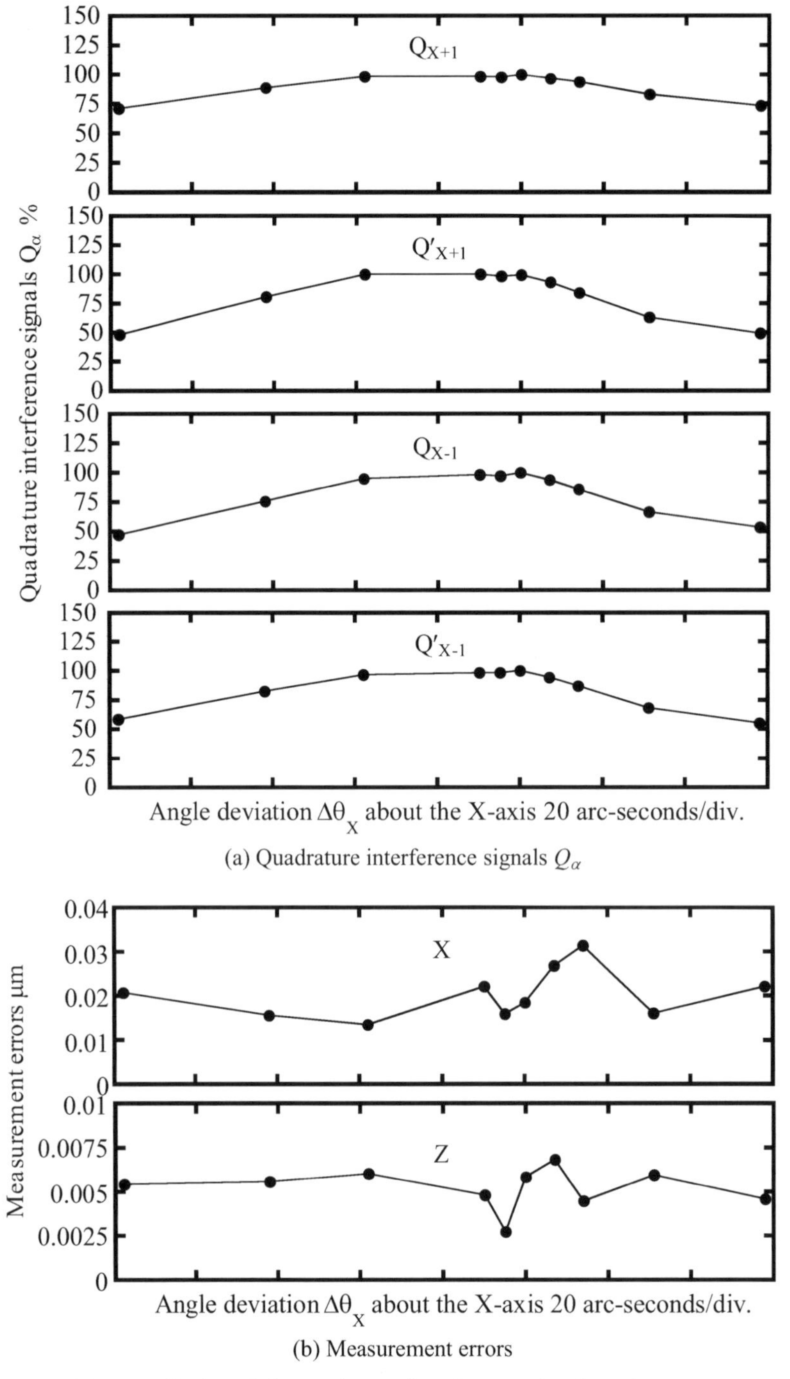

(a) Quadrature interference signals Q_α

(b) Measurement errors

Figure 4.15. Amplitudes of the quadrature interference signals and measurement errors with respect to angular deviation $\Delta\theta_x$ about the *X*-axis

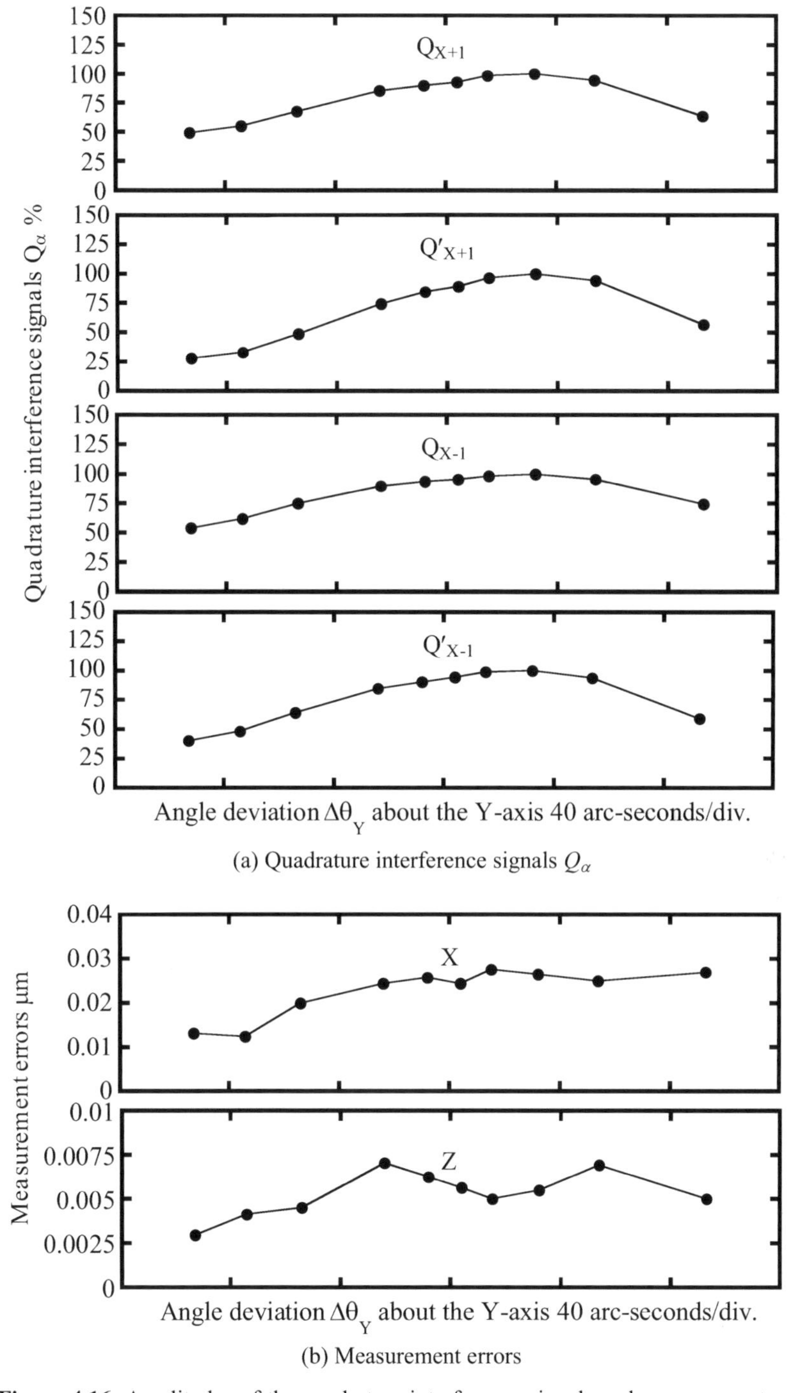

Figure 4.16. Amplitudes of the quadrature interference signals and measurement errors with respect to angular deviation $\Delta\theta_Y$ about the *Y*-axis

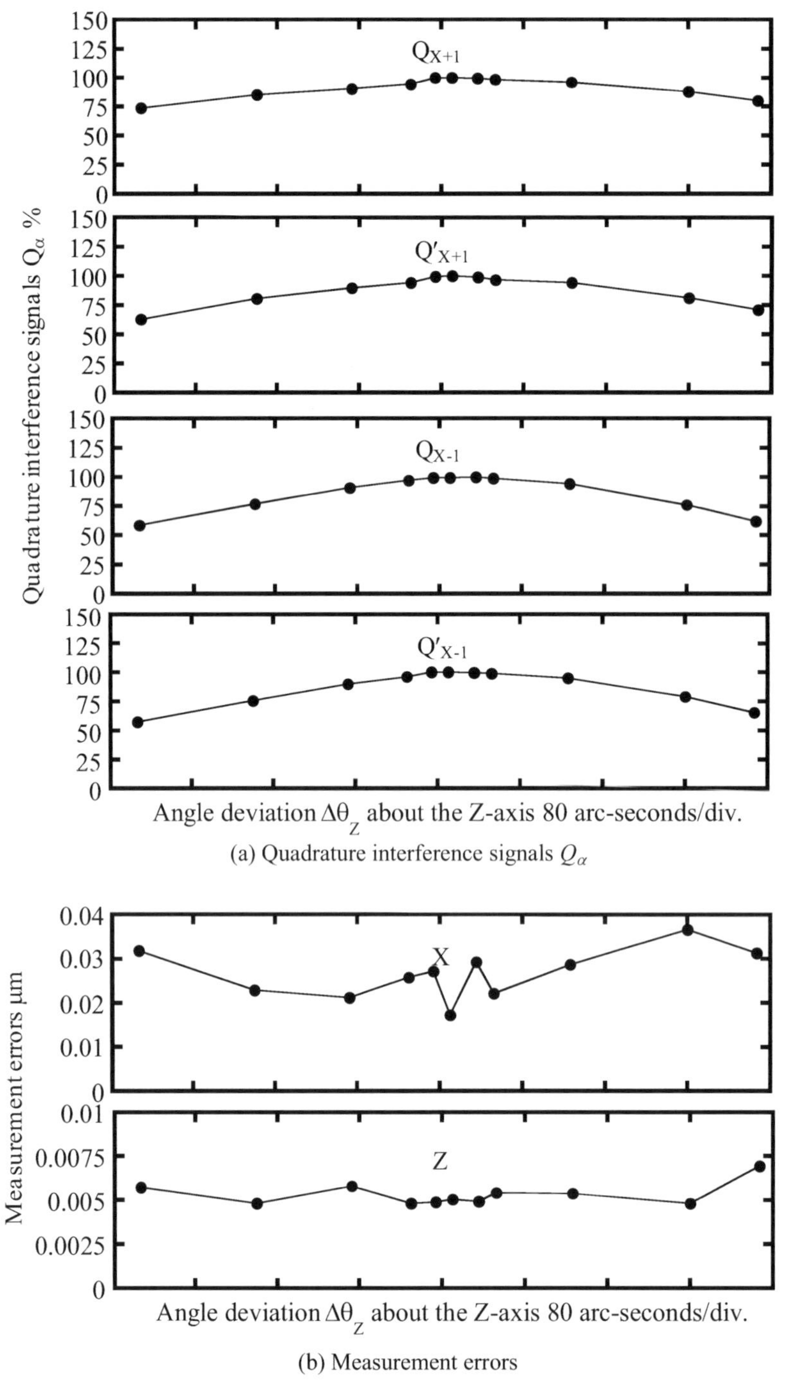

Figure 4.17. Amplitudes of the quadrature interference signals and measurement errors with respect to angular deviation $\Delta\theta_z$ about the Z-axis

4.3 Three-axis Grating Encoder with Sinusoidal Grid

The measurement principle of the 2-DOF linear encoder is expanded to a three-axis grating encoder for displacement measurement along the *X*-, *Y*- and *Z*-axes. Figure 4.18 shows the basic principle for detection of the three-axis displacements. The grating encoder is composed of a read head and a sinusoidal scale grid, which functions as a phase grating. The read head consists of a laser source, a polarizing beam splitter (PBS), two quarter waveplates (QWPs), a sinusoidal reference grid and a photodetector unit. The read head is kept stationary to detect the displacements of the moving scale grid along the *X*-, *Y*- and *Z*-directions, which are denoted as Δx, Δy and Δz, respectively. The scale grid and the reference grid have identical pitches (g) and amplitudes (H).

A collimated beam is output from the laser source with a wavelength of λ. The laser beam is divided into the component of p-polarization and that of s-polarization at the PBS. The p-polarized beam and the s-polarized beam are projected onto the scale grid and the reference grid, respectively. The *X*-directional positive and negative first-order diffracted beams and the *Y*-directional positive and negative first-order diffracted beams generated at each of the grids are employed for measurement of the three-axis displacements. The diffracted beams from the scale grid are bent at the PBS and those from the reference grid pass through the PBS so that the two groups of diffracted beams can interference with each other at the photodetector unit consisting of four photodiodes (A, B, C, D). The polarizer,

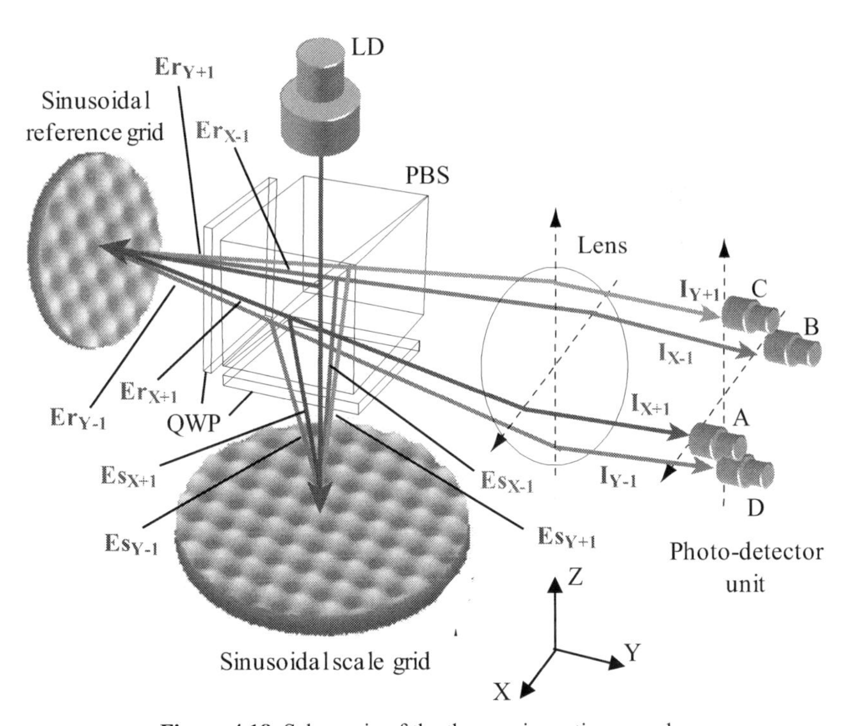

Figure 4.18. Schematic of the three-axis grating encoder

which is placed in front of the photodetector unit in order to obtain the same polarization components from the p- and s-polarizations, is not shown in the figure for simplicity.

Similar to the 2-DOF linear encoder, the interference signals between the corresponding diffracted beams from the scale grating and the reference grating can be expressed as follows [18]:

$$I_{X+1} = (Es_{X+1} + Er_{X+1}) \cdot \overline{Es_{X+1} + Er_{X+1}} = 2E_0{}^2\{1 + \cos(\varphi + \omega)\}, \quad (4.23)$$

$$I_{X-1} = (Es_{X-1} + Er_{X-1}) \cdot \overline{Es_{X-1} + Er_{X-1}} = 2E_0{}^2\{1 + \cos(-\varphi + \omega)\}, \quad (4.24)$$

$$I_{Y+1} = (Es_{Y+1} + Er_{Y+1}) \cdot \overline{Es_{Y+1} + Er_{Y+1}} = 2E_0{}^2\{1 + \cos(\tau + \omega)\}, \quad (4.25)$$

$$I_{Y-1} = (Es_{Y-1} + Er_{Y-1}) \cdot \overline{(Es_{Y-1} + Er_{Y-1})} = 2E_0{}^2\{1 + \cos(-\tau + \omega)\}, \quad (4.26)$$

where φ, ω are the phase differences along the *X*- and *Z*-axes shown in Equations 4.3 and 4.5. The phase difference τ along the *Y*-axis can be expressed by:

$$\tau = \frac{2\pi}{g}\Delta y, \quad (4.27)$$

where *g* is the pitch of the sinusoidal grid.

Figure 4.19 shows the optical arrangement of the grating encoder that can distinguish the moving direction and remove the influence of the variation of E_0, which is proportional to the intensity of the light source. The interference waves of the diffracted beams from the grids are divided by a non-polarization BS and two PBSs to generate four sets of beams. Each set of the beams are detected by a photodetector unit shown in Figure 4.18. Three QWPs are placed in the optical paths between the BS and PBSs to adjust the phase and the polarization direction of each beam. The fast axes of the QWPs are shown in the figure.

The quadrature interference signals Q_α and Q'_α ($\alpha = X + 1,\ X - 1$) can be obtained from I_{X+1} and I_{X-1} as shown in Equations 4.15–4.18. Q_α and Q'_α ($\alpha = Y + 1,\ Y - 1$) can be obtained from I_{Y+1} and I_{Y-1} as follows:

$$Q_{Y+1} = \frac{I_{Y+1}(0^o) - I_{Y+1}(180^o)}{I_{Y+1}(0^o) + I_{Y+1}(180^o)} = \sin(\tau + \omega), \quad (4.28)$$

$$Q'_{Y+1} = \frac{I_{Y+1}(90^o) - I_{Y+1}(270^o)}{I_{Y+1}(90^o) + I_{Y+1}(270^o)} = \cos(\tau + \omega), \quad (4.29)$$

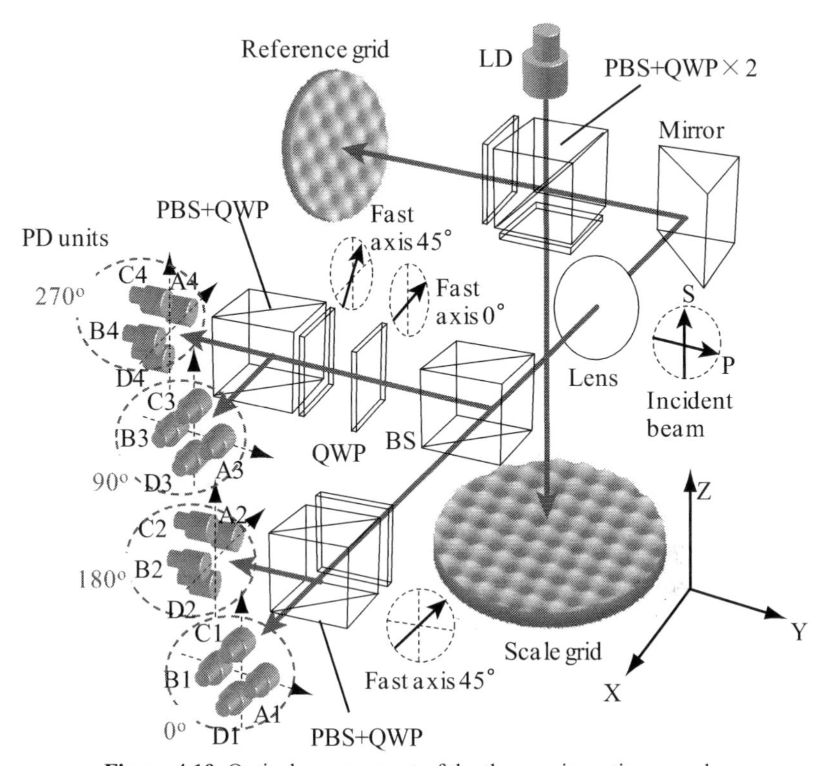

Figure 4.19. Optical arrangement of the three-axis grating encoder

$$Q_{Y-1} = \frac{I_{Y-1}(0^o) - I_{Y-1}(180^o)}{I_{Y-1}(0^o) + I_{Y-1}(180^o)} = \sin(-\tau + \omega), \tag{4.30}$$

$$Q'_{Y-1} = \frac{I_{Y-1}(90^o) - I_{Y-1}(270^o)}{I_{Y-1}(90^o) + I_{Y-1}(270^o)} = \cos(-\tau + \omega). \tag{4.31}$$

The three-axis sensor outputs ($S_{\Delta x}$, $S_{\Delta y}$, $S_{\Delta z}$) can be calculated as follows to obtained the displacements along X-, Y- and Z-axes:

$$S_{\Delta x} = \frac{1}{2} \cdot \frac{g}{2\pi} \cdot \left\{ \arctan\left(\frac{Q_{X+1}}{Q'_{X+1}} \right) - \arctan\left(\frac{Q_{X-1}}{Q'_{X-1}} \right) \right\} = \Delta x, \tag{4.32}$$

$$S_{\Delta y} = \frac{1}{2} \cdot \frac{g}{2\pi} \cdot \left\{ \arctan\left(\frac{Q_{Y+1}}{Q'_{Y+1}} \right) - \arctan\left(\frac{Q_{Y-1}}{Q'_{Y-1}} \right) \right\} = \Delta y, \tag{4.33}$$

$$S_{\Delta y} = \frac{1}{4} \cdot \frac{1}{1+\cos\theta} \cdot \frac{\lambda}{2\pi} \left\{ \arctan\left(\frac{Q_{X+1}}{S'_{X+1}}\right) + \arctan\left(\frac{Q_{X-1}}{S'_{X-1}}\right) \right.$$

$$\left. + \arctan\left(\frac{Q_{Y+1}}{Q'_{Y+1}}\right) + \arctan\left(\frac{Q_{Y-1}}{Q'_{Y-1}}\right) \right\} = \Delta z \,. \tag{4.34}$$

Figure 4.20 shows a photograph of the prototype three-axis grating encoder. A 1.5 mW laser diode with a wavelength of 685 nm is employed as the optical source. The diameter of the collimated laser beam is 6 mm. The size of the sensor is 100 mm (*X*) × 100 mm (*Y*) × 60 mm (*Z*). The sinusoidal *XY*-grids are fabricated in a diamond turning machine equipped with a fast-tool-servo shown in Chapter 3. The pitch of the sine function of the grid is set to be 10 μm based on the ability of the fabrication system. The amplitude of the sine function is designed to be 60 nm based on the result of an optical simulation, at which the first-order diffraction beams have the maximum intensities. The area of the reference grid is 10 mm in diameter, which is required to be larger than the beam diameter. The area of the scale, which determines the measurement range in *XY*-directions, is 20 mm in diameter. It takes approximately 22 h for the fabrication of the sinusoidal grid over such an area. A larger fabrication area can be reached by increasing the fabrication time. Figure 4.21 shows a photograph of the fabrication system and a microscope image of a part of the fabricated grid.

The diffraction angle of the first-order diffraction beams is calculated to be 3.93°. The optical head is designed in such a way that only the first-order diffraction beams can be incident to the optical head.

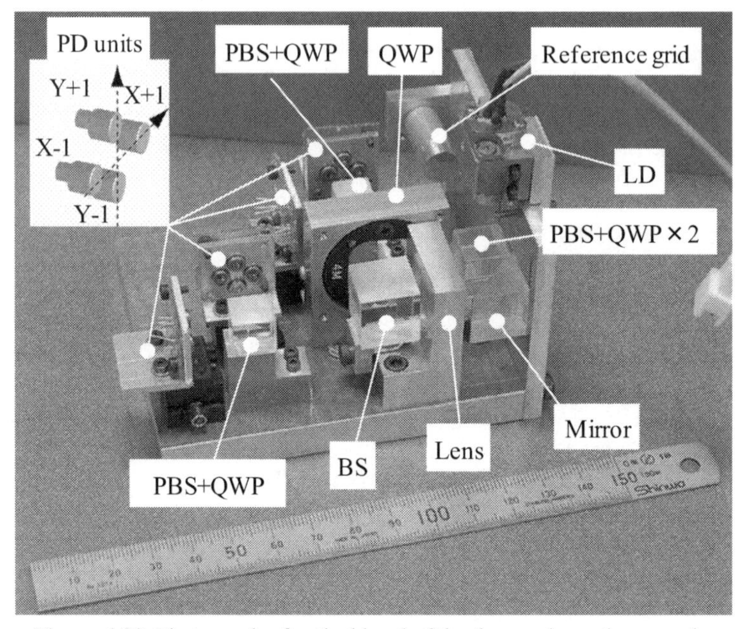

Figure 4.20. Photograph of optical head of the three-axis grating encoder

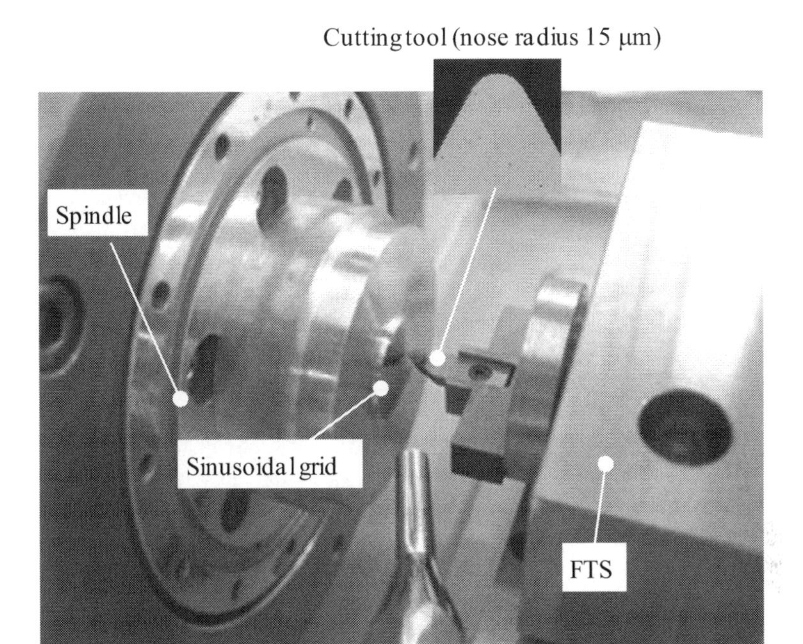

(a) Fabrication system of the sinusoidal grid

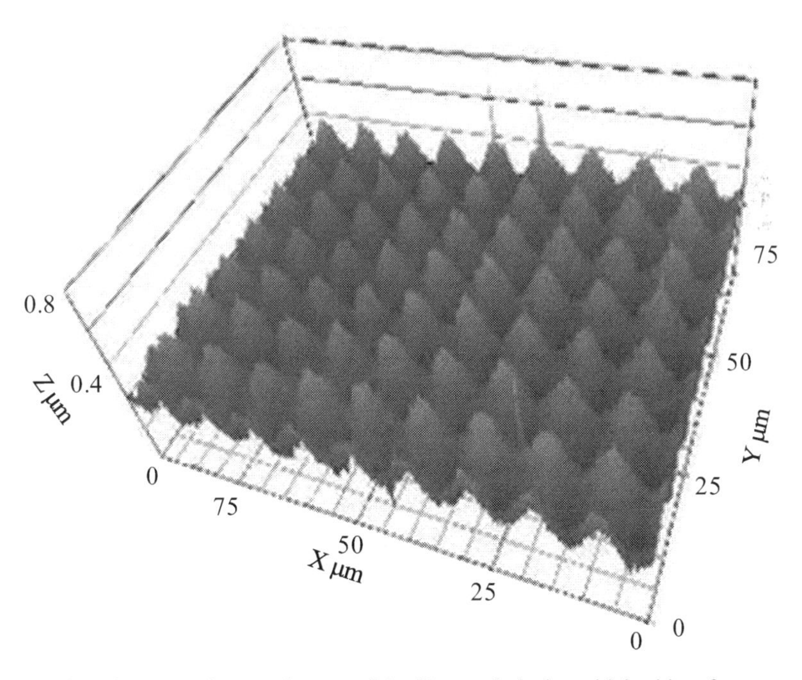

(b) Microscope image of a part of the 10-μm pitch sinusoidal grid surface

Figure 4.21. Fabrication of the 10-μm pitch sinusoidal grid surface

Experiments are carried out to test the basic performance of the prototype three-axis grating encoder. The reading head is kept stationary and the scale grid is mounted on a commercially available *XYZ*-stage system driven by PZT actuators. The bandwidth of the prototype grating encoder is set to be 0–30Hz. The stage system is composed by overlapping three single-axis stages. The bottom is the *X*-stage and the top is the *Z*-stage. Each of the stages is equipped with a capacitive displacement sensor, which makes the stage have a closed-loop linearity of 0.02% over a stroke of 100 μm. Figure 4.22 shows a result of the multi-axis displacement measurement. The scale grid is first moved 100 μm along the *X*-axis, then 100 μm along the *Y*-axis, finally 20 μm along the *Z*-axis, respectively. It can be seen that the grating encoder can measure the *XYZ*-displacement simultaneously. The maximum angular differences between the axes of the results by the grating encoder and the capacitive sensors are 0.011° (*X*-axis), 0.093° (*Y*-axis), and 0.86° (*Z*-axis), respectively. The differences are caused by the misalignment of the stage axes and the sensor axes. Figures 4.23 to 4.25 show the results of resolution test in the *X*-, *Y*- and *Z*-axes, respectively. The scale grid is moved with a step of 5 nm in each of the axes. Both the output data of the grating encoder and the capacitive sensors are plotted in the figure. It can be seen that the grating encoder has the ability to detect the three-axis displacement with a nanometric resolution.

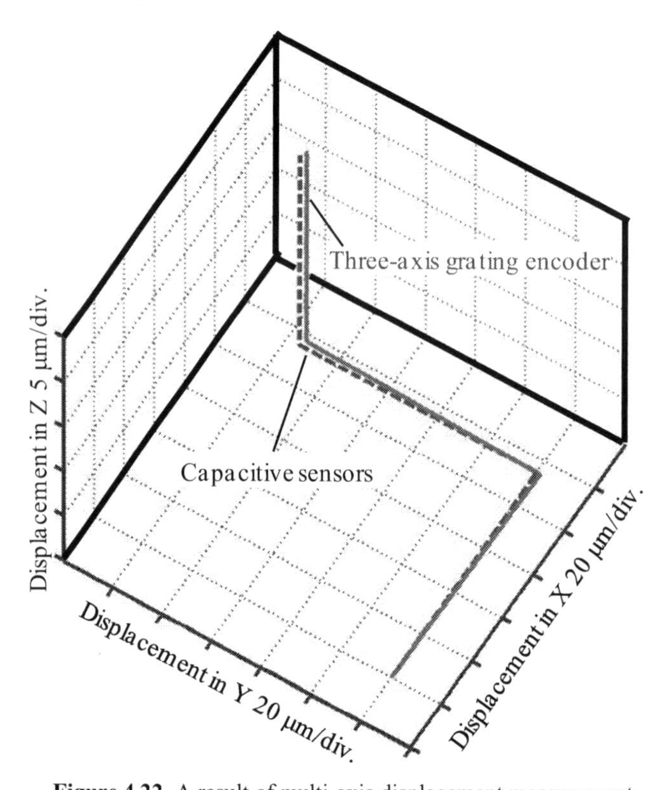

Figure 4.22. A result of multi-axis displacement measurement

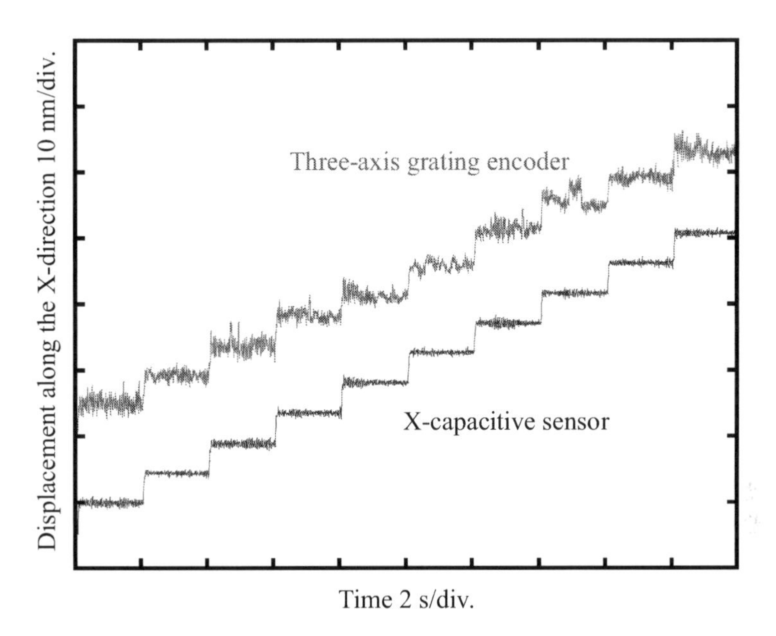

Figure 4.23. Results of resolution test in the *X*-direction

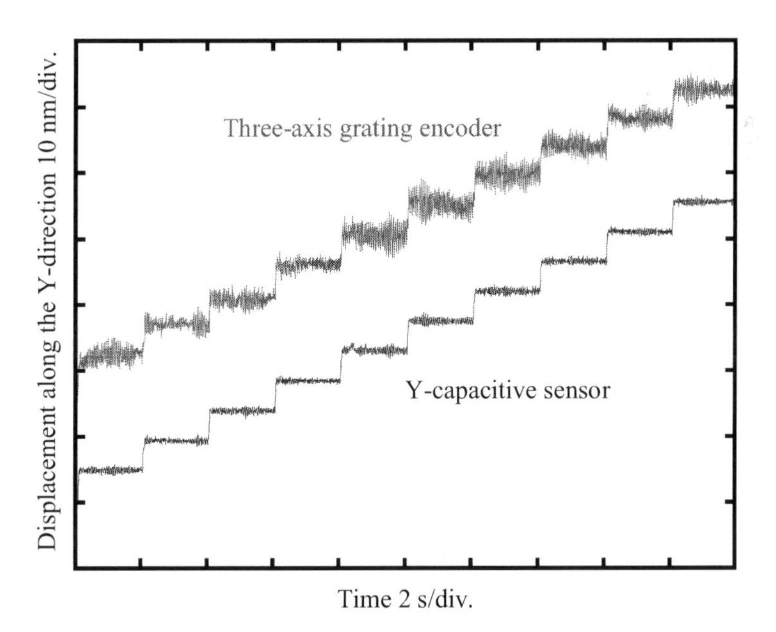

Figure 4.24. Results of resolution test in the *Y*-direction

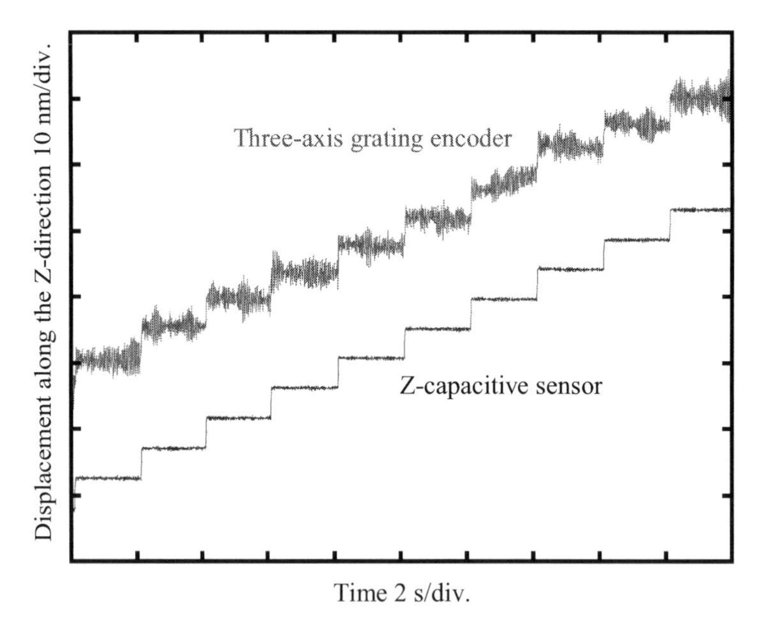

Figure 4.25. Results of resolution test in the *Z*-direction

4.4 Three-axis Grating Encoder with Rectangular *XY*-grid

The resolutions of the three-axis grating encoder in the *X*- and *Y*-directions are determined by the pitch of the *XY*-grid. It is difficult and time-consuming to fabricate a sinusoidal grid with pitches less than 10 μm by mechanical cutting shown in Figure 4.21 (a). In this section, rectangular *XY*-grids with a pitch of 1 μm, which are made by photolithography [19], are employed as the scale grid and the reference grid. Figure 4.26 (a) shows a schematic of the grid. A photograph and a scanning electron microscope image of the grid are shown in Figure 4.26 (b). The grid is designed to have a depth of 0.2 μm and a width of 0.35 μm so that the first-order diffracted beams have large intensities. The size of the scale grid used in the experiment is 30 mm (*X*) × 30 mm (*Y*), and that of the reference grid is 10 mm (*X*) × 10 mm (*Y*). The ability of the photolithography facility allows fabrication of a grid with a maximum size of 130 mm (*X*) × 130 mm (*Y*).

Figure 4.27 shows a schematic of the optical head designed for the 1 μm-pitched *XY*-grid. The principle is the same as that shown in Figures 4.18 and 4.19. A laser diode with a wavelength of 685 nm is employed as the light source. The diffraction angle of the first-order diffracted beams is calculated to be 43.2° with the grid, which is ten times larger than the 10 μm-pitched grid used in Section 4.3. A transparent grid shown in Figure 4.27 (b) is thus employed to bend the diffracted beams from the grid. The grid pitch, which is identical to the scale grid, is 1 μm. The depth of the grid is designed to be 0.6 μm to get high transparent efficiencies for the first-order diffracted beams from the scale grid.

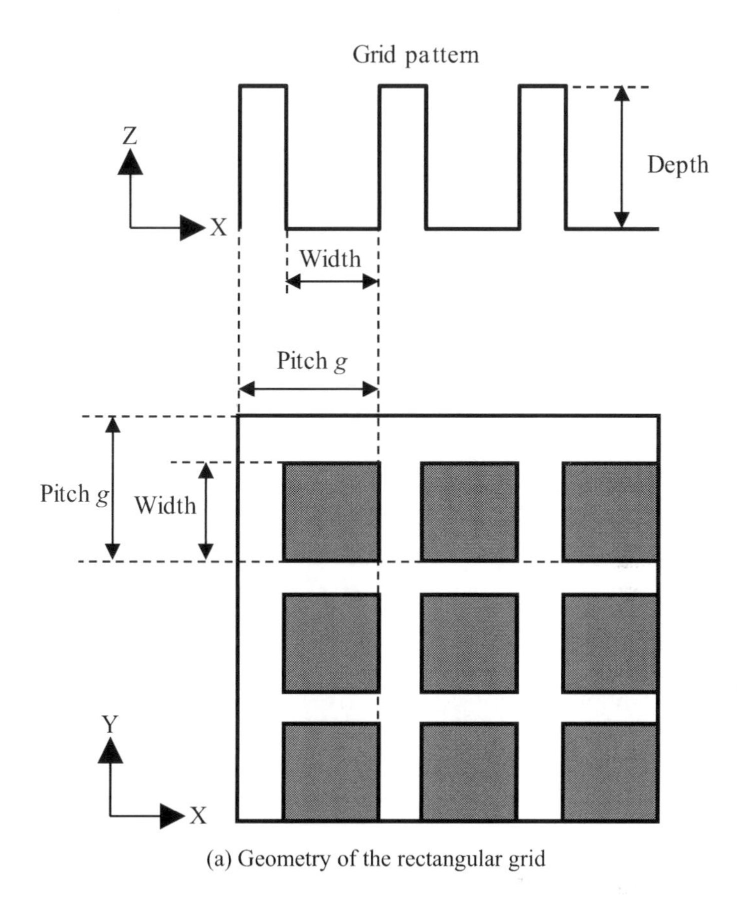

(a) Geometry of the rectangular grid

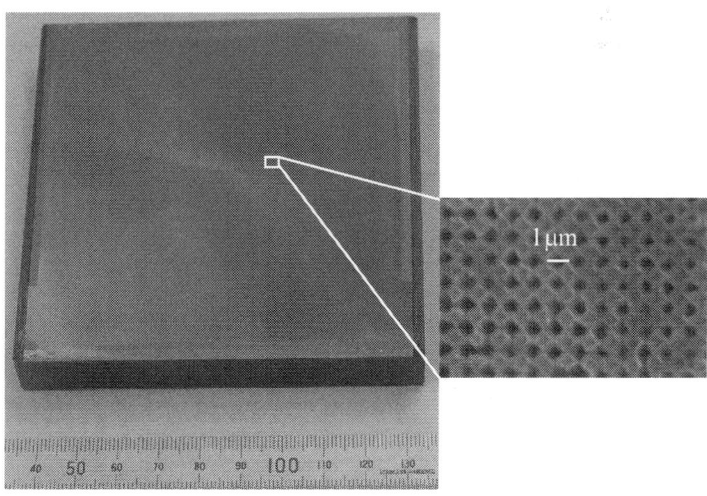

(b) Photograph of the rectangular grid made by photolithography

Figure 4.26. The rectangular *XY*-grid with a pitch of 1 μm

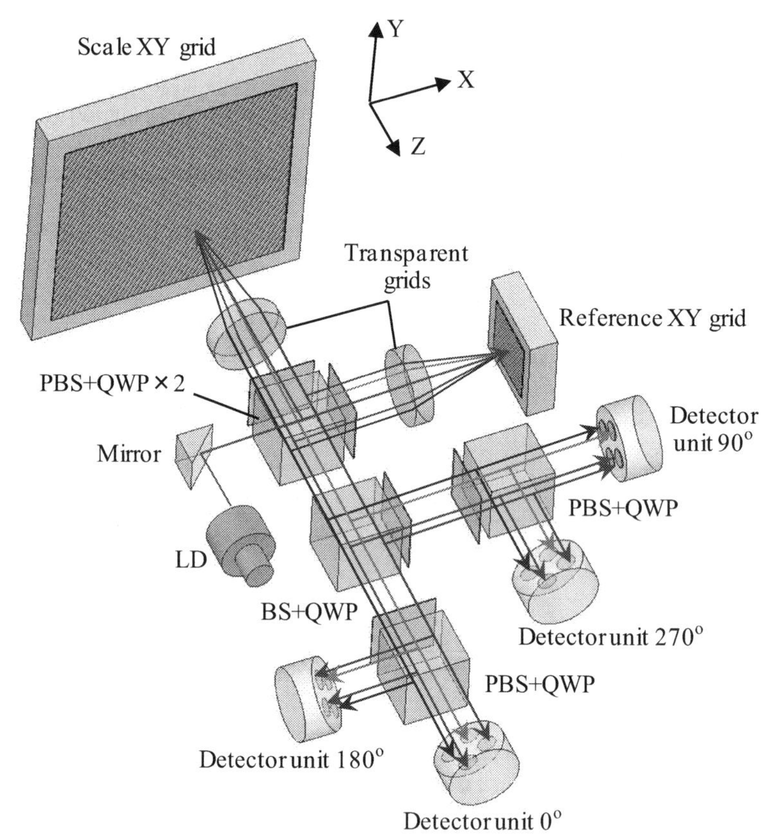

(a) Optical layout of the three-axis grating encoder with rectangular *XY*-grids

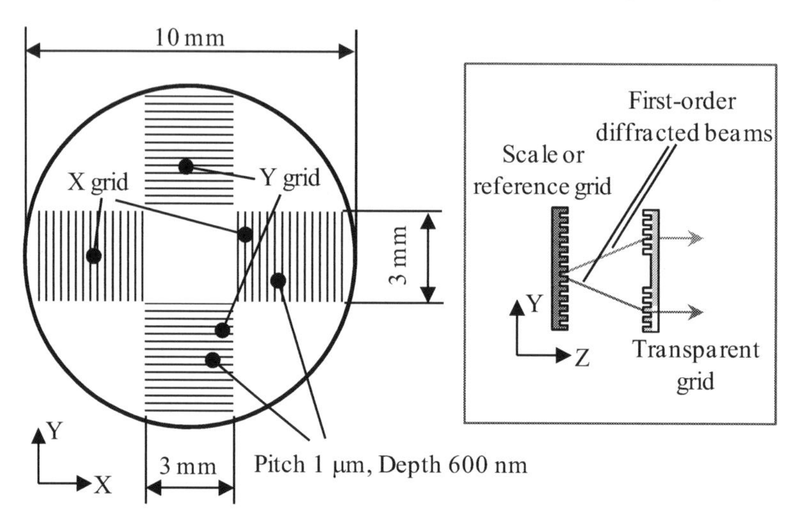

(b) The transparent grid for bending the first-order diffracted beams

Figure 4.27. The optical head for the rectangular *XY*-grids

Figure 4.28 shows the results of the three-axis grating encoder to detect the three-axis vibrations of a table. Three commercial interferometers were also employed to detect the vibrations simultaneously. It can be seen that the three-axis vibrations were detected by the three-axis grating encoder with sub-nanometric resolutions.

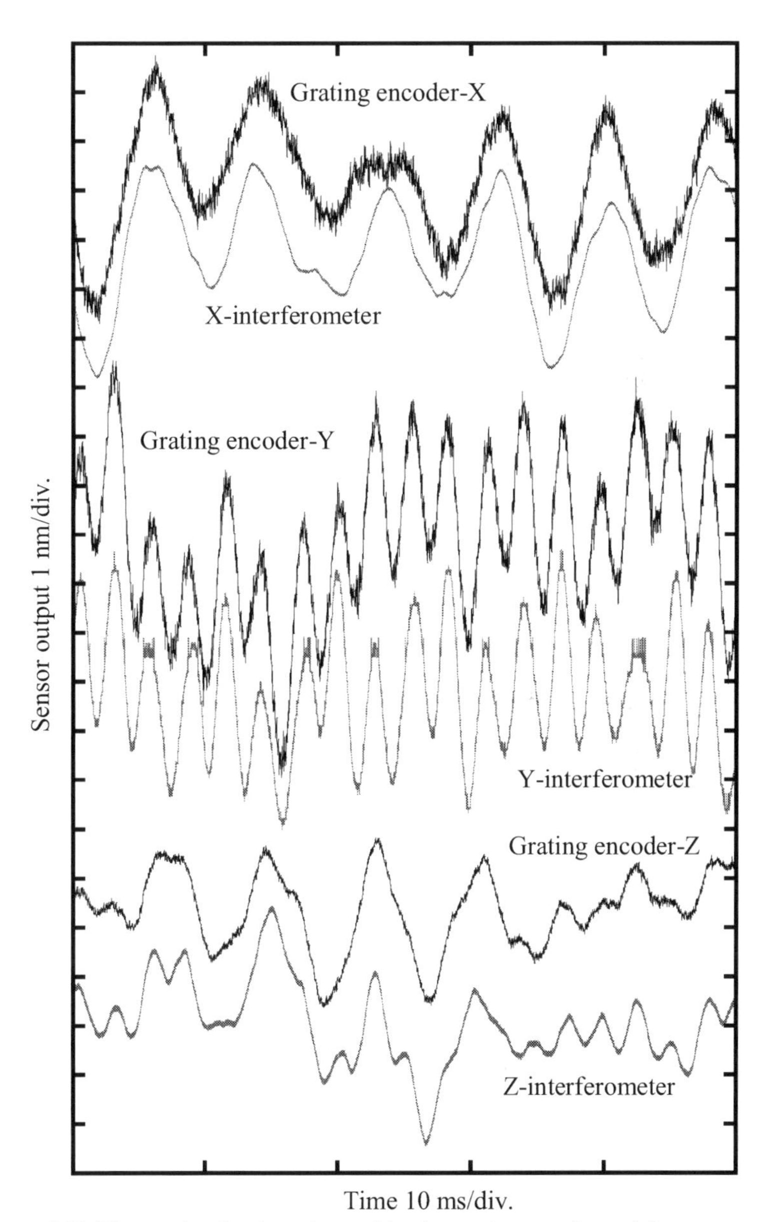

Figure 4.28. Three-axis vibrations detected by the grating encoder and three commercial interferometers

4.5 Summary

A two-degree-of-freedom linear encoder, which can measure the *X*-directional position and the *Z*-directional straightness simultaneously, has been described. In the 2-DOF linear encoder, the positive and negative first-order diffracted beams from a scale grating and a reference grating are superposed with each other to generate interference signals, from which the *X*- and *Z*-directional displacements can be evaluated. An optical head of the 2-DOF linear encoder has been designed and constructed for gratings with pitches of 1.6 μm. It has been verified that the interpolation errors of the quadrature interference signals, which contribute to the *X*-and *Z*-directional sensor outputs as the periodic errors, have amplitudes of 20 and 5 nm, respectively. The resolutions of the sensor were confirmed to be better than 0.5 nm.

The 2-DOF measurement principle has also been expanded for three-axis displacement measurements. Two kinds of *XY*-grids have been fabricated for the three-axis grating encoder. One is a sinusoidal-type *XY*-grid, which has been fabricated by diamond cutting with a fast-tool-servo. The pitch of the sinusoidal grid used in the experiment was 10 μm. The other is a waffle-type *XY*-grid, which was fabricated by photolithography. The pitch of the waffle *XY*-grid used in the experiment was 1 μm. It has been verified that the grating encoder with the sinusoidal grid has nanometric resolutions along the *X*-, *Y*- and *Z*-axes. The grating encoder with the waffle grid has reached sub-nanometric resolutions in all the three-axes.

References

[1] Shamoto E, Murase H, Moriwaki T (2000) Ultra-precision 6-axis table driven by means of walking drive. Ann CIRP 49(1):299–302

[2] Fan KC, Chen MJ (2000) A 6-degree-of-freedom measurement system for the accuracy of X–Y stages. Precis Eng 24:15–23

[3] Jäger G, Manske E, Hausotte T, Büchner HJ (2009) The metrological basis and operation of nanopositioning and nanomeasuring machine NMM-1. Technisches Messen 76(5):227–234

[4] Low KS (2003) Advanced precision linear stage for industrial automation applications. IEEE Trans Instrum Meas 52(3):785–789

[5] Otsuka J, Ichikawa S, Masuda T, Suzuki K (2005) Development of a small ultraprecision positioning device with 5 nm resolution. Meas Sci Technol 16:2186–2192

[6] Matsuda K, Roy M, Eiju T, O'Byrne JW, Sheppard Colin JR (2002) Straightness measurements with a reflection confocal optical system: an experimental study. Appl Opt 41(19):3966–3970

[7] Gao W, Arai Y, Shibuya A, Kiyono S, Park CH (2006) Measurement of multi-degree-of-freedom error motions of a precision linear air-bearing stage. Precis Eng 30(1):96–103

[8] Baldwin RR (1974) Interferometer system for measuring straightness and roll. US Patent 3,790,284, 5 Feb 1974

[9] Agilent Technologies Inc. (2010) Laser interferometer catalogue. Agilent Technologies Inc., Palo Alto

[10] Lin ST (2001) A laser interferometer for measuring straightness. Opt Laser Technol 33:195–199
[11] Estler WT (1985) Calibration and use of optical straightedges in the metrology of precision machines. Opt Eng 24(3):372–379
[12] Gao W, Tano M, Araki T, Kiyono S, Park CH (2007) Measurement and compensation of error motions of a diamond turning machine. Precis Eng 31(3):310–316
[13] Steinmetz CR (1990) Sub-micron position measurement and control on precision machine tools with laser interferometry. Precis Eng 12(1):12–24
[14] Kimura A, Gao W, Arai Y, Zeng LJ (2010) Design and construction of a two-degree-of-freedom linear encoder for nanometric measurement of stage position and straightness. Precis Eng 34(1):145–155
[15] Teimel A (1992) Technology and applications of grating interferometers in high-precision measurement. Precis Eng 14(3):147–154
[16] Renishaw plc (2010) Fibre-optic laser encoder RLE 10. Renishaw plc, Gloucestershire, UK
[17] Canon Inc. (2010) Micro-laser interferometer DS-80. Canon Inc., Tokyo, Japan
[18] Gao W, Kimura A (2007) A three-axis displacement sensor with nanometric resolution. Ann CIRP 56(1):529–532
[19] Zeng LJ, Li L (2007) Optical mosaic gratings made by consecutive, phase-interlocked, holographic exposures using diffraction from latent fringes. Opt Lett 32:1081–1083

5

Scanning Multi-sensor System for Measurement of Roundness

5.1 Introduction

Roundness is one of the most fundamental geometries of precision workpieces. Most of the round-shaped precision workpieces are manufactured by a turning process, in which a spindle is employed to rotate the workpiece. The out-of-roundness of the workpiece is basically determined by the error motion of the spindle. Measurement of the workpiece roundness error and the spindle error is an essential task for assurance of the manufacturing accuracy.

The scanning sensor method [1, 2] is the most widely used method for the measuring the roundness of a workpiece. In this method, a displacement sensor or a slope sensor, which is mounted on a rotary stage, is moved to scan the workpiece surface. The rotational scanning motion of the stage must be good enough so that it can be used as the measurement reference. To nanometrically roundness, however, the required measurement accuracy of the workpiece roundness must be at the same level as that of the scanning motion of the stage. Similar cases can be found in the on-machine measurement where the machine tool spindle is used. In such cases, it is essential to separate the workpiece roundness error from the spindle error [3, 4]. The error separation is also necessary in spindle error measurement, in which a ball or a cylinder is used as the measurement reference.

To perform the error separation, it is necessary to establish simultaneous equations involving the workpiece roundness error and the spindle error. There are two kinds of error separation methods based on how the equations are established [5]. One is known as the multi-step technique [6–8], and the other is the multi-sensor technique [9–13]. The multi-step methods including the reversal method establish the equations by making multiple measurements with one sensor. On the other hand, the multi-sensor methods establish the equations by using multiple sensors to make one measurement. Compared with multi-step methods, multi-sensor methods are more suitable for on-machine measurements because the repeatability of the spindle error is not necessary.

Roundness measurement involves three parameters; roundness error of the workpiece, and the X-directional component and Y-directional component of the spindle error. The three-displacement sensor method [14], which employs three displacement sensors, can realize the error separation to get the accurate workpiece roundness. In this chapter, a new three-sensor method of using three 2D slope sensors, called the three-slope sensor method, is presented to measure not only workpiece roundness but also multi-degree-of-freedom spindle error components.

However, some high-frequency components of roundness error cannot be accurately measured with the three-sensor method due to the problem of harmonic suppression. This problem cannot be completely solved by merely increasing the number of sensors [11, 14]. A new multi-sensor method, called the mixed method, is thus presented in the second part of this chapter to overcome this drawback of the three-sensor method.

5.2 Three-slope Sensor Method

Figure 5.1 shows the principle of the roundness measurement by using one displacement sensor. A displacement sensor, which is fixed spatially, is used to scan a round workpiece while the workpiece is rotating. Let P be a representative point of the workpiece, and the roundness error be described by the function $r(\theta)$, which is the deviation from the circle with an average radius of R_r. θ is the angle between the point P and the sensor. Let α be the angle between the sensor and the Y-axis. If no rotational error of the spindle (spindle error) exists, as shown in Figure 5.1, the roundness error $r(\theta)$ can be obtained correctly from the sensor output $m(\theta)$ as follows:

$$m(\theta) = r(\theta)\,. \tag{5.1}$$

However, if the spindle error exists, as shown in Figure 5.2, the sensor output $m(\theta)$ becomes,

$$m(\theta) = r(\theta) + \Delta m(\theta)\,, \tag{5.2}$$

where

$$\Delta m(\theta) = e_X(\theta)\sin\alpha + e_Y(\theta)\cos\alpha\,. \tag{5.3}$$

Here, $e_X(\theta)$ and $e_Y(\theta)$ are the X-directional and the Y-directional components of the spindle error.

Figure 5.2 shows the principle of the three-displacement sensor method using three displacement sensors schematically. The three displacement sensors are fixed around a cylindrical workpiece. The workpiece is scanned by the sensors while the workpiece is rotating. If the displacement outputs of the sensors are denoted by $m_1(\theta)$, $m_2(\theta)$ and $m_3(\theta)$, respectively, the outputs can be expressed as:

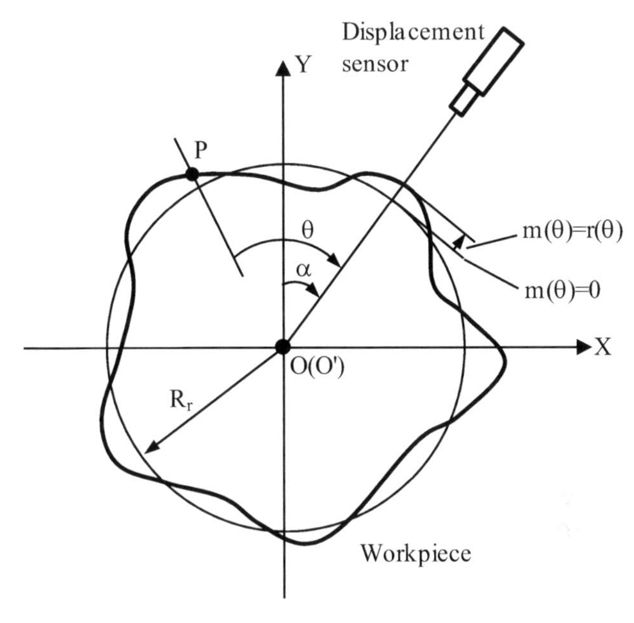

(a) Without spindle error

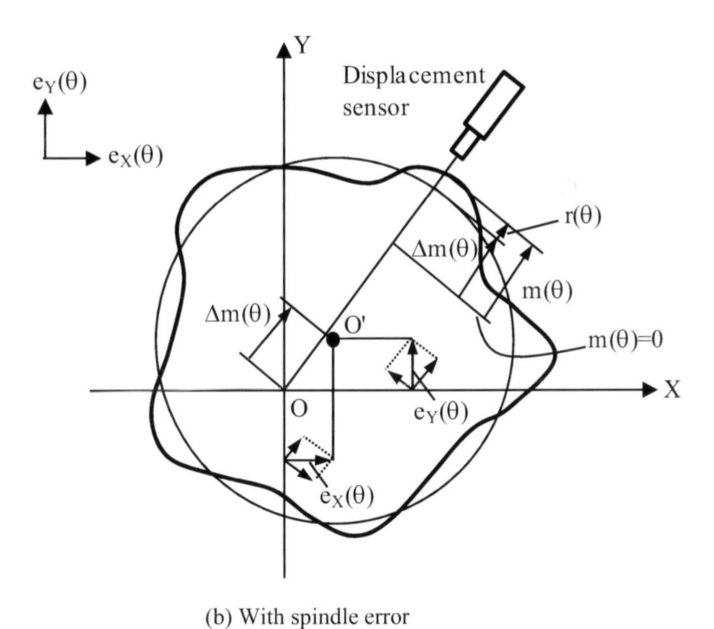

(b) With spindle error

Figure 5.1. Roundness measurement by using a displacement sensor

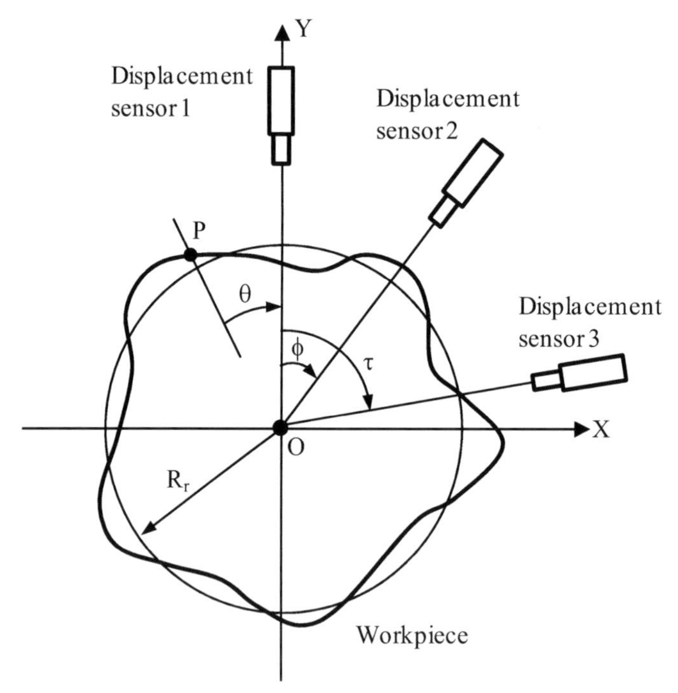

Figure 5.2. The three-displacement sensor method for roundness measurement

$$m_1(\theta) = r(\theta) + e_Y(\theta), \tag{5.4}$$

$$m_2(\theta) = r(\theta + \phi) + e_X(\theta)\sin\phi + e_Y(\theta)\cos\phi, \tag{5.5}$$

$$m_3(\theta) = r(\theta + \tau) + e_X(\theta)\sin\tau + e_Y(\theta)\cos\tau. \tag{5.6}$$

The differential output $m_{3D}(\theta)$ of the three-displacement sensor method, in which the spindle error is canceled, can be denoted as

$$m_{3D}(\theta) = r(\theta) + a_1 r(\theta + \phi) + a_2 r(\theta + \tau), \tag{5.7}$$

where

$$a_1 = -\frac{\sin\tau}{\sin(\tau - \phi)}, \ a_2 = \frac{\sin\phi}{\sin(\tau - \phi)}. \tag{5.8}$$

In Equation 5.7, $r(\theta)$ can be treated as the input and $m_{3D}(\theta)$ as the output. According to the theory of digital filters [15], the relation between the input $r(\theta)$ and the output $m_{3D}(\theta)$ can be defined by the following transfer function of the three-displacement sensor method:

$$H_{3D}(n) = \frac{M_{3D}(n)}{R(n)} = 1 + a_1 e^{-jn\phi} + a_2 e^{-jn\tau}, \tag{5.9}$$

where n is the spatial frequency (the number of undulations per revolution), and $M_{3D}(n)$ and $R(n)$ are the Fourier transforms of $m_{3D}(\theta)$ and $r(\theta)$, respectively. $R(n)$ can be obtained from $M_{3D}(n)$ and $H_{3D}(n)$. $r(\theta)$ can be evaluated by IFFT of $R(n)$.

The amplitude of the transfer function $H_{3D}(n)$ represents the complex harmonic sensitivity of the three-displacement sensor method. The amplitude (harmonic sensitivity) and the phase angle of $H_{3D}(n)$ are expressed as follows:

$$|H_{3D}(n)| = \sqrt{1 + a_1^2 + a_2^2 + 2(a_1 \cos n\phi + a_2 \cos n\tau + a_1 a_2 \cos n(\tau - \phi))}, \tag{5.10}$$

$$\arg[H_{3D}(n)] = \tan^{-1}\left(-\frac{a_1 \sin n\phi + a_2 \sin n\tau}{1 + a_1 \cos n\phi + a_2 \cos n\tau}\right). \tag{5.11}$$

The transfer function $H_{3D}(\omega)$ is shown in Figure 5.3. As can be seen in the figure, the amplitude at some frequencies in the transfer function of the three-displacement sensor method approaches zero. This prevents the three-displacement sensor method from measuring the corresponding frequency components correctly.

As shown in Figure 5.4, we can also use one slope sensor to perform roundness measurement.

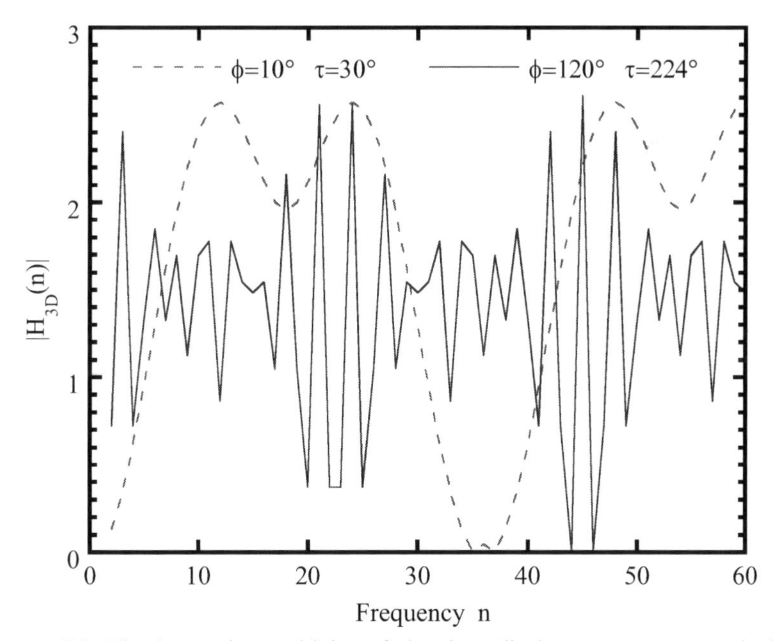

Figure 5.3. The harmonic sensitivity of the three-displacement sensor method for roundness measurement

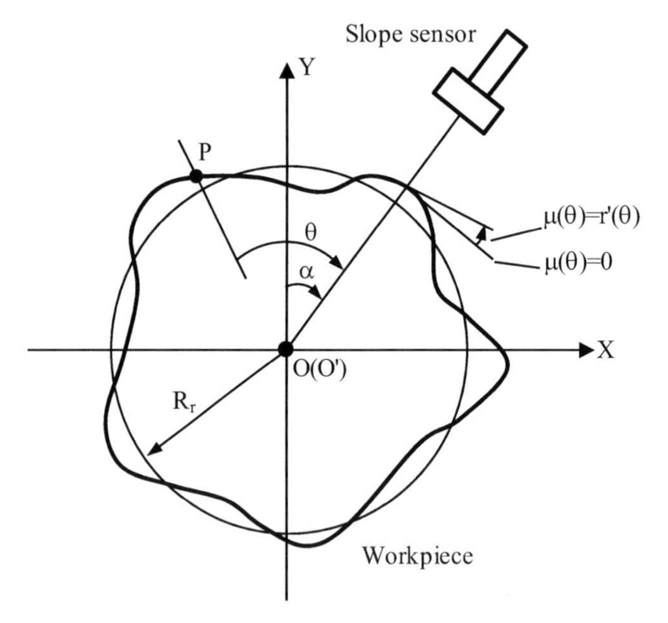

(a) Without spindle error

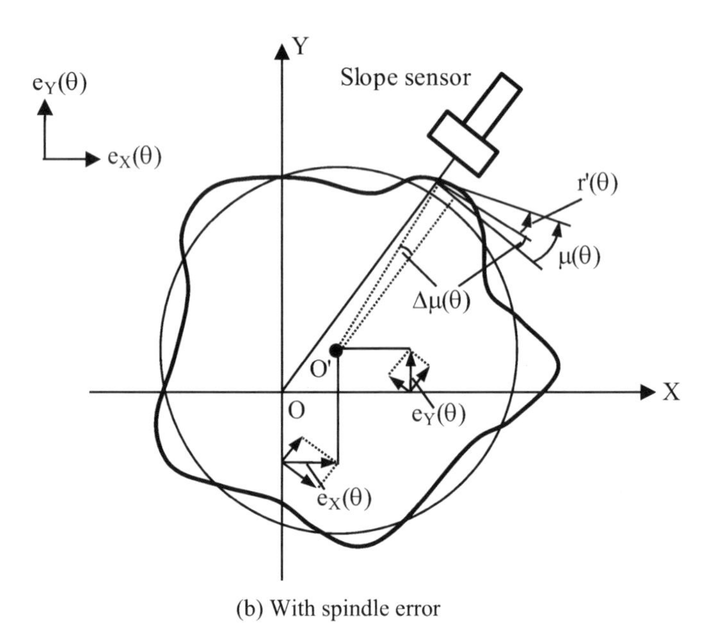

(b) With spindle error

Figure 5.4. Roundness measurement by using a slope sensor

Let the local slope of the surface be described by the function $r'(\theta)$. If no spindle error exists (Figure 5.4 (a)), the local slope $r'(\theta)$ can be measured correctly from the output of the slope sensor as [16]:

$$\mu(\theta) = r'(\theta)\,, \tag{5.12}$$

$$r'(\theta) = \frac{dr(\theta)}{d(R_r \theta)}\,. \tag{5.13}$$

where R_r is the average radius of the workpiece.

The roundness error can then be obtained by integrating $r'(\theta)$. However, if there exists a spindle error as shown in Figure 5.4 (b), because of the shift of the measuring point on the circumference of the workpiece, the spindle error will generate an angle change $\Delta\mu(\theta)$ in the sensor output as follows:

$$\mu(\theta) = r'(\theta) + \Delta\mu(\theta)\,, \tag{5.14}$$

where

$$\Delta\mu(\theta) = \frac{e_X(\theta)\cos\alpha - e_Y(\theta)\sin\alpha}{R_r}\,. \tag{5.15}$$

Figure 5.5 shows the principle of the three-slope sensor method using three-slope sensors schematically. The three-slope sensors are fixed around a cylindrical workpiece, and scan the workpiece while it is rotating. If the slope outputs of the sensors are denoted by $\mu_1(\theta)$, $\mu_2(\theta)$ and $\mu_3(\theta)$, respectively, the outputs can be expressed as:

$$\mu_1(\theta) = r'(\theta) + \frac{e_X(\theta)}{R_r}\,, \tag{5.16}$$

$$\mu_2(\theta) = r'(\theta+\phi) + \frac{e_X(\theta)}{R_r}\cos\phi - \frac{e_Y(\theta)}{R_r}\sin\phi\,, \tag{5.17}$$

$$\mu_3(\theta) = r'(\theta+\tau) + \frac{e_X(\theta)}{R_r}\cos\tau - \frac{e_Y(\theta)}{R_r}\sin\tau\,. \tag{5.18}$$

The differential output $m_{3S}(\theta)$ of the three-slope sensor method can be denoted as

$$m_{3S}(\theta) = r(\theta) + a_1 r(\theta+\phi) + a_2 r(\theta+\tau)\,, \tag{5.19}$$

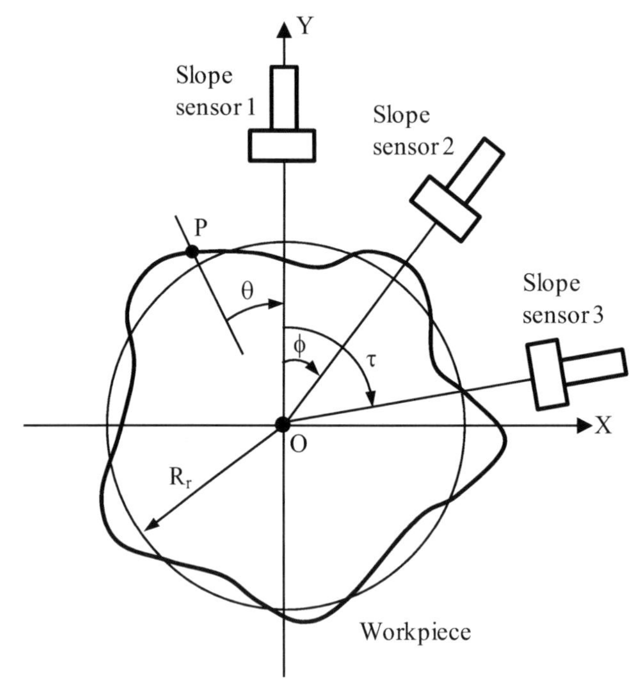

Figure 5.5. The three-slope sensor method for roundness measurement

where

$$a_1 = -\frac{\sin\tau}{\sin(\tau-\phi)}, \quad a_2 = \frac{\sin\phi}{\sin(\tau-\phi)}. \tag{5.20}$$

Consequently, the spindle error is canceled.

The transfer function $H_{3S}(n)$ of the three-slope sensor method can be expressed as follows:

$$H_{3S}(n) = \frac{M_{3S}(n)}{R(n)} = jn(1 + a_1 e^{-jn\phi} + a_2 e^{-jn\tau}), \tag{5.21}$$

where n is the spatial frequency (the number of undulations per revolution), and $M_{3S}(n)$ and $R(n)$ are the Fourier transforms of $m_{3S}(\theta)$ and $r(\theta)$, respectively. $R(n)$ can be obtained from $M_{3D}(n)$ and $H_{3S}(n)$. $r(\theta)$ can be evaluated by IFFT of $R(n)$. The amplitude (harmonic sensitivity) and the phase angle of $H_{3S}(n)$ are expressed as follows:

$$|H_{3S}(n)| = n\sqrt{1 + a_1^2 + a_2^2 + 2(a_1\cos n\phi + a_2\cos n\tau + a_1 a_2 \cos n(\tau-\phi))}, \tag{5.22}$$

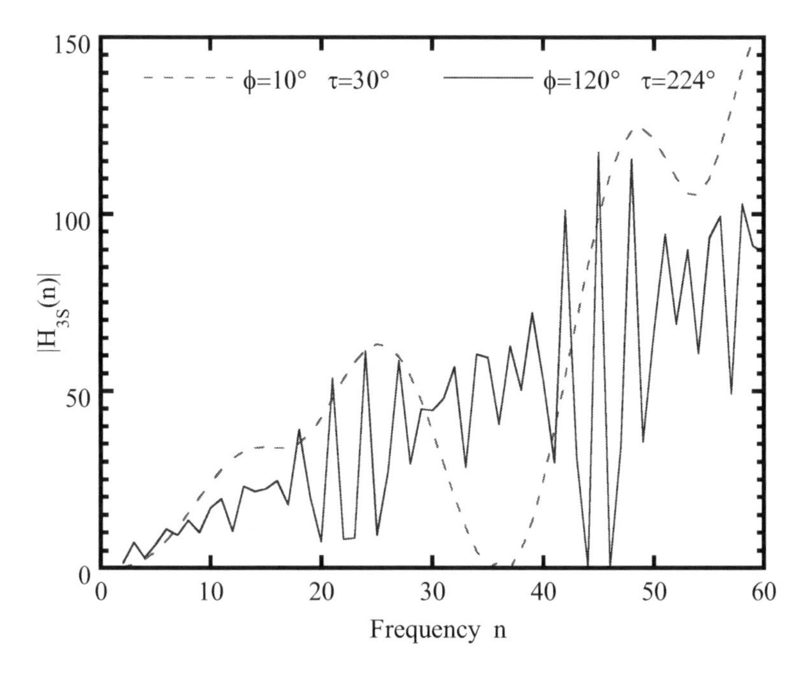

Figure 5.6. The harmonic sensitivity of the three-slope probe method for roundness measurement

$$\arg[H_{3S}(n)] = \tan^{-1}\left(\frac{1 + a_1 \cos n\phi + a_2 \cos n\tau}{a_1 \sin n\phi + a_2 \sin n\tau}\right). \tag{5.23}$$

The amplitude of the transfer function $H_{3S}(n)$ is shown in Figure 5.6. Similar to the three-displacement sensor method, the amplitude at some frequencies in the transfer function of the three-slope sensor method approaches zero. The frequencies at which the amplitude is zero are the same as those of the three-displacement sensor method with the same sensor arrangement.

The three-slope sensor method can also be used for measurement of the surface slope along the *Z*-direction and the spindle tilt error motions by using two-dimensional slope sensors. Figure 5.8 shows the schematic of the measurement principle where $r(\theta, z)$ is the surface profile of the cylindrical workpiece. Each sensor detects the two-dimensional local slopes of a point on the workpiece surface. The local slope component along the circumference direction $r'_r(\theta, z)$ is the same as that shown in Equation 5.13 and can be separated from the spindle radial error component $e_r(\theta, z)$ by using the radial output components of the slope sensors as shown in Equations 5.16–5.20. On the other hand, the local slope component along the *Z*-direction $r'_Z(\theta, z)$ is defined as:

$$r'_z(\theta) = \frac{dr(\theta, z)}{dz}. \tag{5.24}$$

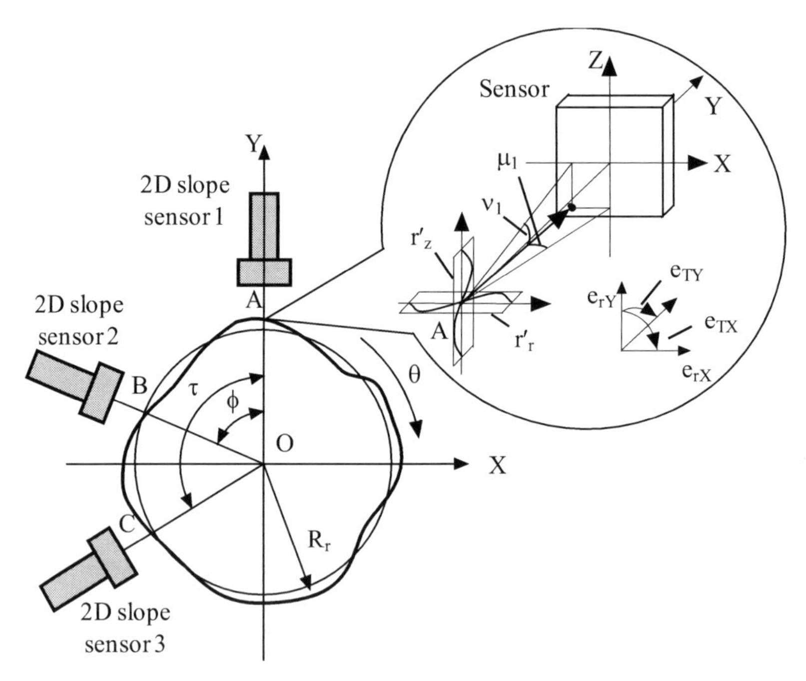

Figure 5.7. The three-slope sensor method with two-dimensional slope sensors

The $r'_Z(\theta, z)$ and the tilt error motion $e_T(\theta, z)$ can be similarly separated from each other by using the angular output components of the slope sensor shown below:

$$\nu_1(\theta) = r_z'(\theta) - e_{TY}(\theta), \tag{5.25}$$

$$\nu_2(\theta) = r_z'(\theta+\phi) + e_{TX}(\theta)\cdot\sin\phi - e_{TY}(\theta)\cdot\cos\phi, \tag{5.26}$$

$$\nu_3(\theta) = r_z'(\theta+\tau) + e_{TX}(\theta)\cdot\sin\tau - e_{TY}(\theta)\cdot\cos\tau. \tag{5.27}$$

The differential outputs for calculating the $e_{TX}(\theta)$ and $e_{TY}(\theta)$ are expressed by:

$$\begin{aligned}\Delta\nu_{TX}(\theta) &= a_3\cdot\nu_1(\theta+\phi) - a_3\cdot\nu_2(\theta) + a_4\cdot\nu_3(\theta) - a_4\cdot\nu_1(\theta+\tau)\\ &\quad + a_5\cdot\nu_2(\theta+\tau) - a_5\cdot\nu_3(\theta+\phi)\\ &= e_{TX}(\theta) + a_1\cdot e_{TX}(\theta+\phi) + a_2\cdot e_{TX}(\theta+\tau),\end{aligned} \tag{5.28}$$

$$\begin{aligned}\Delta\nu_{TY}(\theta) &= a_1\cdot\nu_2(\theta) - a_1\cdot\nu_1(\theta+\phi) + a_2\cdot\nu_3(\theta) - a_2\cdot\nu_1(\theta+\tau)\\ &= e_{TY}(\theta) + a_1\cdot e_{TY}(\theta+\phi) + a_2\cdot e_{TY}(\theta+\tau),\end{aligned} \tag{5.29}$$

where a_1 and a_2 are those defined in Equation 5.20. a_3, a_4 and a_5 are defined by

$$a_3 = \frac{\cos\tau}{\sin(\tau-\phi)}, \quad a_4 = \frac{\cos\phi}{\sin(\tau-\phi)}, \quad a_5 = \frac{1}{\sin(\tau-\phi)}. \tag{5.30}$$

Figure 5.8 shows a schematic of the 2D slope sensor unit designed and constructed for experiments. The sensor unit consists of three 2D slope sensors. The angles ϕ and τ between sensors were designed to be 60 and 96°, respectively. This sensor arrangement makes the three-slope sensor method sensitive to undulations per revolution of up to 28, which is the highest calculation number in the experiments. It should be pointed out that the sensor arrangement is needed to be optimized if it is necessary to measure higher undulations per revolution. The sensors utilize the principle of autocollimation for slope detection. As can be seen from the schematic of sensor 1, a collimated beam from a laser diode is projected onto point A on the workpiece surface after passing through a polarization beam splitter (PBS) and a quarter waveplate. The reflected beam is reflected again at the PBS and then accepted by the autocollimation unit consisting of an objective lens and a quadrant photodiode (QPD) placed at the focal position of the lens. The 2D

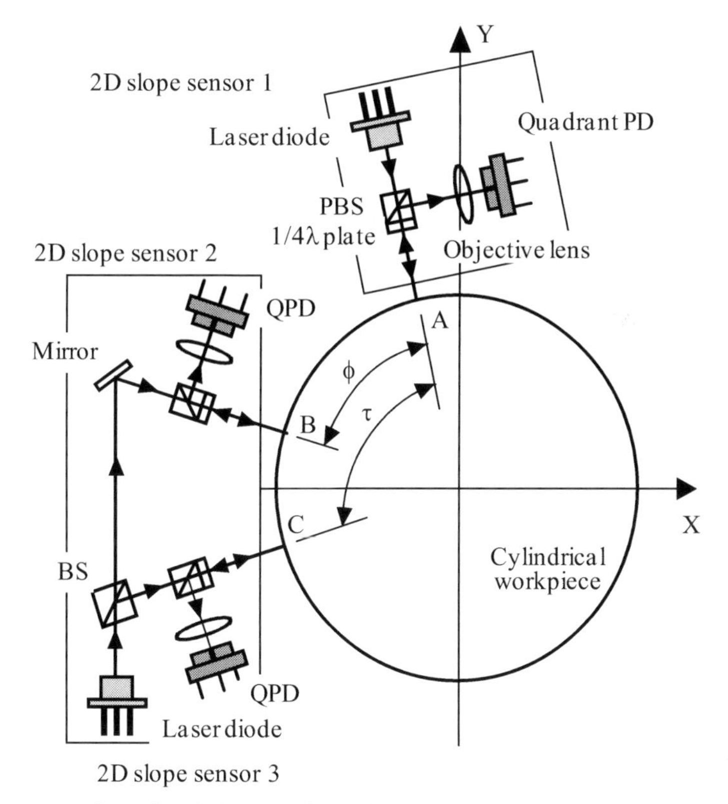

Figure 5.8. Schematic of the two-dimensional slope sensor unit for the three-slope sensor method

slope information at point A can be obtained from the photoelectric currents of the QPD. Sensors 2 and 3 share the same optical source for the purpose of compactness. The laser beam is divided into two beams by a beam splitter (BS). The beams are projected onto points B and C on the workpiece surface. The reflected beams are received by two autocollimation units so that the 2D slope information at points B and C can be detected, respectively. Diameters of the laser beams were set to be 0.5 mm and focal distances of the lenses were 30 mm. The slope sensitivity of the sensors was approximately 0.1 arcsec. The size of sensor A was 90 mm (L) × 80 mm (W) × 40 mm (H) and that of sensors B and C was 160 mm (L) × 100 mm (W) × 40 mm (H).

Figure 5.9 shows a schematic of the experimental setup for roundness and spindle error measurement. A diamond turned cylindrical workpiece with a diameter of 80 mm was employed as the artifact. The workpiece was mounted on an air-spindle. The outputs of the slope sensors were sampled simultaneously by a personal computer via a 12-bit AD converter. The rotational angle of the spindle was measured by an optical encoder. The roundness of the workpiece was also measured by the conventional three-displacement sensor method employing three capacitive-type displacement sensors for comparison. The measured spindle error and workpiece roundness are shown in Figures 5.10 and 5.11, respectively.

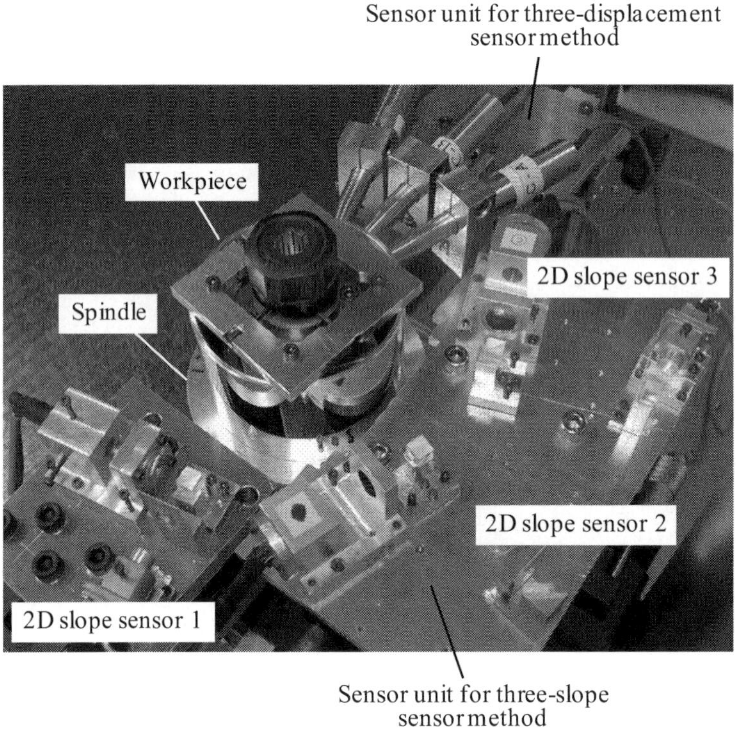

Sensor unit for three-displacement sensor method

Workpiece

Spindle

2D slope sensor 3

2D slope sensor 2

2D slope sensor 1

Sensor unit for three-slope sensor method

Figure 5.9. Photograph of the experimental setup for roundness and spindle error measurement by the three-slope sensor method

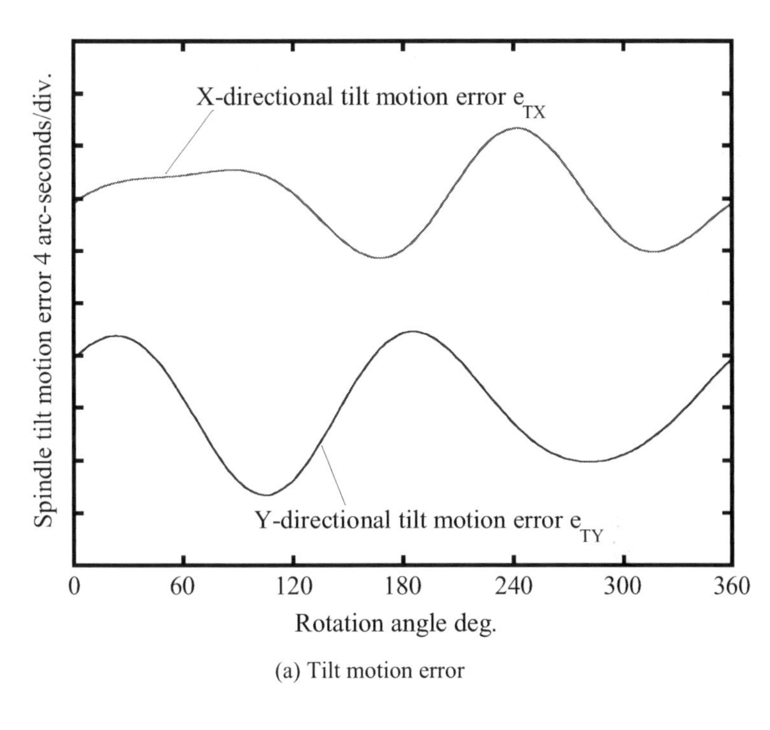

(a) Tilt motion error

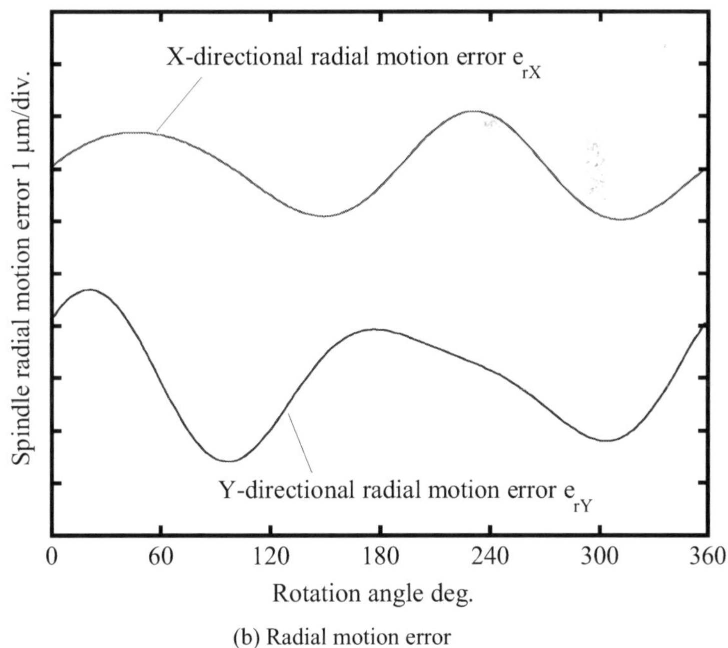

(b) Radial motion error

Figure 5.10. Measured spindle motion error

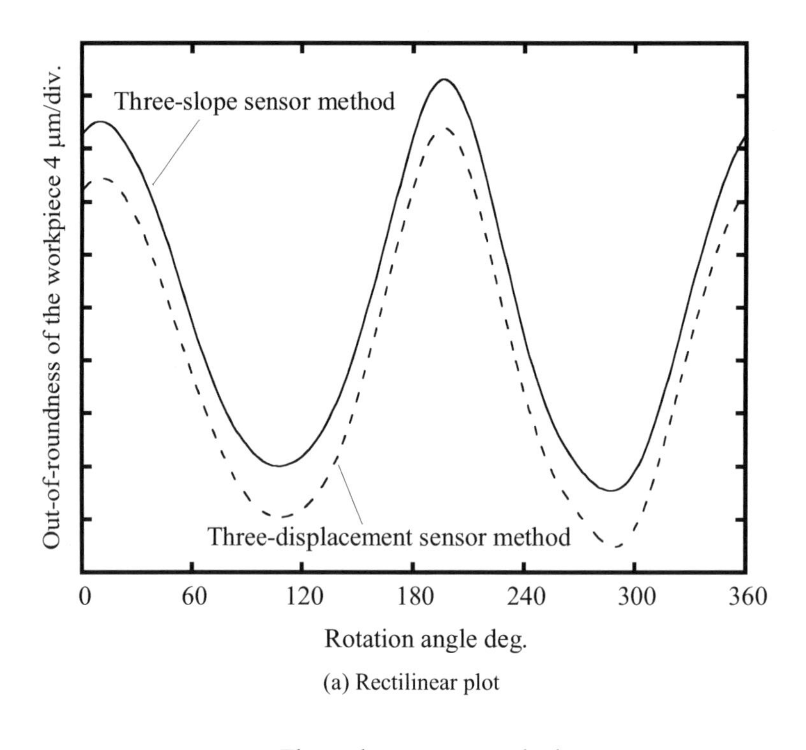

(a) Rectilinear plot

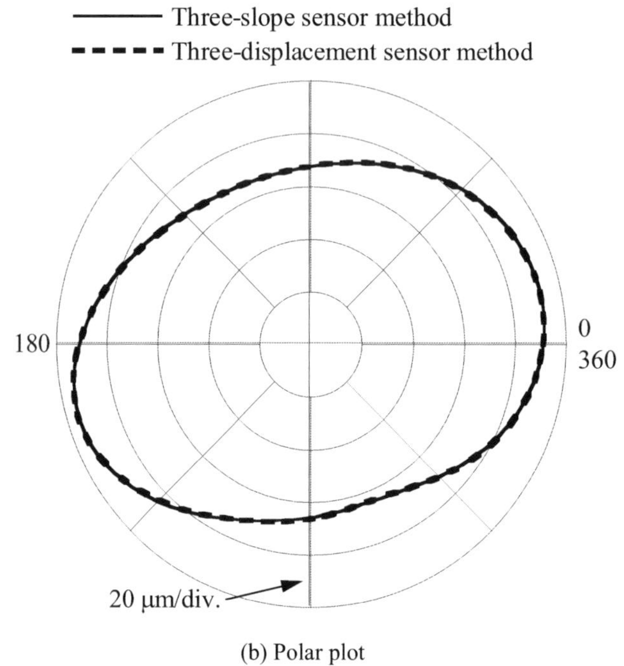

(b) Polar plot

Figure 5.11. Measured workpiece roundness

In the experiment, the spindle speed was approximately 60 rpm and the sampling number over one revolution was 200. Figure 5.10 (a) shows the two-directional components of the tilt error motion, which were approximately 9.9 arcsec in the *X*-direction and 12.0 arcsec in the *Y*-direction, respectively. The two-directional components of the radial error motion of the spindle are plotted in Figure 5.10 (b). The *X*-directional component was approximately 2.0 μm and the *Y*-directional component was approximately 3.3 μm. Figure 5.11 shows the measured roundness of the cylindrical workpiece. The roundness was approximately 31.1 μm. The same workpiece was also measured by the conventional three-displacement sensor method and the result was shown in the figure. It can be seen that the two results corresponded well with each other.

5.3 Mixed Method

The three-sensor method described above can separate the workpiece roundness error from the spindle error motion. However, some frequency components of roundness cannot be accurately measured using this method because of the problem of harmonic suppression. This problem cannot be completely solved by merely increasing the number of sensors [11, 14]. In this section, a method of mixing the displacement sensors and the slope sensors, which is called the mixed method, is described. This method can separate the roundness error from the spindle error completely, and capture high-frequency components.

Figure 5.12 shows the schematics of the two-displacement/one-slope (2D1S) mixed method and the one-displacement/two-slope (1D2S) mixed method. In the 2D1S mixed method, two displacement sensors (sensors 1 and 3) and one slope sensor (sensor 2) are employed. Let ϕ and τ be the angles between the sensors. The sensor outputs in the 2D1S mixed method are expressed by:

$$m_1(\theta) = r(\theta) + e_Y(\theta), \tag{5.31}$$

$$\mu_2(\theta) = r'(\theta+\phi) + \frac{e_X(\theta)}{R_r}\cos\phi - \frac{e_Y(\theta)}{R_r}\sin\phi, \tag{5.32}$$

$$m_3(\theta) = r(\theta+\tau) + e_X(\theta)\sin\tau + e_Y(\theta)\cos\tau. \tag{5.33}$$

The differential output $m_{2D1S}(\theta)$ to cancel the spindle error can be denoted as:

$$\begin{aligned} m_{2D1S}(\theta) &= m_1(\theta)(\tan\phi + \cot\tau) - \frac{m_3(\theta)}{\sin\tau} + \frac{R_r\mu_2(\theta)}{\cos\phi} \\ &= r(\theta)(\tan\phi + \cot\tau) - \frac{r(\theta+\tau)}{\sin\tau} + \frac{R_r r'(\theta+\phi)}{\cos\phi}. \end{aligned} \tag{5.34}$$

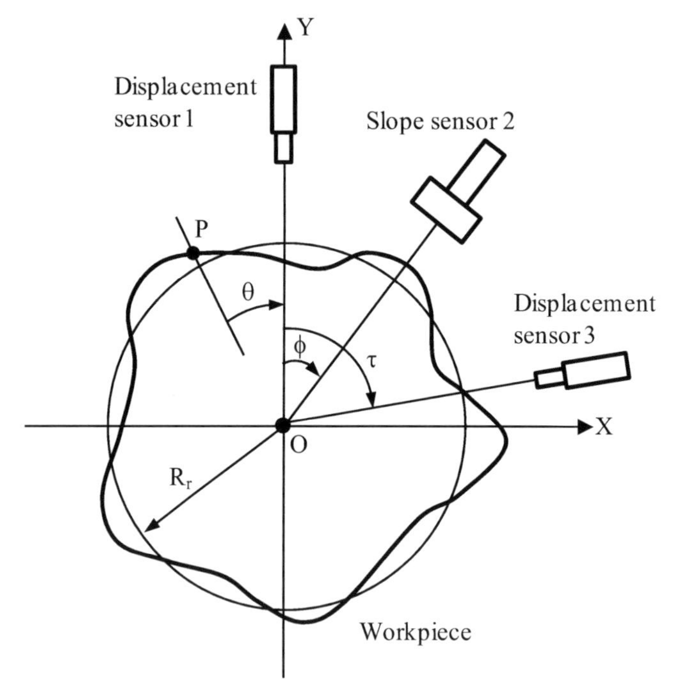

(a) The two-displacement/one-slope (2D1S) mixed method

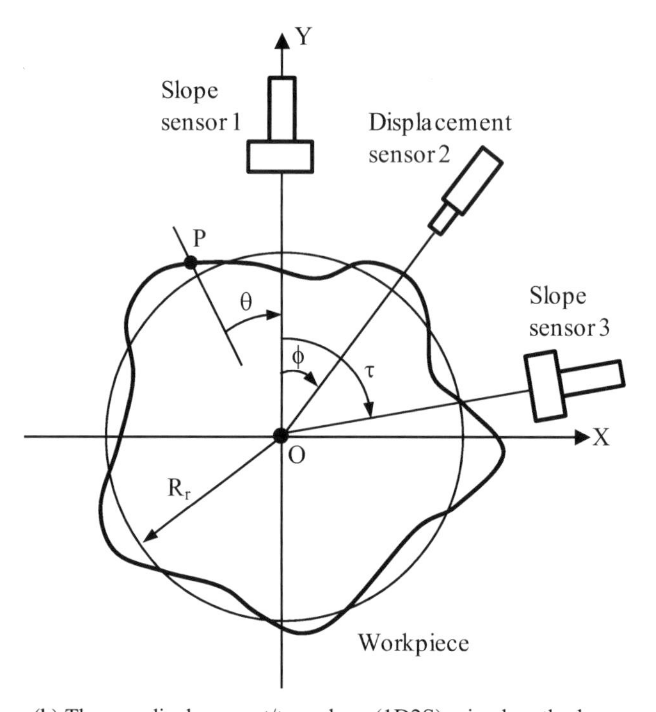

(b) The one-displacement/two-slope (1D2S) mixed method

Figure 5.12. The mixed method mixing displacement probes and slope probes for roundness measurement

Let the Fourier transforms of $r(\theta)$ and $m_{2D1S}(\theta)$ be $R(n)$ and $M_{2D1S}(n)$, respectively, the transfer function of the 2D1S mixed method can be defined by:

$$H_{2D1S}(n) = \frac{M_{2D1S}(n)}{R(n)} = \tan\phi + \cot\tau - \frac{e^{jn\tau}}{\sin\tau} + \frac{jne^{jn\phi}}{\cos\phi}, \tag{5.35}$$

where n is the frequency. The amplitude (harmonic sensitivity) and the phase angle of $H_{2D1S}(n)$ are expressed as:

$$|H_{2D1S}(n)| = \frac{1}{\cos\phi\sin\tau}[(\sin\phi\sin\tau + \cos\phi\cos\tau - \cos\phi\cos n\tau$$

$$- n\sin n\phi\sin\tau)^2 + (n\cos n\phi\sin\tau - \cos\phi\sin n\tau)^2]^{1/2}, \tag{5.36}$$

$$\arg[H_{2D1S}(n)] = \tan^{-1}[(n\cos n\phi\sin\tau - \cos\phi\sin n\tau)/(\sin\phi\sin\tau$$

$$+\cos\phi\cos\tau - \cos\phi\cos n\tau - n\sin n\phi\sin\tau)]. \tag{5.37}$$

We can also employ one displacement sensor and two slope sensors to construct the mixed method as shown in Figure 5.12 (b). The sensor outputs of the 1D2S mixed method can be expressed as:

$$\mu_1(\theta) = r'(\theta) + \frac{e_X(\theta)}{R_r}, \tag{5.38}$$

$$m_2(\theta) = r(\theta+\phi) + e_X(\theta)\sin\phi + e_Y(\theta)\cos\phi, \tag{5.39}$$

$$\mu_3(\theta) = r'(\theta+\tau) + \frac{e_X(\theta)}{R_r}\cos\tau - \frac{e_Y(\theta)}{R_r}\sin\tau. \tag{5.40}$$

The differential output, in which the effect of the spindle error is cancelled, can be given by:

$$m_{1D2S}(\theta) = R_r\mu_1(\theta)(\tan\phi + \cot\tau) - \frac{m_2(\theta)}{\cos\phi} - \frac{R_r\mu_3(\theta)}{\sin\tau}. \tag{5.41}$$

The transfer function of the 1D2S mixed method can be defined as follows:

$$H_{1D2S}(n) = \frac{M_{1D2S}(n)}{R(n)} = jn(\tan\phi + \cot\tau) - \frac{e^{jn\phi}}{\cos\phi} - \frac{jne^{jn\tau}}{\sin\tau}, \tag{5.42}$$

$$|H_{1D2S}(n)| = \frac{1}{\cos\phi\sin\tau}[(n\cos\phi\sin n\tau - \cos n\phi\sin\tau)^2$$

$$+(n\sin\phi\sin\tau + n\cos\phi\cos\tau - \sin n\phi\sin\tau$$

$$-n\cos\phi\cos n\tau)^2]^{1/2}, \tag{5.43}$$

$$\arg[H_{1D2S}(n)] = \tan^{-1}[(n\sin\phi\sin\tau + n\cos\phi\cos\tau - \sin n\phi\sin\tau$$

$$-n\cos\phi\cos n\tau)/(n\cos\phi\sin\tau - \cos n\phi\sin\tau)], \tag{5.44}$$

where $M_{1D2S}(n)$ is the Fourier transform of $m_{1D2S}(\theta)$.

With the angular distances ϕ and τ acting as parameters, the amplitudes of the transfer functions (harmonic sensitivities) of the two methods are plotted in Figure 5.13, respectively. As can be seen in the figures, there is no frequency at which the harmonic sensitivity approaches zero, indicating that both the two mixed methods can measure high-frequency components correctly. In comparison with the 2D1S mixed method, the 1D2S mixed method is more sensitive when $n \leq 7$. On the other hand, the 2D1S mixed method is superior to the 1D2S mixed method in the higher-frequency range. The harmonic sensitivity of the 2D1S mixed method in the low-frequency range, and that of 1D2S mixed method in the high-frequency range can be seen to be closely related to the sensor arrangement. As shown in Figure 5.13 (a), the minimum harmonic sensitivity $|H_{2D1S}(2)|$ is only 0.16 for the symmetrical sensor arrangement $(\phi, \tau) = (22.5^\circ, 45^\circ)$. This value can be increased by changing the sensor arrangement to one that is asymmetrical. For a fixed τ, the largest $|H_{2D1S}(2)|$ is obtained when ϕ is equal to 0° or τ. In this sensor arrangement, the slope sensor and one of the displacement sensors are placed at the same position. $|H_{2D1S}(2)|$ also improves as β increases. When $(\phi, \tau) = (0^\circ, 90^\circ)$ or $(\phi, \tau) = (90^\circ, 90^\circ)$, $|H_{2D1S}(2)|$ reaches its limiting value. In this sensor arrangement, the 2D1S mixed method yields the most well-balanced harmonic response. The same can be said with regard to the 1D2S mixed method. The mixed method with this sensor arrangement is called the orthogonal mixed method.

Figure 5.14 shows the principle of the orthogonal mixed method. The output $m_1(\theta)$ of the displacement sensor and the output $\mu_2(\theta)$ of the slope sensor can be expressed as follows:

$$m_1(\theta) = r(\theta) + e_Y(\theta), \tag{5.45}$$

$$\mu_2(\theta) = r'(\theta + \frac{\pi}{2}) - \frac{e_Y(\theta)}{R_r}. \tag{5.46}$$

Therefore, the differential output $m_{om}(\theta)$, in which the roundness error is separated from the spindle error, can be denoted as:

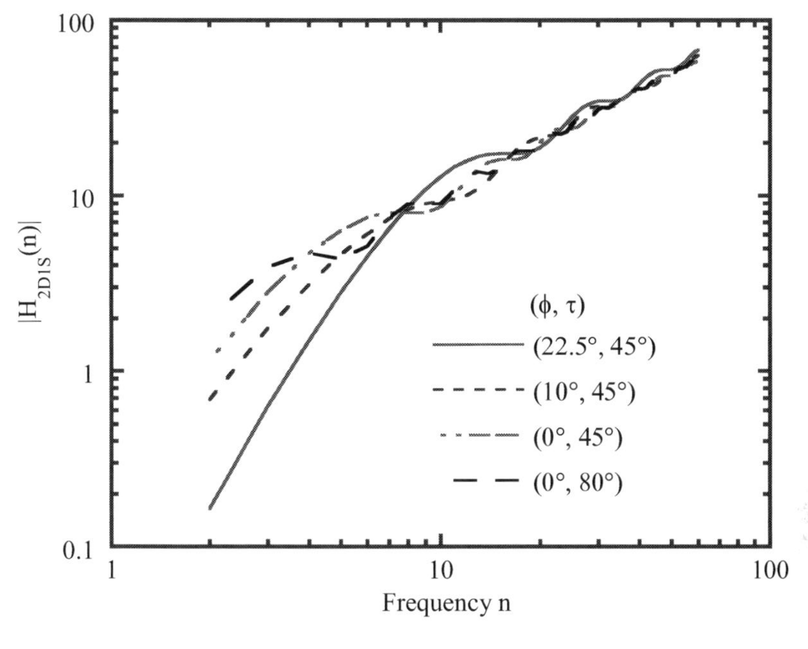

(a) The 2D1S mixed method

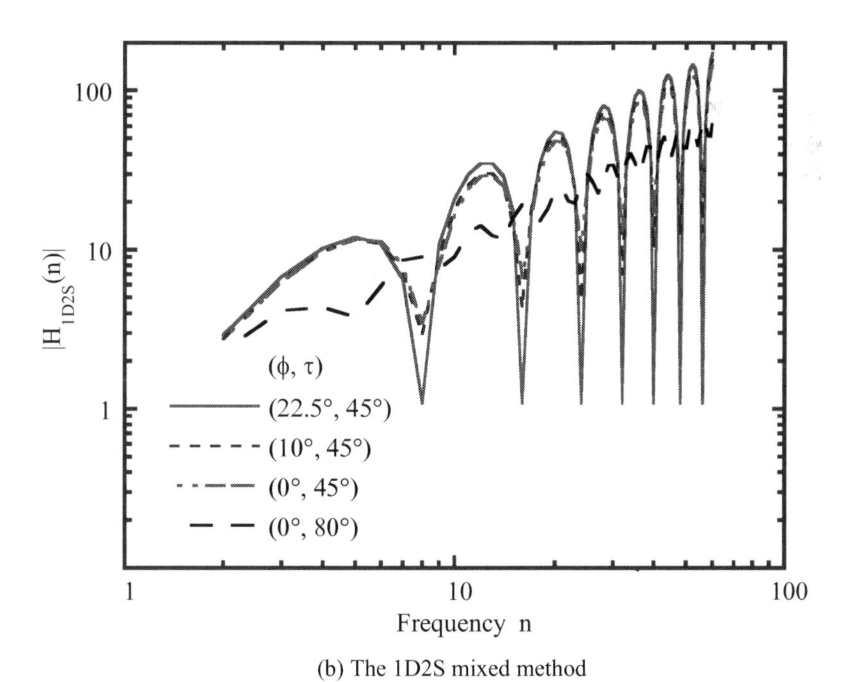

(b) The 1D2S mixed method

Figure 5.13. The harmonic sensitivities of the mixed method

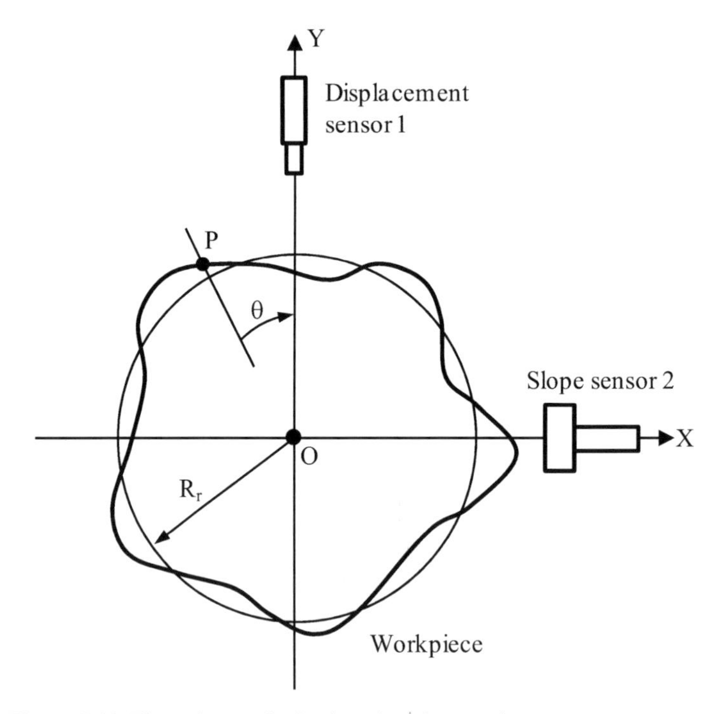

Figure 5.14. The orthogonal mixed method for roundness measurement

$$m_{om}(\theta) = m_1(\theta) + R_r \mu_2(\theta) = r(\theta) + R_r r'(\theta + \frac{\pi}{2}). \tag{5.47}$$

The transfer function of the orthogonal mixed method can be expressed by:

$$H_{om}(n) = \frac{M_{om}(n)}{R(n)} = 1 + jne^{jn\frac{\pi}{2}}, \tag{5.48}$$

$$|H_{om}(n)| = [(1 - n\sin n\frac{\pi}{2})^2 + (n\cos n\frac{\pi}{2})^2]^{1/2}, \tag{5.49}$$

$$\arg[H_{om}(n)] = \tan^{-1}[(n\cos n\frac{\pi}{2})/(1 - n\sin n\frac{\pi}{2})]. \tag{5.50}$$

where $M_{om}(n)$ is the Fourier transform of $m_{om}(\theta)$.

Figure 5.15 shows the amplitude of the transfer function (harmonic sensitivity) of the orthogonal mixed method. As can be seen from thc figure, $H_{om}(n)$ yields good characteristics. The minimum harmonic sensitivity $|H_{om}(2)|$ is 2.24.

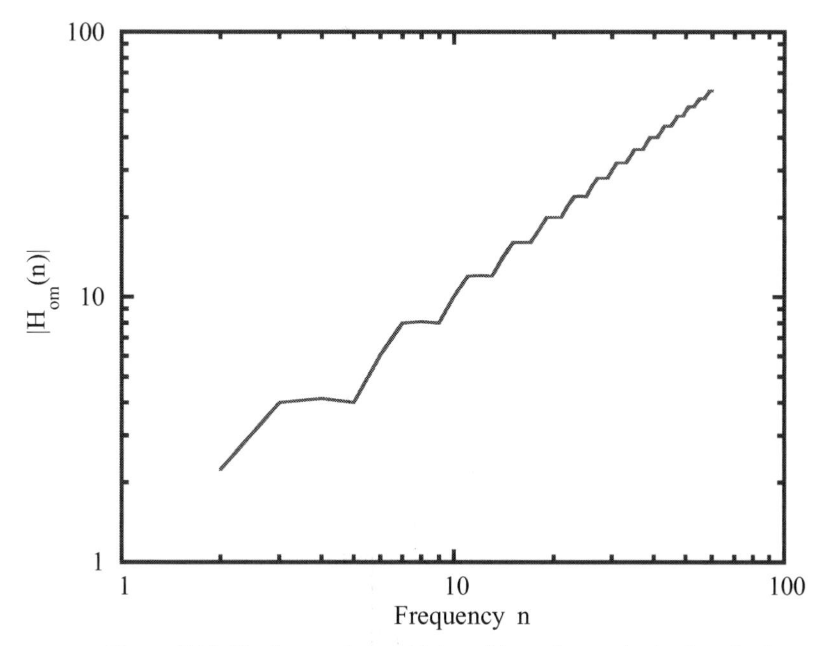

Figure 5.15. The harmonic sensitivity of the orthogonal mixed method

As shown in Figure 5.16, if a setting error $\Delta\phi$ in the angular distance between the two sensors exists, the differential output of the orthogonal mixed method becomes

$$m_e(\theta) = r(\theta + \Delta\phi) + R_r r(\theta + \frac{\pi}{2})$$

$$\approx r(\theta) + R_r r'(\theta + \frac{\pi}{2}) + \Delta\phi r'(\theta) . \tag{5.51}$$

An error $\Delta m_{om}(\theta)$ in the differential output occurs as:

$$\Delta m_{om}(\theta) = \Delta\phi r'(\theta) . \tag{5.52}$$

The evaluated Fourier transform $R_e(n)$ of the roundness error then becomes

$$R_e(n) = \frac{M_e(n)}{H_{om}(n)}$$

$$= R(n) + \frac{\Delta M_e(n)}{H_{om}(n)}$$

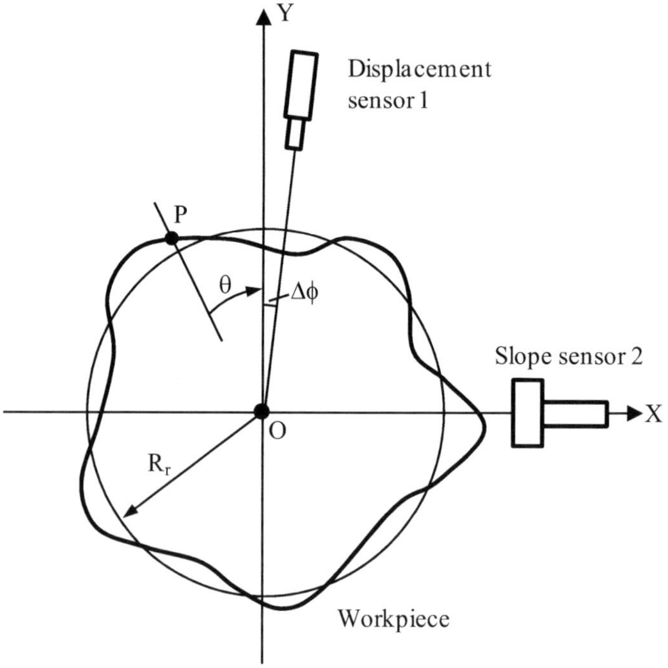

Figure 5.16. Setting error of angular distance

$$= R(n) + \Delta R(n), \tag{5.53}$$

where

$$R_e(n) = \frac{\Delta M_e(n)}{H_{om}(n)}. \tag{5.54}$$

Here, $R(n)$ is the real Fourier transform of the roundness error. $M_e(n)$ and $\Delta M_e(n)$ are the Fourier transforms of $m_e(\theta)$ and $\Delta m_e(\theta)$, respectively. The relative error of $|R_e(n)|$ to $|R(n)|$ can then be evaluated as:

$$\Delta E(n) = 1 - \left|\frac{R_r(n)}{R(n)}\right| = \left|\frac{\Delta R_r(n)}{R(n)}\right|$$

$$= \frac{n\Delta\phi}{|H_{om}(n)|}. \tag{5.55}$$

Figure 5.17 shows $\Delta E(n)$ plotted versus n when $\Delta\phi = 0.5°$. It can be seen that the largest error occurs at $n = 5$ ($\Delta E(5) = 1.1\%$).

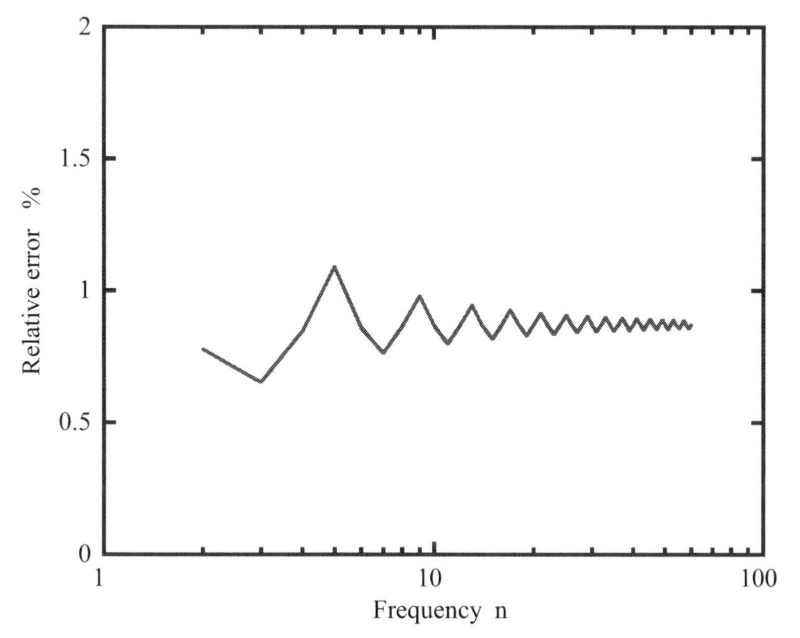

Figure 5.17. Evaluation error of roundness caused by the setting error

An optical sensor system consisting of one displacement sensor [17] and one slope sensor [18] with an angular distance of 90º was constructed to realize the orthogonal mixed method. Both the displacement sensor and the slope sensor utilize the principle of the critical angle method of total reflection.

Figure 5.18 shows the experimental setup for displacement sensor calibration. A capacitive sensor was used as the reference. The surface was moved by using a PZT actuator, and the displacement of the surface was simultaneously measured by the developed displacement sensor and the reference sensor. Figure 5.19 shows calibration results obtained in two separate measurements. The residual error from fitting with a third-order polynomial is also plotted in this figure, and can be seen to be approximately 0.5% of the calibration range. Figure 5.20 shows the experimental setup for the slope sensor calibration. A photoelectric autocollimator was used as the reference. A lever system was used to introduce the angular displacement. The lever was driven by using a PZT actuator so that the lever could rotate about its fulcrum. The angular displacement of the lever was measured simultaneously by the developed slope sensor and the autocollimator. Two separate calibration results are plotted in Figure 5.21. The residual error from fitting with a third-order polynomial is approximately 0.5% of the calibration range.

The experimental setup shown in Figure 5.22 was used to investigate the feasibility of canceling the spindle error in the differential output of the orthogonal mixed method defined in Equation 5.47. A precision ball with a diameter of 1 inch was used as the target. The spindle error $e_Y(\theta)$ was introduced by moving the ball in the Y-direction by using a PZT. As shown in Figure 5.23, the spindle error $e_Y(\theta)$ is canceled in the differential output of the orthogonal mixed method.

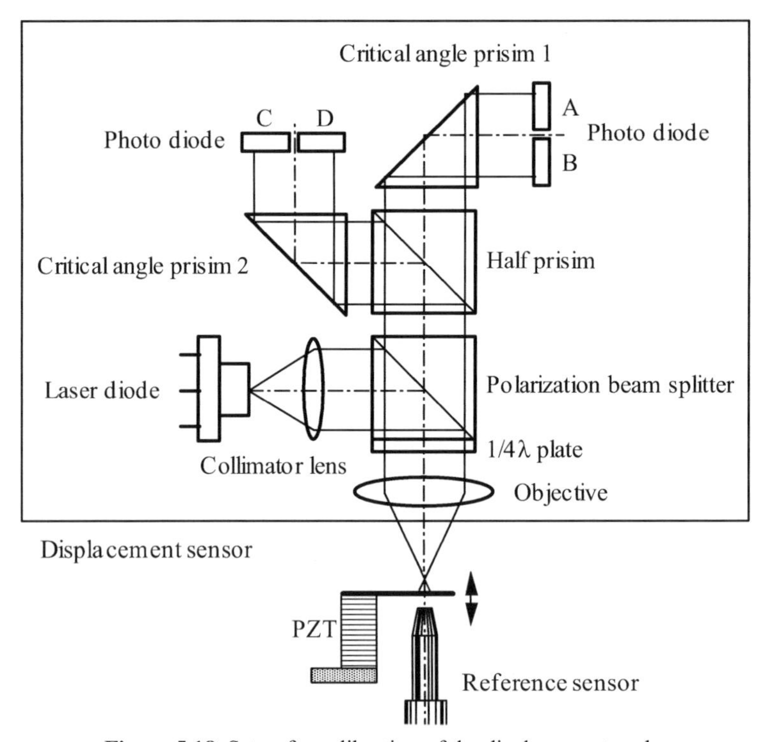

Figure 5.18. Setup for calibration of the displacement probe

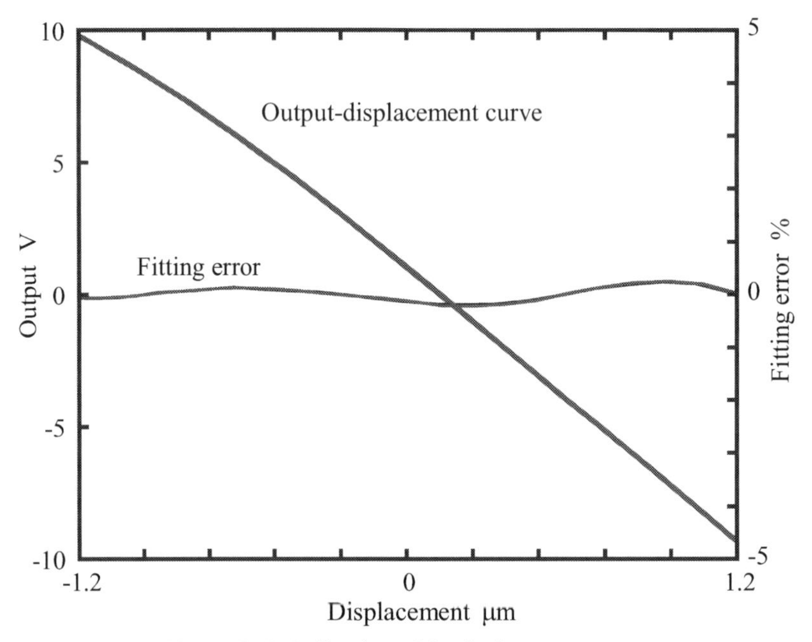

Figure 5.19. Calibration of the displacement sensor

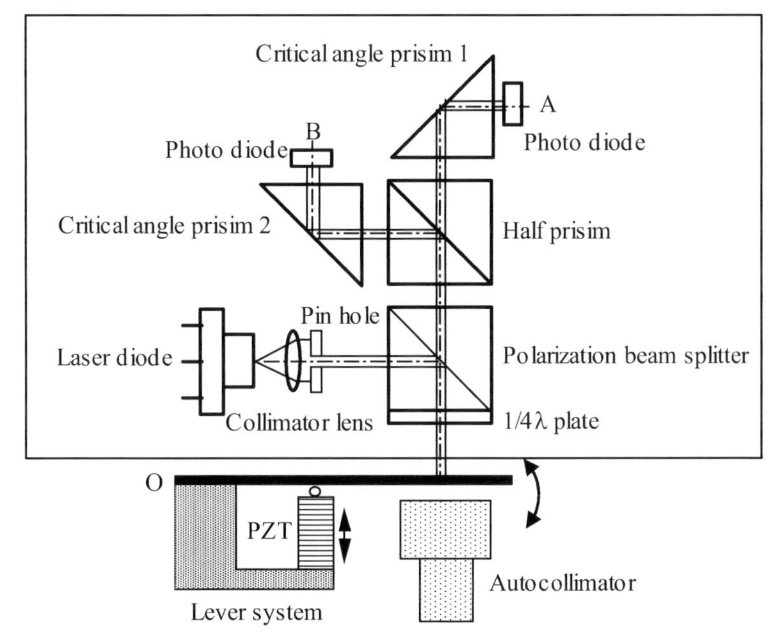

Figure 5.20. Setup for calibration of the slope probe

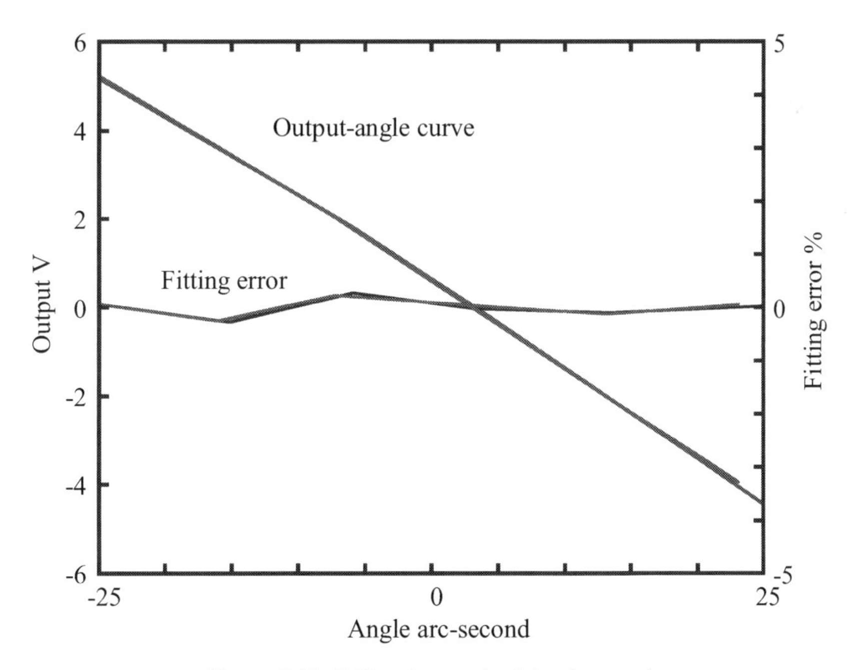

Figure 5.21. Calibration result of the slope probe

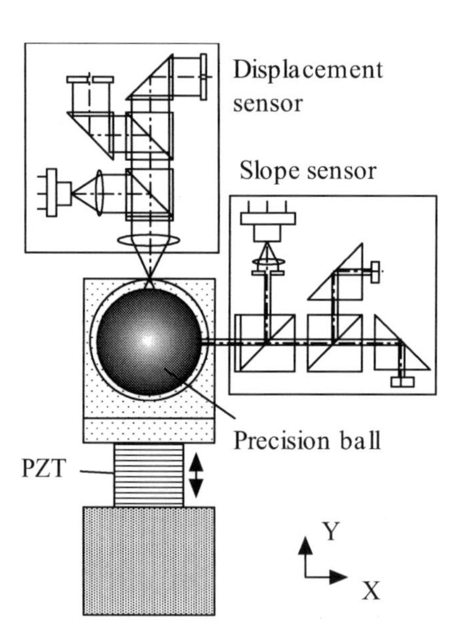

Figure 5.22. Setup for testing of the differential output of the orthogonal mixed method

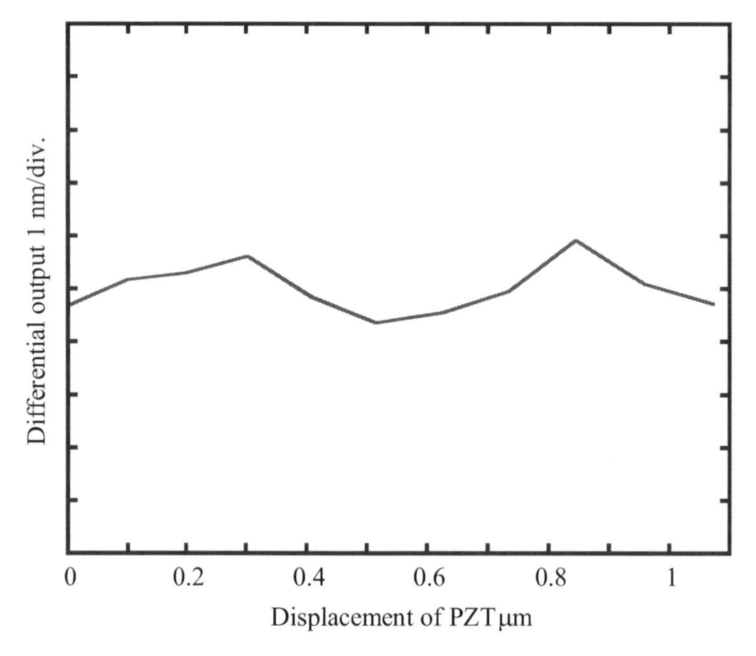

Figure 5.23. Differential output of the orthogonal mixed method

Figure 5.24 shows a photograph of the experimental measurement system constructed for the roundness measurement based on the orthogonal mixed method. The measurement system consists of the developed displacement and slope sensors, the precision ball shown in Figure 5.22, an air-spindle, and an optical rotary encoder. The rotational angle of the spindle was measured by the optical encoder. The positional signal of the optical encoder is sent to the AD converter as a trigger signal. The output signals are sampled simultaneously in order to avoid errors attributable to the sampling time delay. The ball can be adjusted in the *X*- and *Y*-directions by using adjustment screws so that the eccentric error can be adjusted to fall within the measurement ranges of the sensors. The sensors are mounted on *XYZ* micro-stages and the positions of the sensors relative to the ball can be adjusted in the *X*-, *Y*- and *Z*-directions.

Figure 5.25 shows the result of stability test of the optical sensor. In the test, the output signals were sampled without rotating the ball. The displacement sensor and the slope sensor can be seen to have the stabilities of 1 nm and 0.01 arc-second, respectively in a test term of 20 s.

Figure 5.26 shows the measured roundness errors of two separate measurements and the repeatability error between the two measured results. Figure 5.26 (a) shows the polar plot of the roundness error and the repeatability error, and Figure 5.26 (b) shows the corresponding rectilinear plot. The sampling number was 512. It can be seen that the roundness error is approximately 60 nm, and the repeatability error is approximately 5 nm. The measured spindle errors of the two repeated measurements and the difference between them are shown in Figure 5.27. The spindle error was approximately 800 nm, and the difference was approximately 140 nm. Vibration components, which were caused by the improper coupling between the spindle and the encoder, were found in the spindle errors. Comparison of the results plotted in Figures 5.26 and 5.27 shows that the roundness error is separated from large spindle errors with high repeatability. This confirmed the effectiveness of the orthogonal mixed method.

5.4 Summary

A multi-sensor method using three two-dimensional slope sensors for roundness and spindle error measurement, called the three-slope sensor method, has been described. This method can simultaneously measure the workpiece roundness error, the spindle radial error motion and the spindle tilt error motion accurately.

A multi-sensor method mixing displacement sensors and slope sensors for roundness and spindle error measurement, known as the mixed method, has also been presented. This method can separate roundness and spindle error completely, and is well suited for measuring profiles that include high-frequency components. It was verified that well-balanced harmonic response can be achieved over the entire frequency range when the angular distance between the slope sensor and the displacement sensor is set to be 90°. The mixed method employing this sensor arrangement is called the orthogonal mixed method. This sensor arrangement is also the simplest one because the separation of the roundness error from the spindle error requires only one displacement sensor and one slope sensor.

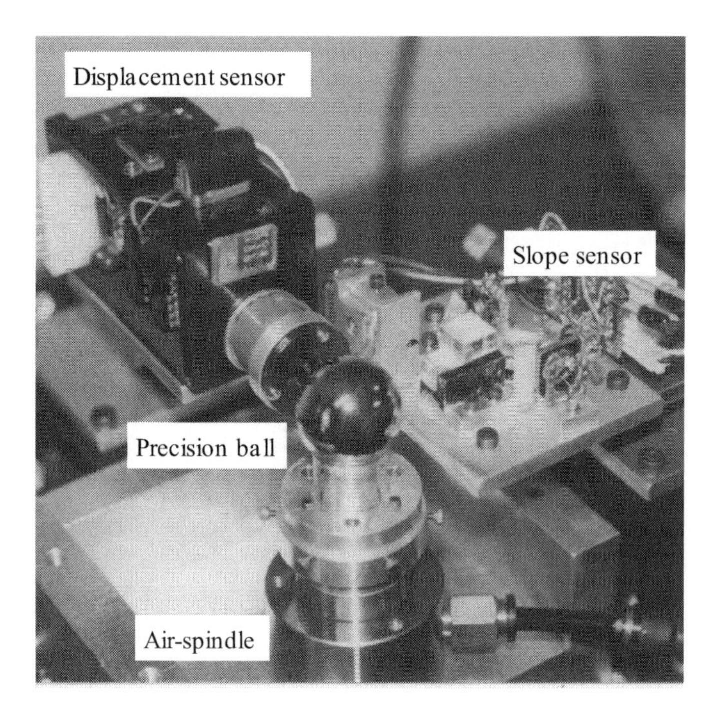

Figure 5.24. Setup for roundness and spindle error measurement with the orthogonal mixed method

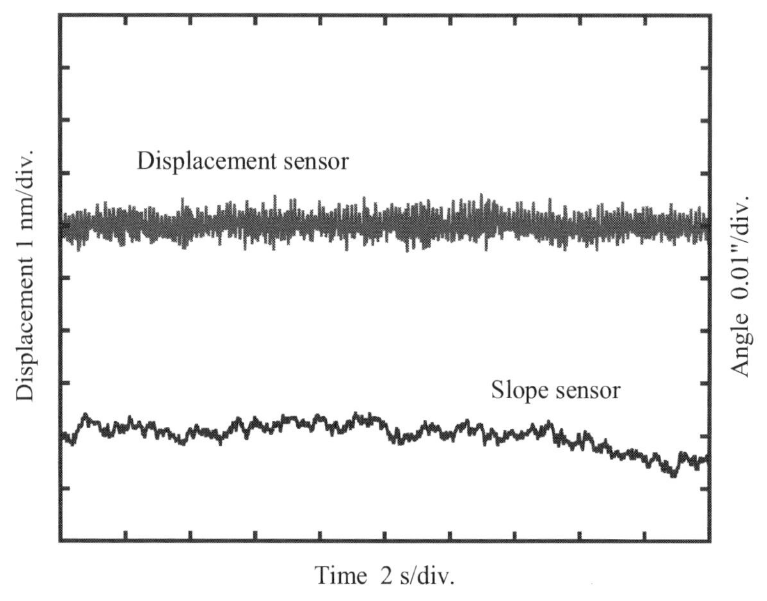

Figure 5.25. Results of stability test

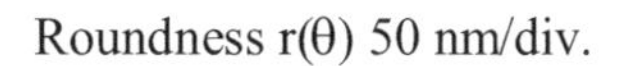

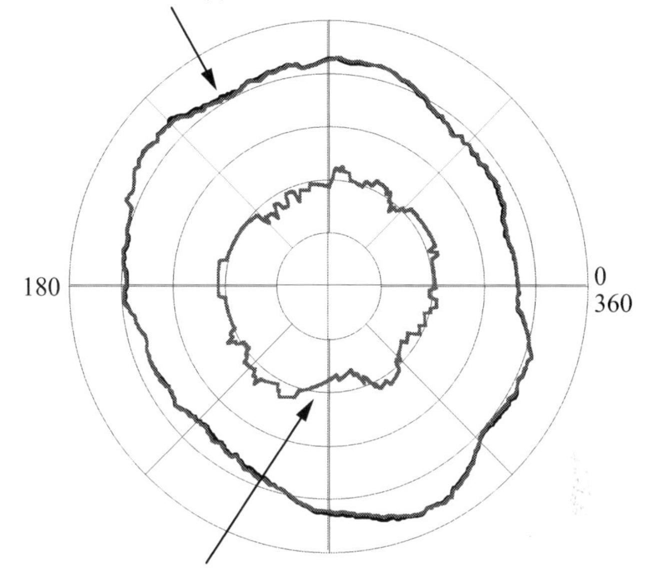

(a) Polar plot

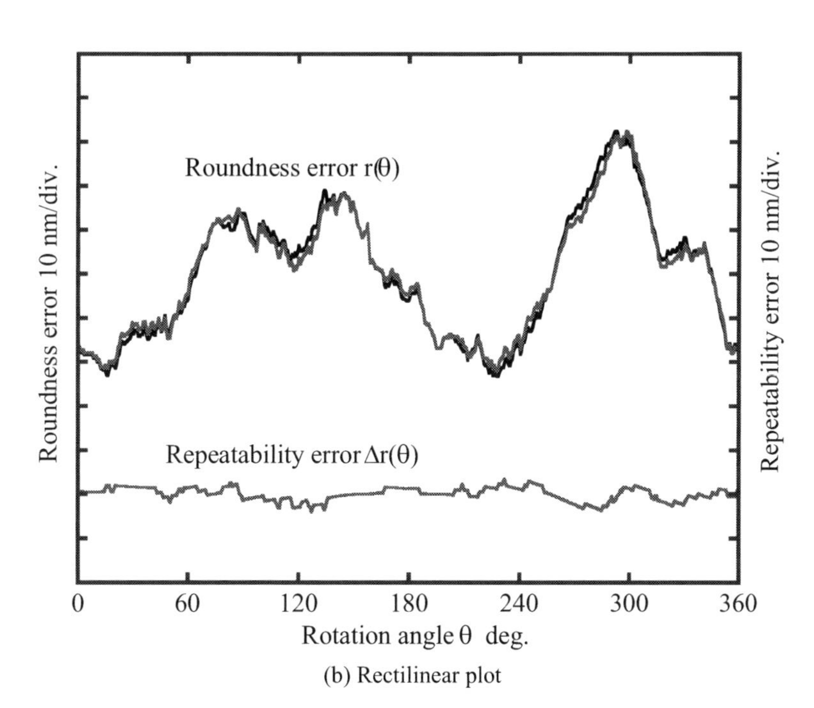

(b) Rectilinear plot

Figure 5.26. Measurement result of ball roundness

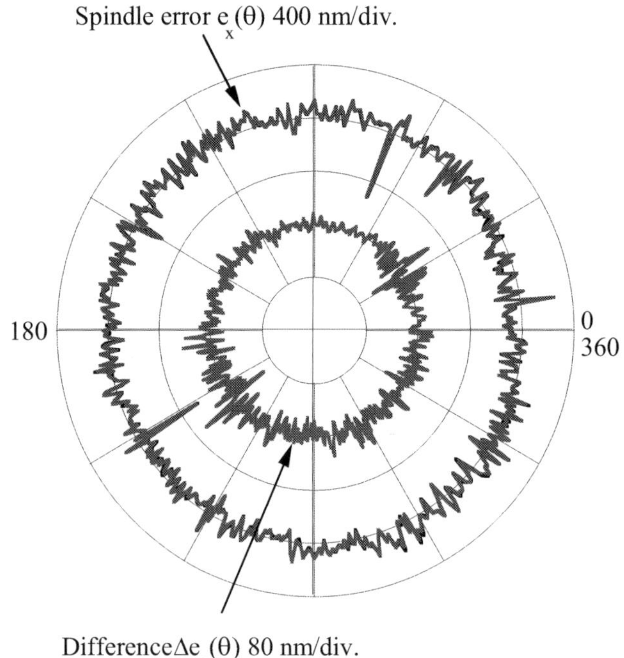

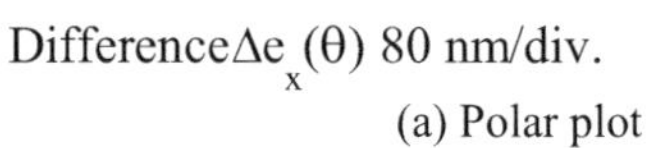

(a) Polar plot

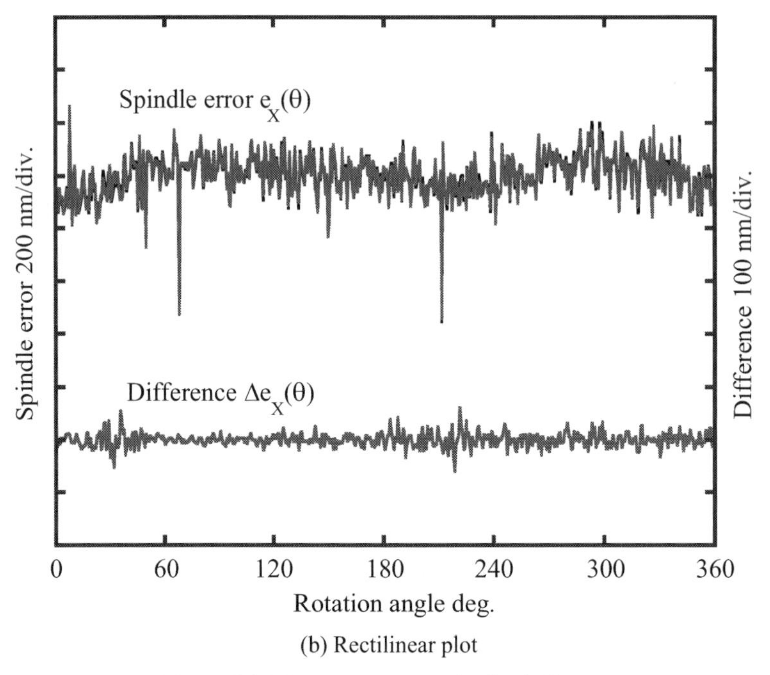

(b) Rectilinear plot

Figure 5.27. Measurement result of spindle error

References

[1] Bryan J, Clouser, R, Holland E (1967) Spindle accuracy. Amer Machinist 612:149–164

[2] CIRP STC Me (1976) Unification document Me: axes of rotation. Ann CIRP 25(2):545–564

[3] Kakino Y, Kitazawa J (1978) In situ measurement of cylindricity. Ann CIRP 27(1):371–375

[4] Shinno H, Mitsui K, Tanaka N, Omino T, Tabata T (1987) A new method for evaluating error motion of ultra-precision spindle. Ann CIRP 36(1):381–384

[5] Whitehouse DJ (1976) Some theoretical aspects of error separation techniques in surface metrology. J Phys E Sci Instrum 9:531–536

[6] Donaldson RR (1972) A simple method for separating spindle error from test ball roundness error. Ann CIRP 21(1):125–126

[7] Evans CJ, Hocken RJ, Estler WT (1996) Self-calibration: reversal, redundancy, error separation and "absolute testing". Ann CIRP 45(2):617–634

[8] Estler WT, Evans CJ, Shao LZ (1997) Uncertainty estimation for multi-position form error metrology. Precis Eng 21(2/3):72–82

[9] Ozono S (1974) On a new method of roundness measurement based on the three points method. In: Proceedings of the International Conference on Production Engineering, Tokyo, Japan, pp 457–462

[10] Moore D (1989) Design considerations in multi-probe roundness measurement. J Phys E Sci Instrum 9:339–343

[11] Zhang GX, Wang RK (1993) Four-point method of roundness and spindle error measurements. Ann CIRP 42(1):593–596

[12] Gao W, Kiyono S, Nomura T (1996) A new multi-probe method of roundness measurements. Precis Eng 19(1):37–45

[13] Gao W, Kiyono S, Sugawara T (1997) High accuracy roundness measurement by a new error separation method. Precis Eng 21(2/3):123–133

[14] Ozono S, Hamano Y (1976) On a new method of roundness measurement based on the three-point method, 2nd report. Expanding the measurable maximum frequency. In: Proceedings of the Annual Meeting of JSPE, pp 503–504

[15] Hanmming RW (1989) Digital filters. Prentice Hall, Upper Saddle River, NJ

[16] Gao W, Kiyono S, Satoh E (2002) Precision measurement of multi-degree-of-freedom spindle errors using two-dimensional slope sensors. Ann CIRP 52(2):447–450

[17] Kohno T, Ozawa N, Miyamoto K, Musha T (1988) High precision optical surface sensor. Appl Opt 27(1):103–108

[18] Huang P, Kiyono S, Kamada O (1992) Angle measurement based on the internal-reflection effect: a new method. Appl Opt 31(28):6047–6055

6

Scanning Error Separation System for Measurement of Straightness

6.1 Introduction

The straightness is another fundamental geometric parameter of precision workpieces. The straightness of a workpiece surface can be measured by scanning a displacement sensor or a slope sensor over the workpiece surface by a linear stage (slide). Because the axis of motion of the linear stage functions as the reference for the measurement, any out-of-straightness error motion of the slide will cause a measurement error. Because the out-of-straightness error of a precision linear slide (slide error) is typical on the order of 100 nm over a 100 mm moving stroke [1], it is necessary to separate the error motion for precision nanometrology of the workpiece straightness. The influence of the straightness error of a straightedge surface, which is employed as the reference for measurement of slide error, should also be removed.

Similar to the roundness measurement described in Chapter 5, error separation can be carried out by the multi-sensor method and the reversal method. This chapter provides solutions to some key issues inherent in conventional error separation methods for measurement of workpiece straightness and slide error.

6.2 Three-displacement Sensor Method with Self Zero-adjustment

6.2.1 Three-displacement Sensor Method and Zero-adjustment Error

The three-displacement sensor method [2–4], which uses three displacement sensors, is the most typical multi-sensor method for the straightness measurement. A double integration operation of the differential output of the three-displacement sensor method, in which the error motions are cancelled, is used to evaluate the workpiece straightness. However, if the zero-values of the sensors are not adjusted or measured (zero-adjustment) precisely, the difference between the zero-values will introduce an offset in the differential output of the three-displacement sensor method. The double integration of the offset will thus yield a parabolic error term in the profile evaluation result of the three-displacement sensor method [5–7]. The zero-adjustment can be made by measuring an accurate reference flat surface. However this method is not efficient for measurement of a long workpiece since the parabolic error term is proportional to the square of the measurement length. Even a small zero-adjustment error introduced by the flatness error of the reference surface will cause a large profile evaluation error. For this reason, the zero-adjustment error is the largest error source for the straightness measurement of a long workpiece by the three-displacement sensor method. This is the most significant difference between the straightness measurement and the roundness measurement by using multi-sensor methods.

Figure 6.1 shows the schematic of the three-displacement sensor method for measurement of straightness. In this method, a sensor unit (sensor unit-A) consisting of three displacement sensors is mounted on a scanning stage moving along the X-direction. The straightness profile of the workpiece side 1 surface along the X-axis is scanned by the sensor unit. Assume that the profile height of side 1 is $f(x_i, 0°)$ at sampling position x_i. Let the corresponding sensor outputs be $m_1(x_i)$, $m_2(x_i)$ and $m_3,(x_i)$, respectively; they can be expressed as:

$$m_1(x_i) = f(x_i - d, 0°) + e_Z(x_i) - de_{yaw}(x_i), \tag{6.1}$$

$$m_2(x_i) = f(x_i, 0°) + e_Z(x_i), \tag{6.2}$$

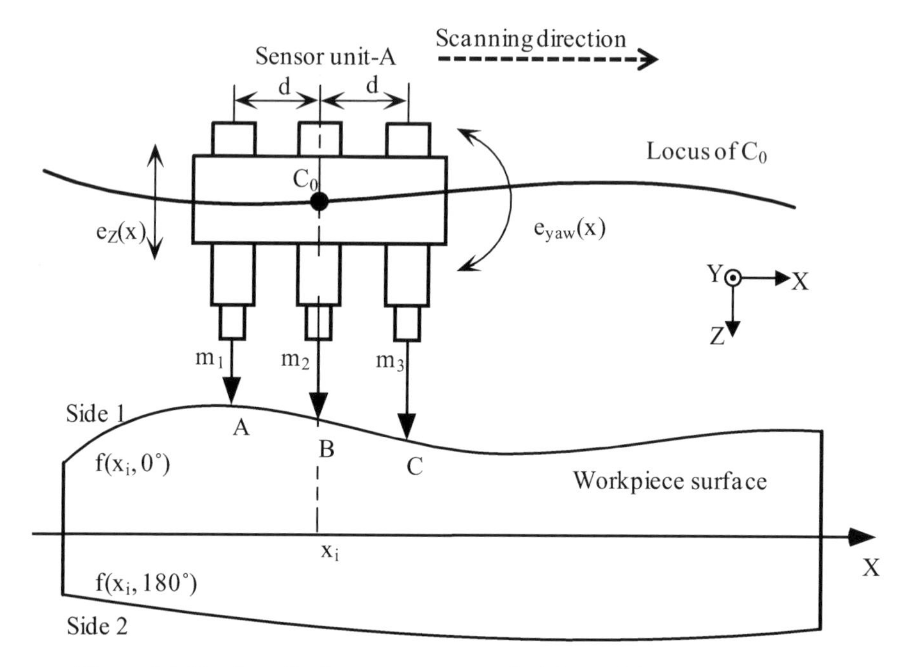

Figure 6.1. The three-displacement sensor method for straightness measurement

$$m_3(x_i) = f(x_i + d, 0°) + e_Z(x_i) + de_{yaw}(x_i), \ (i = 1, \ldots, N), \tag{6.3}$$

where d is the sensor interval, and N is the sampling number over the scanning length L. The sampling is assumed to be conducted at an equal sampling period s ($= L/N$). $e_Z(x_i)$ and $e_{yaw}(x_i)$ are the Z-directional translation error and yaw error of the scanning stage at x_i, respectively. A differential output $m_s(x_i)$ is calculated to cancel the error motions as follows:

$$m_s(x_i) = \frac{m_3(x_i) - 2m_2(x_i) + m_1(x_i)}{d^2}$$

$$= \left[\frac{(f(x_i + d, 0°) - f(x_i, 0°)) - (f(x_i, 0°) - f(x_i - d, 0°))}{d} \right] \frac{1}{d}$$

$$\approx f''(x_i, 0°), \ (i = 1, \ldots, N). \tag{6.4}$$

An approximated straightness profile can thus be evaluated from double integration of $m_s(x_i)$ without the influence of the error motions.

$$z(x_i) = \sum_{k=1}^{i} \left(\sum_{j=1}^{k} (m_s(x_j) \cdot s) \cdot s \right), \ (i = 2, \ldots N) \tag{6.5}$$

The difference between $z(x_i)$ and $f(x_i, 0°)$, the data processing error, is mainly caused by the discrete derivative and integration operations in the data processing procedure shown above. The error is a function of the sensor interval d and the sampling period s. It is also related to the spatial wavelength components of the surface profile. The data processing error is very small in the long spatial wavelength range and relatively large in the short wavelength range [8, 9]. The error will also become zero in the entire spatial wavelength range when d is equal to s. In this case, however, the profile can only be evaluated with very limited data points. In practice, we set s to be smaller than d to get more data points based on the fact that the interesting spatial wavelength components of the surface profile of most precision cylinders are limited in the long wavelength range. In this chapter, the data processing error is ignored and $z(x_i)$ is assumed equal to $f(x_i, 0°)$.

Most of displacement sensors can only perform relative measurement. It means the absolute zero-value of a displacement sensor is unknown. As shown in Figure 6.2, let the unknown zero-values of the three sensors in the sensor unit be e_{m1}, e_{m2} and e_{m3}, respectively. Equations 6.1–6.3 can be rewritten as:

$$m_1(x_i) = f(x_i - d, 0°) + e_Z(x_i) - de_{yaw}(x_i) + e_{m1}, \tag{6.6}$$

$$m_2(x_i) = f(x_i, 0°) + e_Z(x_i) + e_{m2}, \tag{6.7}$$

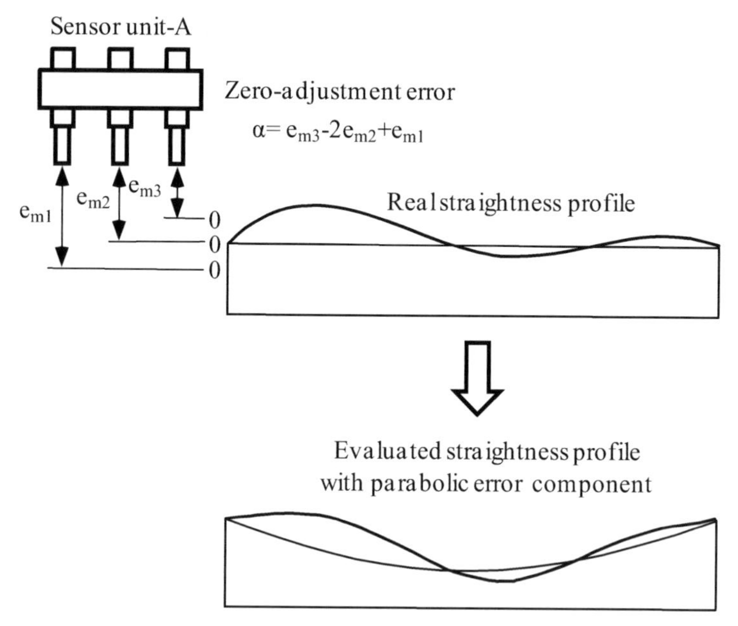

Figure 6.2. The zero-adjustment error

$$m_3(x_i) = f(x_i + d, 0°) + e_Z(x_i) + de_{yaw}(x_i) + e_{m3}, \ (i = 1, \ldots, N). \tag{6.8}$$

The corresponding profile evaluation result in Equation 6.5 becomes:

$$z_1(x_i) = \sum_{k=1}^{i}\left(\sum_{j=1}^{k}(m_s(x_j)\cdot s)\cdot s\right) + \frac{\alpha}{2d^2}x_i^2$$

$$= f(x_i, 0°) + \frac{\alpha}{2d^2}x_i^2, \ (i = 2, \ldots, N), \tag{6.9}$$

where α is the difference between the zero-values, which is defined as:

$$\alpha = (e_{m3} - e_{m2}) + (e_{m1} - e_{m2}) = e_{m3} - 2e_{m2} + e_{m1}. \tag{6.10}$$

It can be seen that a parabolic error term in the profile evaluation result is caused by α. Since this profile evaluation error term is proportional to the square of the measurement length, large profile evaluation errors will be caused in the measurement of long workpieces. To realize the precision straightness profile measurement, α has to be measured accurately. Here, we call α the zero-difference of the sensor unit of the three-displacement sensor method, and refer to the process of measuring the zero-difference as zero-adjustment.

As can be seen in Equation 6.10, α is related to the relative differences of zero-values between sensors. The absolute zero-value of each sensor is not an issue. Theoretically, α can be measured through targeting the sensor unit to a reference flat surface as shown in Figure 6.3. In practice, however, it is difficult and expensive for this zero-adjustment method to achieve high accuracy. As can be seen in Figure 6.4, a 10 nm zero-adjustment error ($\Delta\alpha$) will cause profile evaluation errors of 4.5 μm and 0.18 μm when the sensor interval is set to be 10 and 50 mm, respectively, for a measurement length of 600 mm. In other words, we

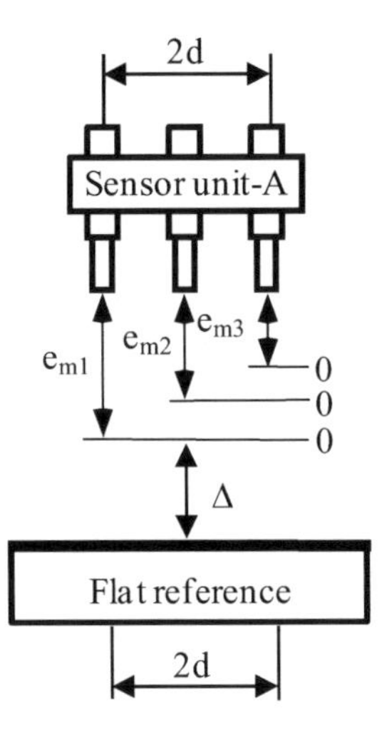

Figure 6.3. Zero-adjustment by using flat reference

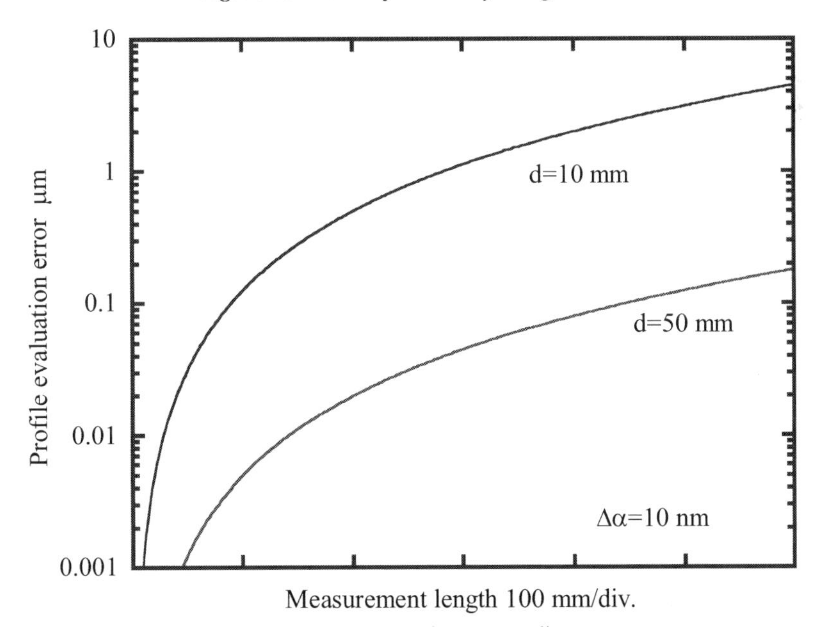

Figure 6.4. Influence of the zero-adjustment error

need to set the sensor interval to be 50 mm and use a reference flat surface with a flatness of 10 nm over a length of 100 mm to realize a profile measurement accuracy of 0.18 μm. It is not easy to get such a reference flat surface in practice.

6.2.2 Three-displacement Sensor Method with Self Zero-adjustment

This section presents a self zero-adjustment method to accurately carry out the zero-adjustment for the three-displacement sensor method. In addition to the sensor unit-A in Figure 6.1, another sensor unit-B with the same sensor interval d is employed (Figure 6.5 (a)). The two sensor units are placed on the two sides of the workpiece to simultaneously scan $f(x, 0°)$ and $f(x, 180°)$. The outputs of unit-B, corresponding to those of unit-A shown in Equations 6.6–6.8, are denoted as:

$$n_1(x_i) = f(x_i - d, 180°) - e_Z(x_i) + de_{yaw}(x_i) + e_{n1}, \quad (6.11)$$

$$n_2(x_i) = f(x_i, 180°) - e_Z(x_i) + e_{n2}, \quad (6.12)$$

$$n_3(x_i) = f(x_i + d, 180°) - e_Z(x_i) - de_{yaw}(x_i) + e_{n3}, \ (i = 1, \ldots, N). \quad (6.13)$$

where e_{n1}, e_{n2}, e_{n3} are the zero-values of sensors in sensor unit-B. The zero-difference of sensor unit-B can be expressed as follows:

$$\beta = e_{n3} - 2e_{n2} + e_{n1}. \quad (6.14)$$

Note that the two sensor units are mounted on the same scanning stage, and sense the same error motions $e_Z(x_i)$ and $e_{yaw}(x_i)$. After the first scanning, as shown in Figure 6.5 (b), the workpiece is rotated 180° about the X-axis, and scanned by the sensor units again. The sensor outputs during the second scanning can be written as:

$$m_{1r}(x_i) = f(x_i - d, 180°) + e_{Zr}(x_i) - de_{yawr}(x_i) + e_{m1}, \quad (6.15)$$

$$m_{2r}(x_i) = f(x_i, 180°) + e_{Zr}(x_i) + e_{m2}, \quad (6.16)$$

$$m_{3r}(x_i) = f(x_i + d, 180°) + e_{Zr}(x_i) + de_{yawr}(x_i) + e_{m3}, \quad (6.17)$$

$$n_{1r}(x_i) = f(x_i - d, 0°) - e_{Zr}(x_i) + de_{yawr}(x_i) + e_{n1}, \quad (6.18)$$

$$n_{2r}(x_i) = f(x_i, 0°) - e_{Zr}(x_i) + e_{n2}, \quad (6.19)$$

$$n_{3r}(x_i) = f(x_i + d, 0°) - e_{Zr}(x_i) - de_{yawr}(x_i) + e_{n3}, \ (i = 1, \ldots, N). \quad (6.20)$$

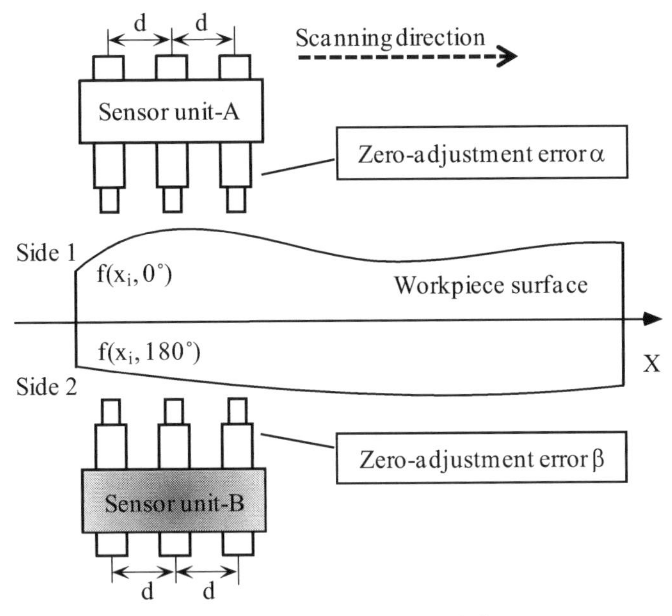

(a) Step 1: First scan of the workpiece

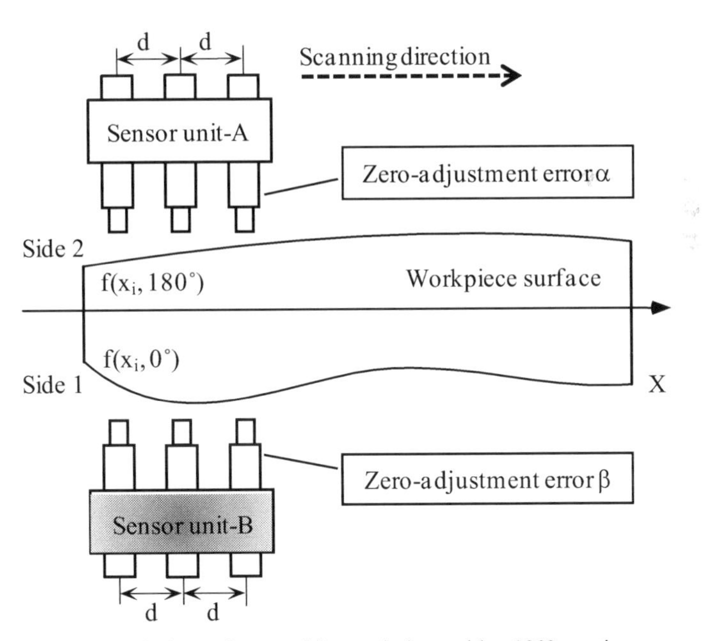

(b) Step 2: Second scan of the workpiece with a 180° rotation

Figure 6.5. Scans for straightness measurement and zero-adjustment with rotation of the workpiece

Here, $e_{Zr}(x_i)$ and $e_{yawr}(x_i)$ are the error motions of the stage during the second scanning, which are generally different from $e_Z(x_i)$ and $e_{yaw}(x_i)$ during the first scanning. From the sensor outputs of the two scanning steps, the zero-differences α and β can then be calculated from the following equations:

$$\beta+\alpha = \frac{\sum_{i=1}^{N-N_d}\left\{\begin{array}{l}(m_3(x_i) - m_2(x_i))-(m_2(x_i+d) - m_1(x_i+d))\\ +(m_{3r}(x_i)-m_{2r}(x_i))-(m_{2r}(x_i+d)-m_{1r}(x_i+d))\\ +(n_3(x_i) - n_2(x_i))-(n_2(x_i+d) - n_1(x_i+d))\\ +(n_{3r}(x_i)-n_{2r}(x_i))-(n_{2r}(x_i+d)-n_{1r}(x_i+d))\end{array}\right\}}{2(N-N_d)}, \tag{6.21}$$

$$\beta-\alpha = \frac{\sum_{i=1}^{N}\left\{\begin{array}{l}(n_3(x_i) - 2n_2(x_i) + n_1(x_i))\\ +(n_{3r}(x_i) - 2n_{2r}(x_i) + n_{1r}(x_i))\\ -(m_3(x_i) - 2m_2(x_i) + m_1(x_i))\\ -(m_{3r}(x_i)-2m_{2r}(x_i)+m_{1r}(x_i))\end{array}\right\}}{2N}, \tag{6.22}$$

where $N_d = d/s$.

Once the zero-difference is obtained, the parabolic error term in Equation 6.9 can be compensated and an accurate straightness profile can be evaluated. It can be seen that no accurate reference surfaces or auxiliary artifacts are used in the zero-adjustment. The averaging operations in Equations 6.21 and 6.22 also greatly reduce the influence of random errors in the sensor outputs. The influence of the positioning error of sampling can also be reduced. The disadvantage of the zero-adjustment method is that an extra sensor unit is necessary.

The zero-adjustment can also be realized by exchanging the positions of the sensor units. As can be seen in Figure 6.6, the sensor units are reversed after the first scanning. Zero-differences of the sensor units can be measured exactly in the same way shown above. Another modified zero-adjustment method is shown in Figure 6.7. The sensor units are rotated around the Y-axis after the first scanning. The sensor outputs of the second scanning can be expressed by:

$$m_{1rr}(x_i) = f(x_i+d, 180°) + e_{Zr}(x_i) - de_{yawr}(x_i) + e_{m1}, \tag{6.23}$$

$$m_{2rr}(x_i) = f(x_i, 180°) + e_{Zr}(x_i) + e_{m2}, \tag{6.24}$$

$$m_{3rr}(x_i) = f(x_i-d, 180°) + e_{Zr}(x_i) + de_{yawr}(x_i) + e_{m3}, \tag{6.25}$$

$$n_{1rr}(x_i) = f(x_i+d, 0°) - e_{Zr}(x_i) + de_{yawr}(x_i) + e_{n1}, \tag{6.26}$$

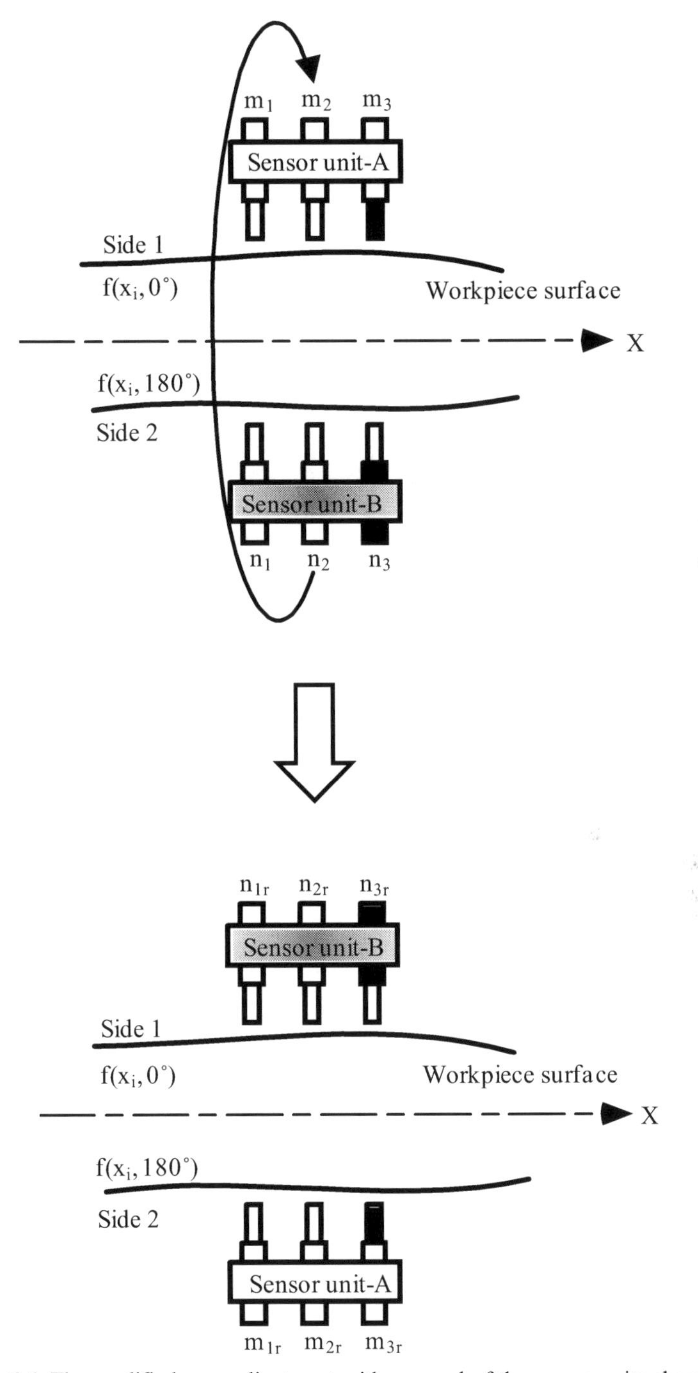

Figure 6.6. The modified zero-adjustment with reversal of the sensor units about the *X*-axis

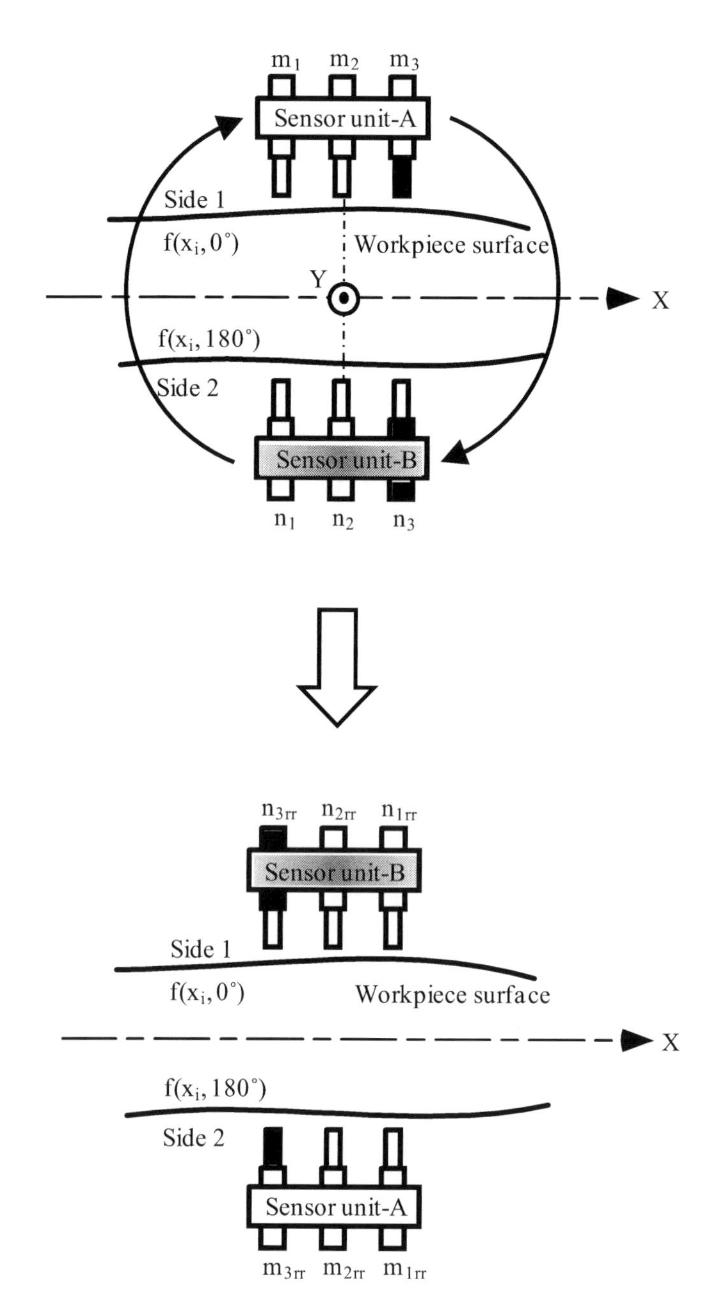

Figure 6.7. The modified zero-adjustment method with rotation of the sensor units about the *Y*-axis

$$n_{2rr}(x_i) = f(x_i, 0°) - e_{Zr}(x_i) + e_{n2}, \tag{6.27}$$

$$n_{3rr}(x_i) = f(x_i - d, 0°) - e_{Zr}(x_i) - de_{yawr}(x_i) + e_{n3}, \; (i = 1, \ldots, N). \tag{6.28}$$

From Equations 6.6–6.8, Equations 6.11–6.13 and Equations 6.23–6.28, the following equations can be derived for obtaining the zero-differences:

$$\beta + \alpha = \frac{\sum_{i=1}^{N-Nd} \left\{ \begin{array}{l} (m_3(x_i) - m_2(x_i)) - (m_2(x_i + d) - m_1(x_i + d)) \\ + (m_{3rr}(x_i + d) - m_{2rr}(x_i + d)) - (m_{2rr}(x_i) - m_{1rr}(x_i)) \\ + (n_3(x_i) - n_2(x_i)) - (n_2(x_i + d) - n_1(x_i + d)) \\ + (n_{3rr}(x_i + d) - n_{2rr}(x_i + d)) - (n_{2rr}(x_i) - n_{1rr}(x_i)) \end{array} \right\}}{2(N - N_d)}, \tag{6.29}$$

$$\beta - \alpha = \frac{\sum_{i=1}^{N} \left\{ \begin{array}{l} (n_3(x_i) - 2n_2(x_i) + n_1(x_i)) \\ + (n_{3rr}(x_i) - 2n_{2rr}(x_i) + n_{1rr}(x_i)) \\ - (m_3(x_i) - 2m_2(x_i) + m_1(x_i)) \\ - (m_{3rr}(x_i) - 2m_{2rr}(x_i) + m_{1rr}(x_i)) \end{array} \right\}}{2N}. \tag{6.30}$$

It should be pointed out that the zero-differences calculated above are mean values of those during the two scanning steps. Variations of zero-differences during the measurement will cause profile evaluation errors. An improved algorithm for calculation of the zero-difference, which can be used to obtain the variation of zero-difference during the scanning, is specially developed for measurement of cylinder workpieces. Figure 6.8 shows the sampling positions of the improved zero-adjustment method. The sampling way is different from that of Figure 6.5, in which the cylinder is fixed during each sampling step. As can be seen in Figure 6.8, the cylinder is being rotated by a motor-driven spindle while the sensor units are moved along the X-axis. At each position x_i along the X-axis, the sensors sample one revolution of the cylinder surface profile. Assume that the sampling positions along the circumference are θ_j ($j = 1, \ldots, K$). The sensor outputs at the sampling position of (x_i, θ_j) can be expressed as:

$$m_1(x_i, \theta_j) = f(x_i - d, \theta_j) + e_Z(x_i, \theta_j) - de_{yaw}(x_i, \theta_j) + e_{m1}, \tag{6.31}$$

$$m_2(x_i, \theta_j) = f(x_i, \theta_j) + e_Z(x_i, \theta_j) + e_{m2}, \tag{6.32}$$

$$m_3(x_i, \theta_j) = f(x_i + d, \theta_j) + e_Z(x_i, \theta_j) + de_{yaw}(x_i, \theta_j) + e_{m3}, \tag{6.33}$$

$$n_1(x_i, \theta_j) = f(x_i - d, \theta_{j+K/2}) - e_Z(x_i, \theta_j) + de_{yaw}(x_i, \theta_j) + e_{n1}, \tag{6.34}$$

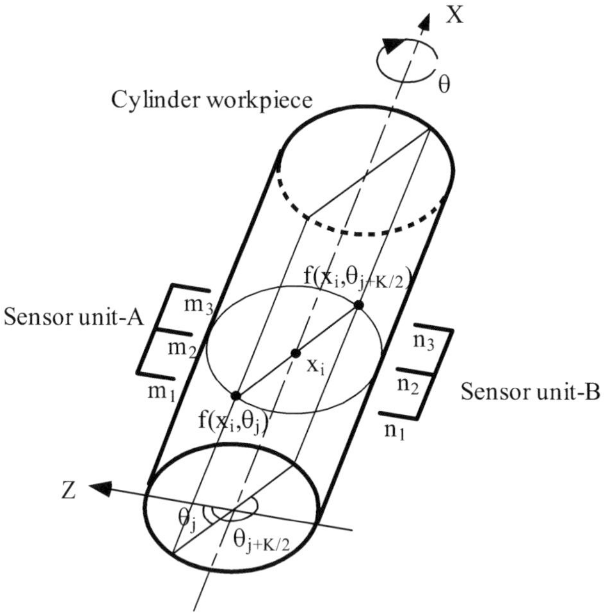

Figure 6.8. The improved zero-adjustment method with a rotating cylinder workpiece

$$n_2(x_i,\theta_j) = f(x_i,\theta_{j+K/2}) - e_Z(x_i,\theta_j) + e_{n2}\,, \tag{6.35}$$

$$n_3(x_i,\theta_j) = f(x_i+d,\theta_{j+K/2}) - e_Z(x_i,\theta_j) - de_{yaw}(x_i,\theta_j) + e_{n3}\,,$$

$(i = 1, \ldots, N, j = 1, \ldots, K)$.(6.36)

The term $e_Z(x_i, \theta_j)$ is a combination of the Z-directional translational error motion of the stage to move the sensor units and the Z-directional radial error motion of the spindle to rotate the cylinder. $e_{yaw}(x_i, \theta_j)$ is a combination of the yaw error motion of the stage and the angular error motion of the spindle. The zero-differences $\alpha(x_i)$ and $\beta(x_i)$ at position x_i along X-axis can then be obtained from the next equations:

$$\beta(x_i)+\alpha(x_i) = \frac{\sum_{j=1}^{K/2}\left\{\begin{array}{l} \left(m_3(x_i,\theta_j)-m_2(x_i,\theta_j)\right)-\left(m_2(x_i+d,\theta_j)-m_1(x_i+d,\theta_j)\right) \\ +\left(m_3(x_i,\theta_{j+K/2})-m_2(x_i,\theta_{j+K/2})\right) \\ -\left(m_2(x_i+d,\theta_{j+K/2})-m_1(x_i+d,\theta_{j+K/2})\right) \\ +\left(n_3(x_i,\theta_j)-n_2(x_i,\theta_j)\right)-\left(n_2(x_i+d,\theta_j)-n_1(x_i+d,\theta_j)\right) \\ +\left(n_{3r}(x_i,\theta_{j+K/2})-n_{2r}(x_i,\theta_{j+K/2})\right) \\ -\left(n_{2r}(x_i+d,\theta_{j+K/2})-n_{1r}(x_i+d,\theta_{j+K/2})\right) \end{array}\right\}}{K}, \tag{6.37}$$

$$\beta(x_i)-\alpha(x_i)=\frac{\sum_{j=1}^{K/2}\left\{\begin{array}{l}\left(n_3(x_i,\theta_j)-2n_2(x_i,\theta_j)+n_1(x_i,\theta_j)\right)\\+\left(n_3(x_i,\theta_{j+K/2})-2n_2(x_i,\theta_{j+K/2})+n_1(x_i,\theta_{j+K/2})\right)\\-\left(m_3(x_i,\theta_j)-2m_2(x_i,\theta_j)+m_1(x_i,\theta_j)\right)\\-\left(m_3(x_i,\theta_{j+K/2})-2m_2(x_i,\theta_{j+K/2})+m_1(x_i,\theta_{j+K/2})\right)\end{array}\right\}}{K}. \tag{6.38}$$

As can be seen from above, the variation of the zero-difference caused by thermal drift during the scanning can be monitored by this improved zero-adjustment method. This will improve the accuracy of the straightness profile measurement through compensating for the variation of the zero-difference.

Figure 6.9 shows a schematic of the experimental system based on the method in Figure 6.5. A cylinder workpiece of 80 mm in diameter was mounted on a spindle and can be rotated around its center axis, which was along the *X*-axis. The sensor units were placed on the table of a linear stage. The stage was driven by a servo motor and had a travel range of 1 m in the *X*-direction. The height positions of the sensor centers were carefully aligned to be consistent with that of the cylinder center. Six capacitive-type displacement sensors were employed. Each sensor had a measurement range of 100 μm, and the non-linearity was of up to 0.4% of the measurement range. The footprint size of the sensor was 1.7 mm. The sensor interval was set to be 50 mm in both sensor units.

Figure 6.10 shows a result of testing the sensor stability. Both the stage and the cylinder were kept stationary during the test. Some vibration signals with peak-to-valley values of up to approximately 400 nm can be found in the output of each sensor. The experimental setup was set in a machine shop where some vibration sources existed. The vibrations were reduced to 100 nm in the differential output, which was close to the resolution of the AD converter to acquire the sensor outputs. This indicates the excellent anti-vibration ability of the three-displacement sensor method. Figure 6.11 (a) shows the sensor outputs of the first scanning shown in Figure 6.5 (a). The measurement length was 600 mm. The sampling period was 1 mm. Since the outputs of sensors in each sensor unit showed almost the same phase variations, and the two sensor units showed a 180° phase difference, it can be said that most of the sensor outputs were associated with error motions of the stage. Figure 6.11 (b) shows the sensor outputs of the second scanning shown in Figure 6.5 (b) after rotating the cylinder 180°. The measurement time for the two scanning steps was approximately 10 min. An additional tilt component, which was caused by the tilt of the cylinder axis when rotating the cylinder, can be found in the sensor outputs.

As can be seen in Figures 6.5 (a) and 6.5 (b), the same profile ($f(x, 0°)$ and/or $f(x, 180°)$) is measured by both the sensor units, and the profile evaluation results by the two sensor units should agree with each other. In the following, the difference between the results of $f(x, 0°)$ evaluated by the two sensor units is used to estimate the reliability of zero-adjustment. The result evaluated by unit-A is referred to as $f_1(x)$ and that by unit-B as $f_2(x)$. The reproducibility error, which is the difference between $f_1(x)$ and $f_2(x)$, is referred to as $\Delta f(x)$.

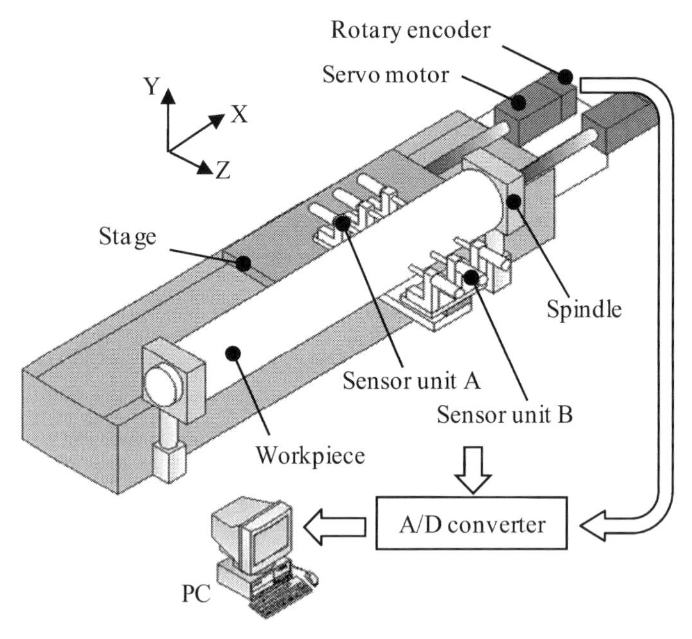

(a) Schematic view

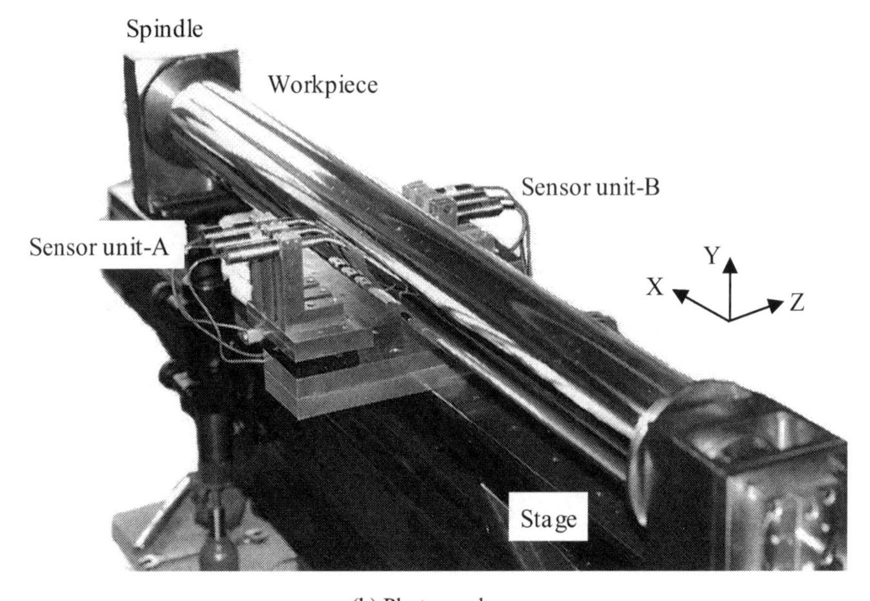

(b) Photograph

Figure 6.9. The experimental system for straightness measurement by using the three-displacement probe method with self zero-adjustment

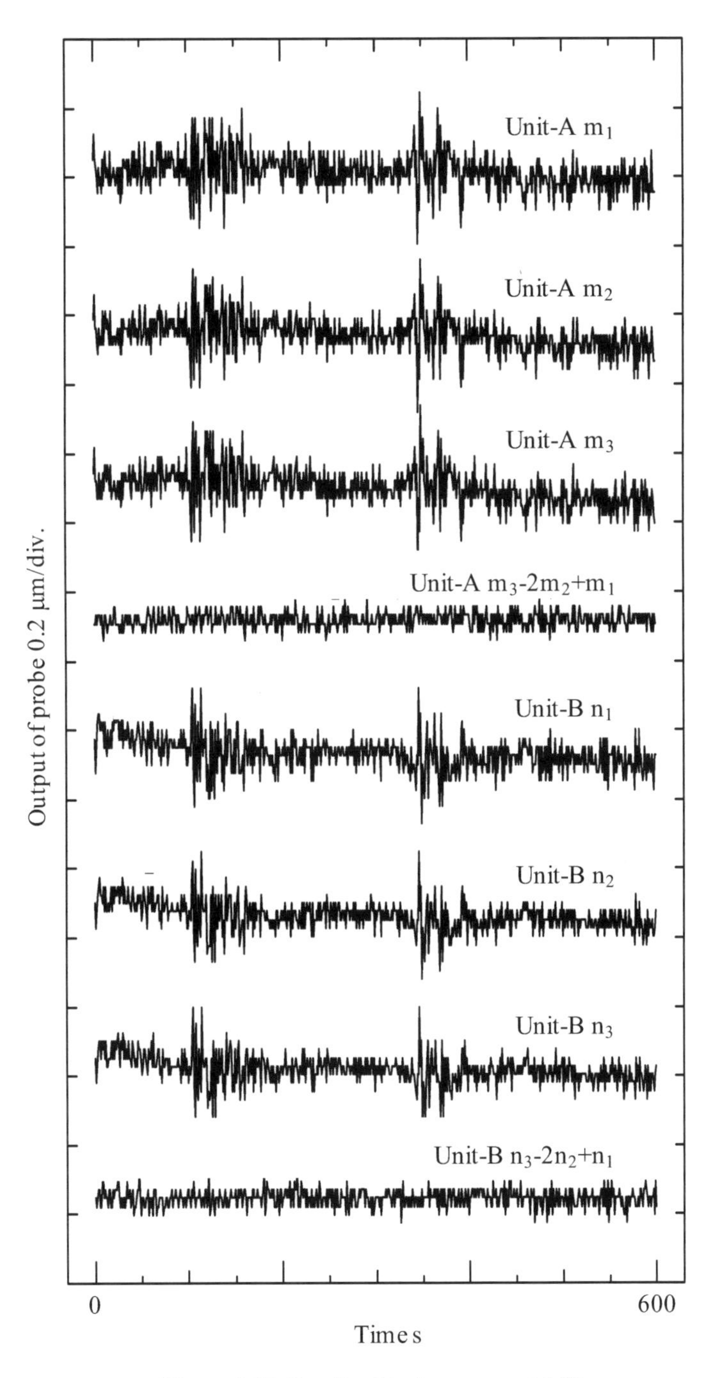

Figure 6.10. Results of testing sensor stability

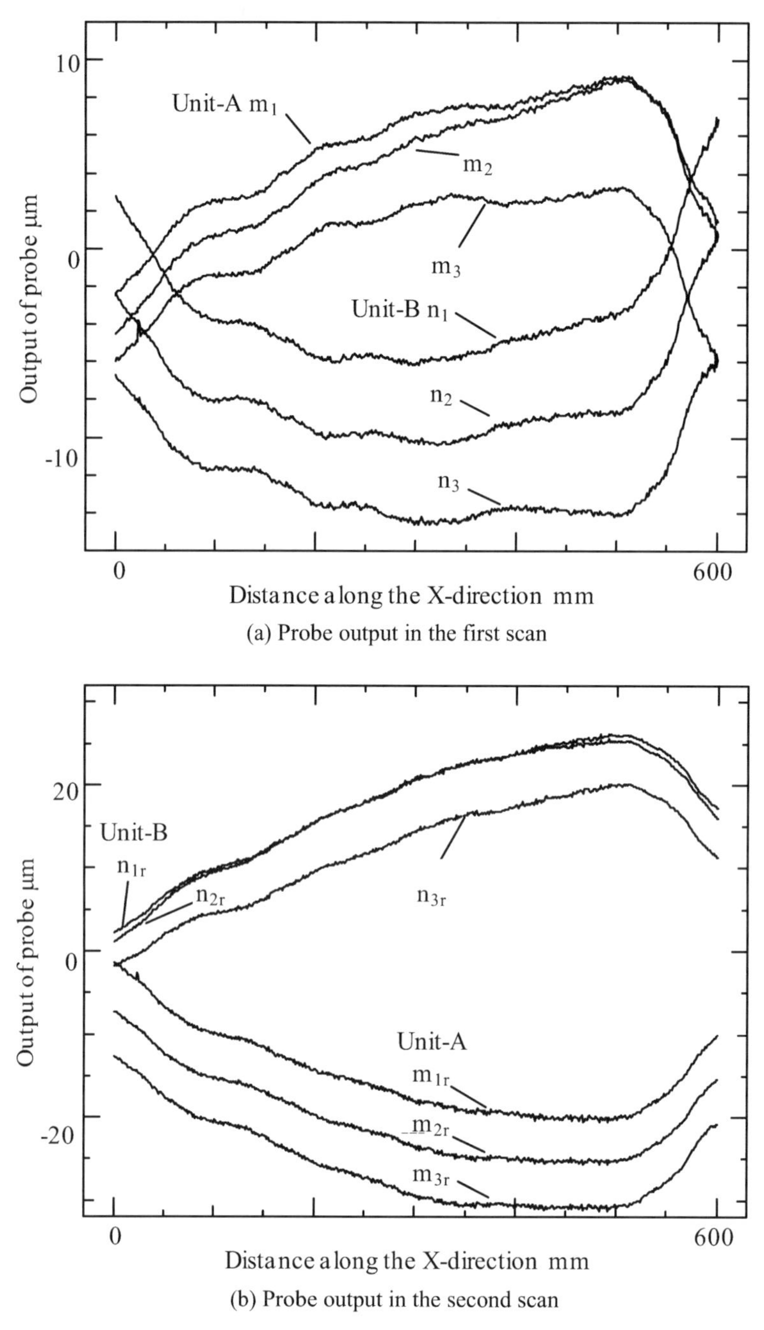

(a) Probe output in the first scan

(b) Probe output in the second scan

Figure 6.11. Probe outputs when scanning the workpiece

For comparison, the straightness profile was first evaluated by using the three-displacement sensor method without accurate zero-adjustment. The sensor outputs at the position of x = 0 mm were used as the sensor zero-values (e_{m1}, e_{m2}, e_{m3}, e_{n1}, e_{n2}, e_{n3}) to calculate the zero-differences (α and β) of the sensor units. This is similar to the conventional zero-adjustment method shown in Figure 6.3, and the accuracy of the zero-adjustment is limited by the straightness error of the cylinder surface. The calculated zero-difference was then used to compensate the parabolic error term in Equation 6.9 to get the straightness profile. Figure 6.12 shows the calculated zero-difference and the profile evaluation result of each sensor unit. The difference between the two profile evaluation results $f_1(x)$ and $f_2(x)$ was approximately 15 μm. This amount corresponds to a zero-adjustment error of approximately 0.83 μm, which was caused by the straightness error of the cylinder surface. The result indicates that the precision straightness profile measurement by using the three-displacement sensor method cannot be realized without accurate zero-adjustment of sensors.

Figure 6.13 shows the results of zero-adjustment using the proposed method shown in Figure 6.5. The sensor outputs obtained in Figure 6.11 are used to calculate the zero-differences (α and β) based on Equations 6.21 and 6.22. As can be seen from Figure 6.13 (a), the profile evaluation results by the two sensor units agree with each other quite well. The reproducibility error is approximately 0.4 μm, corresponding to a residual zero-adjustment error of approximately 20 nm. The residual zero-adjustment error is also much smaller than the stability level of the sensor unit shown in Figure 6.10, indicating the effect of the averaging operations in Equations 6.21 and 6.22.

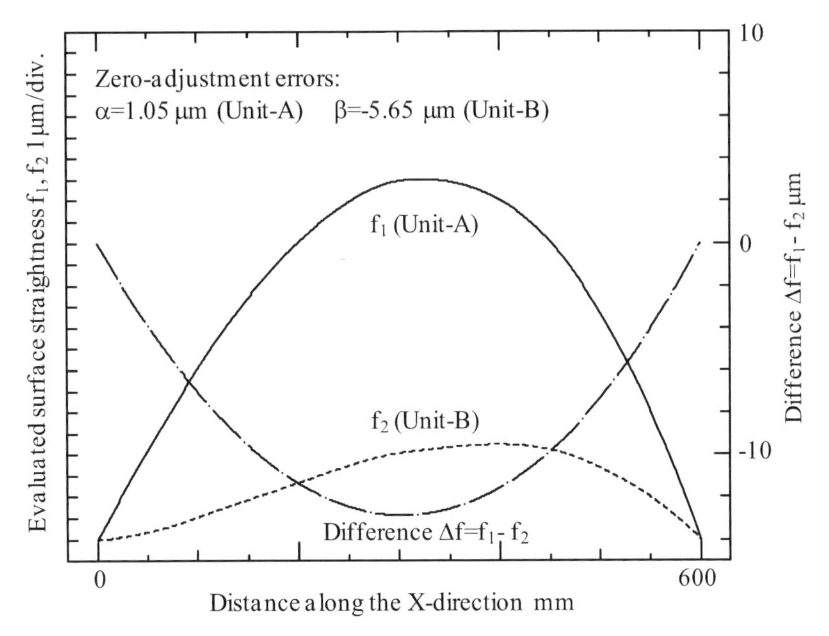

Figure 6.12. Evaluated straightness profile by the three-displacement sensor method without zero-adjustment

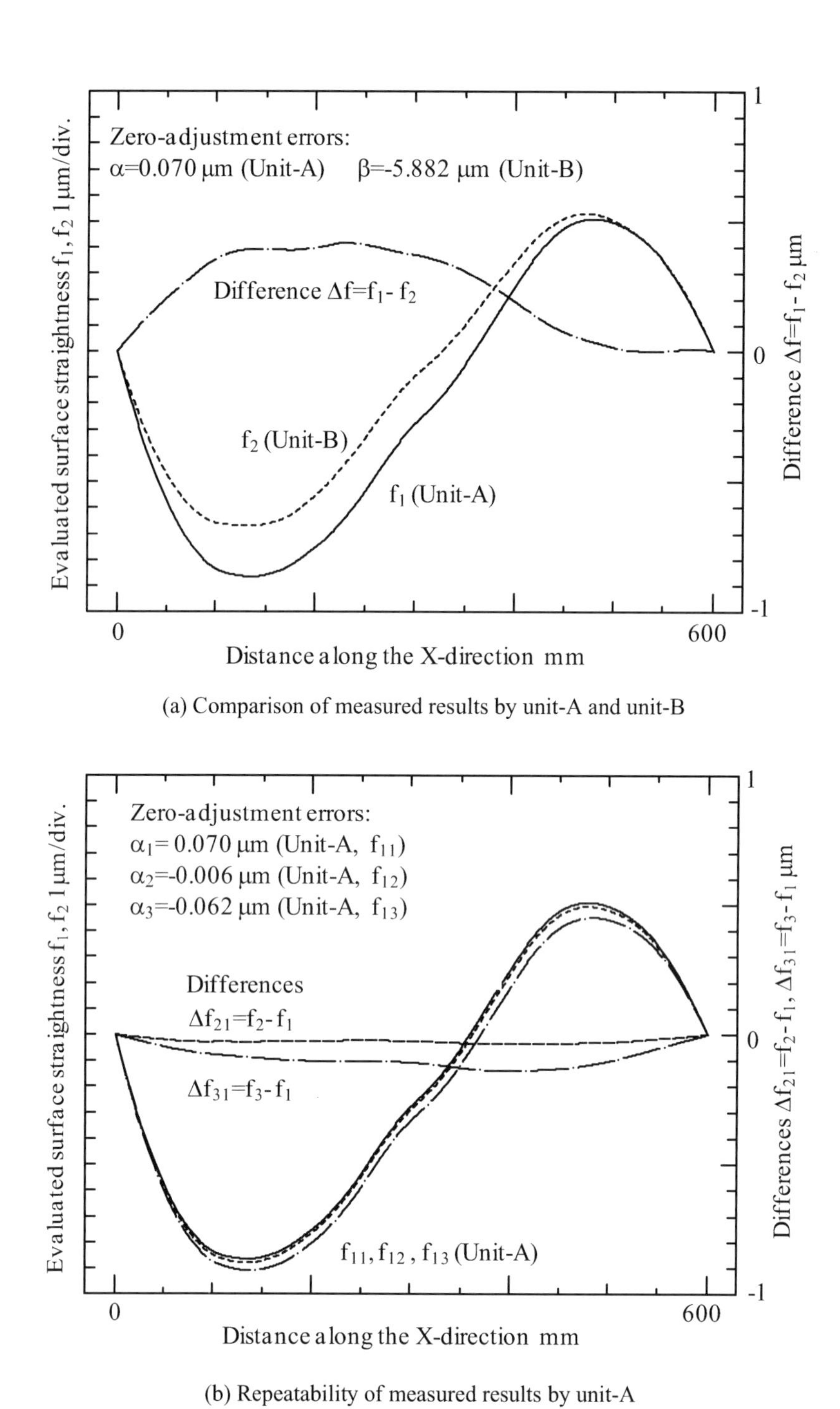

(a) Comparison of measured results by unit-A and unit-B

(b) Repeatability of measured results by unit-A

Figure 6.13. Evaluated straightness profile by the three-displacement sensor method with self zero-adjustment

Figure 6.13 (b) shows the results of three repeated measurements. It took approximately 30 min for the three measurements, each for approximately 10 min. Only the results from sensor unit-A are shown in the figure. It can be seen that the zero-difference changed from 0.070 μm (α_1) to –0.006 μm (α_2), then to –0.062 μm (α_3), which was caused by thermal drift during the measurement. The difference between α_1 and α_3 is approximately 0.13 μm, which could cause a repeatability error of approximately 2.3 μm in the profile evaluation result if the zero-adjustment is not carried out. As can be seen in the Figure 6.13 (b), however, the repeatability errors are less than 0.2 μm with the zero-adjustment.

To verify the feasibility of the three-displacement sensor method with self zero-adjustment, the two-displacement sensor method with angle sensor compensation shown in Figure 6.14 is employed to measure the same workpiece. Two displacement sensors are mounted on the scanning stage to scan the workpiece surface profile $f(x_i)$. An angle sensor is employed to measure the yaw error $e_{yaw}(x_i)$ of the scanning stage. Assume that the outputs of the displacement sensors are $m_1(x_i)$, $m_2(x_i)$ and the output of the angle sensor is $\mu(x_i)$. The outputs can be expressed by:

$$m_1(x_i) = f(x_i - d) + e_Z(x_i) - de_{yaw}(x_i), \tag{6.39}$$

$$m_2(x_i) = f(x_i) + e_Z(x_i), \tag{6.40}$$

$$\mu(x_i) = e_{yaw}(x_i), (i = 1, \ldots, N). \tag{6.41}$$

The differential output $m_{s2}(x_i)$ to cancel the error motions is as follows:

$$\begin{aligned} m_{s2}(x_i) &= \frac{m_2(x_i) - m_1(x_i)}{d} - \mu(x_i) \\ &= \frac{f(x_i) - f(x_i - d)}{d} \\ &\approx f'(x_i), (i = 1, \ldots, N). \end{aligned} \tag{6.42}$$

Let the sampling period be s. The straightness profile can be evaluated from:

$$z_2(x_i) = \sum_{k=1}^{i}(m_s(x_k) \cdot s), (i = 2, \ldots, N). \tag{6.43}$$

Figure 6.15 shows the measurement results of the cylinder workpiece in Figure 6.9 by different methods. The result from the reversal method [10], whose principles will be shown in the next section, is also plotted in the figure. It can be seen that the results by the three different methods are consistent with each other.

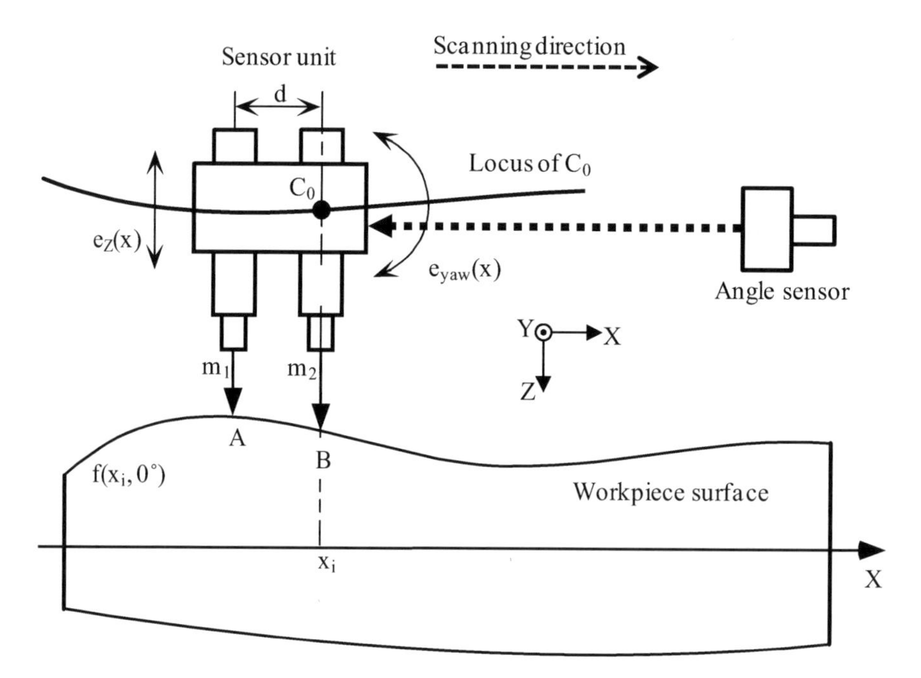

Figure 6.14. The two-displacement sensor method with angle sensor compensation

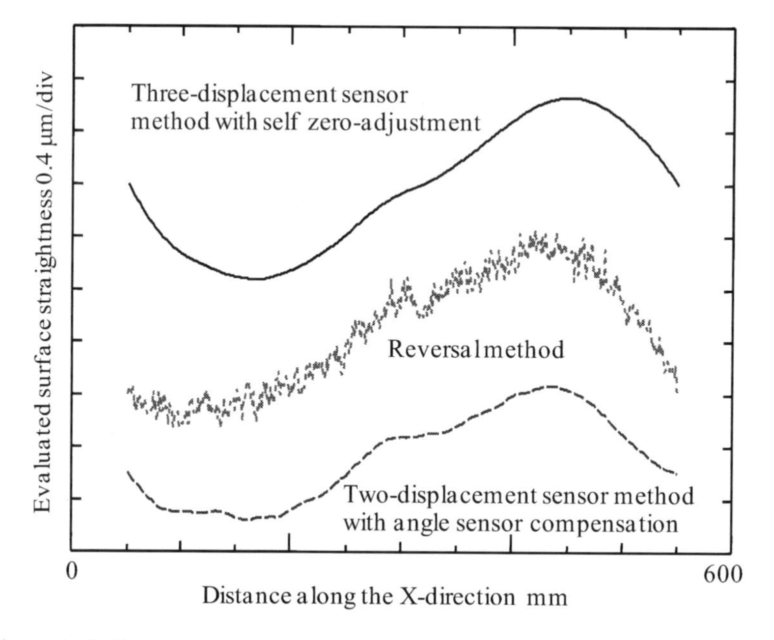

Figure 6.15. The workpiece straightness measured by different error separation methods

6.3 Error Separation Method for Machine Tool Slide

6.3.1 Slide Straightness Error of Precision Machine Tool

Slides are important elements of machine tools. For example, in a precision lathe, the cylinder workpiece rotated by the spindle is cut by a cutting tool that is mounted on the slide of the lathe moving along the axis of the spindle [11–12]. The surface form of the cylinder is basically a transfer of the motion of the tool relative to the spindle axis, which combines the rotational motion of the spindle and the translational motion of the slide [13–15]. In addition to the measurement of the spindle error, the measurement of the slide error is also important not only for the inspection/evaluation of the diamond turning machine but also for quality control of the machined cylinder [16–18]. Compared with the spindle error, the slide error has a larger influence on the machining accuracy when the travel range of the slide is long [1].

The slide straightness error includes the out-of-straightness component, which is defined as the deviation from the slide moving axis, and the out-of-parallelism component, which is defined as the misalignment angle between the slide moving axis and the spindle rotating axis. Employing an accurate straightedge is the most popular method for the measurement of out-of-straightness of a slide [19]. In addition to the disadvantage of high cost, however, the measurement accuracy of the straightedge method is limited by the out-of-straightness of the straightedge. The conventional reversal method can only separate the out-of-straightness of the straightedge under the condition that the slide error is repeatable. Moreover, the out-of-parallelism component of the slide error with respect to the spindle axis [20, 21], which results in a taper-shaped form error on the cylinder workpiece, cannot be measured by both the reversal method and the multi-sensor method.

This section presents improved error separation methods for measurement of slide straightness errors of a precision lathe. The methods can overcome the shortcomings of the conventional reversal method. A cylinder workpiece turned by the precision lathe is employed in the measurement instead of the straightedge in the presented methods.

6.3.2 Error Separation Method for Slide Straightness Measurement

Figure 6.16 shows the precision lathe for measurement, which has a T-base structure. The spindle, which rotates about the *Z*-axis, is mounted on the *Z*-slide to move along the *Z*-axis. The *X*-slide on which the tool is mounted moves along the *X*-axis. The lathe can be used to cut the end surface and the outer surface of the workpiece by moving the cutting tool along the *X*-axis and the *Z*-axis, respectively. Figure 6.17 shows a photograph of a precision lathe with the T-base structure.

Figure 6.18 shows a simplified model for the spindle and the *Z*-slide. The spindle axis is treated as the reference axis, which is assumed to be stationary during the measurement. The slide error $e_{z\text{-slide}}(z)$ and the spindle error $e_{\text{spindle}}(z, \theta)$ are defined as the deviations with respect to the spindle axis.

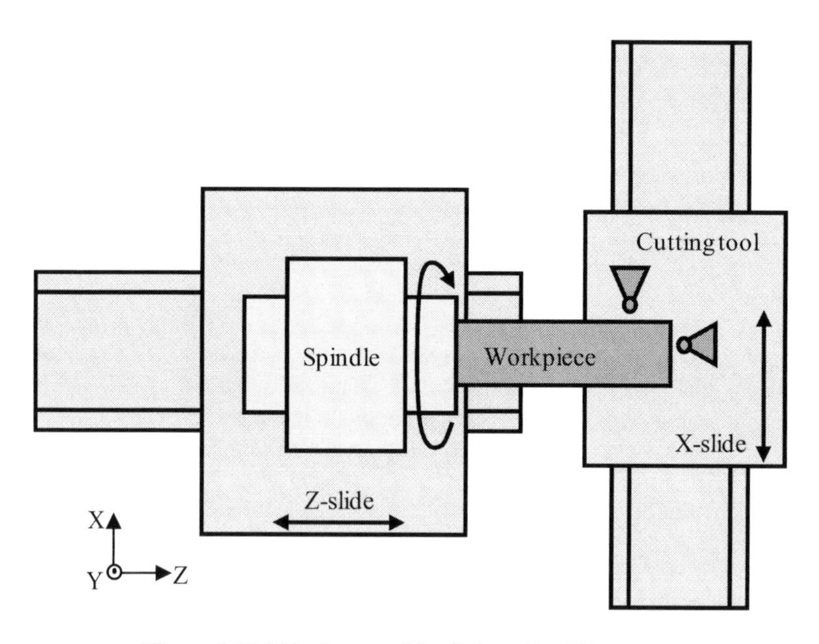

Figure 6.16. Sides in a precision lathe with a T-base structure

Figure 6.17. Photograph of a precision lathe with a T-base structure

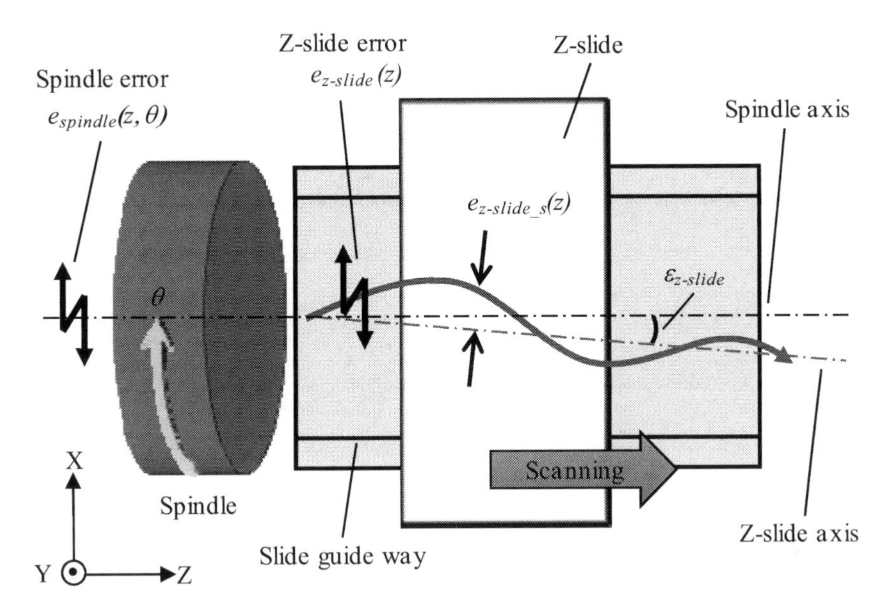

Figure 6.18. Schematic model of the *Z*-slide and the spindle of precision lathe

The spindle error, which combines the radial error component and tilt error component with respect to the spindle axis, is a function of the spindle rotation angle θ and the slide position *z*. The *Z*-slide moves along the slide axis, which has a misalignment angle $\varepsilon_{z\text{-slide}}$ with respect to the spindle axis. $\varepsilon_{z\text{-slide}}$ is referred to as the out-of-parallelism error component of the slide error $e_{z\text{-slide}}(z)$. In addition to $\varepsilon_{z\text{-slide}}$, the out-of-straightness $e_{z\text{-slide\_s}}(z)$, which is defined as the deviation from the slide axis, is the other error component of the slide error. For simplicity, the relationship between $e_{z\text{-slide}}(z)$, $e_{z\text{-slide\_s}}(z)$ and $\varepsilon_{z\text{-slide}}$ is defined as follows:

$$e_{z-slide}(z) = e_{z-slide\_s}(z) - z\varepsilon_{z-slide}\,. \tag{6.44}$$

To identify the slide error $e_{z\text{-slide}}(z)$, it is necessary to measure both $e_{z\text{-slide\_s}}(z)$ and $\varepsilon_{z\text{-slide}}$. In the following discussion, the spindle error and the slide error are assumed to be repeatable. In the conventional reversal method for slide error measurement, a displacement sensor is moved by the slide to scan side 1 of a stationary straightedge as shown in Figure 6.19 (a). The sensor output m_1 can be expressed by:

$$m_1(z) = e_{z-slide}(z) + f(x) + z\psi_1\,, \tag{6.45}$$

where $f(x)$ is the out-of-straightness component of the surface form error of side 1 of the straightedge. ψ_1 is the mount error of the straightedge axis with respect to *Z*-axis. It can be seen that the measurement of the slide error $e_{z\text{-slide}}(z)$ is influenced by $f(x)$ and ψ_1.

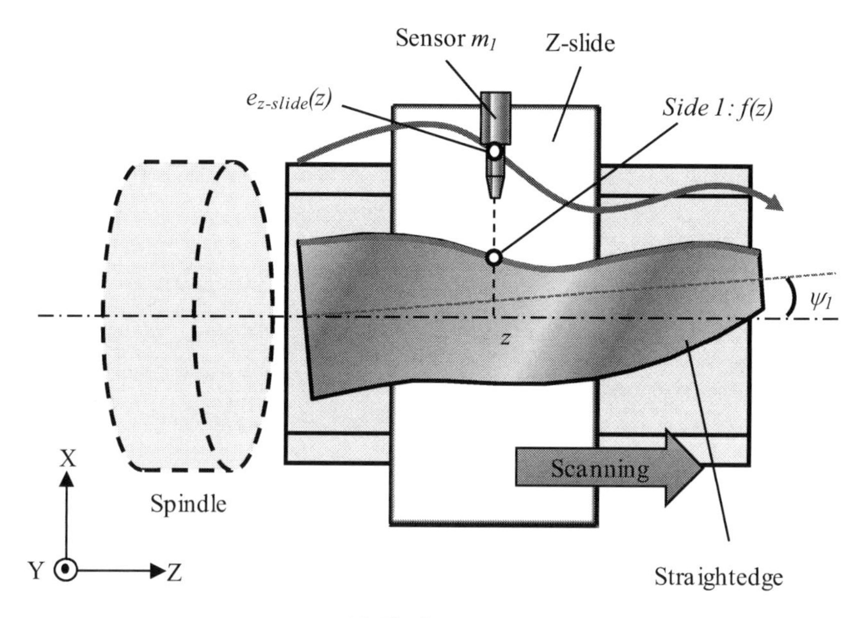

(a) The first scan

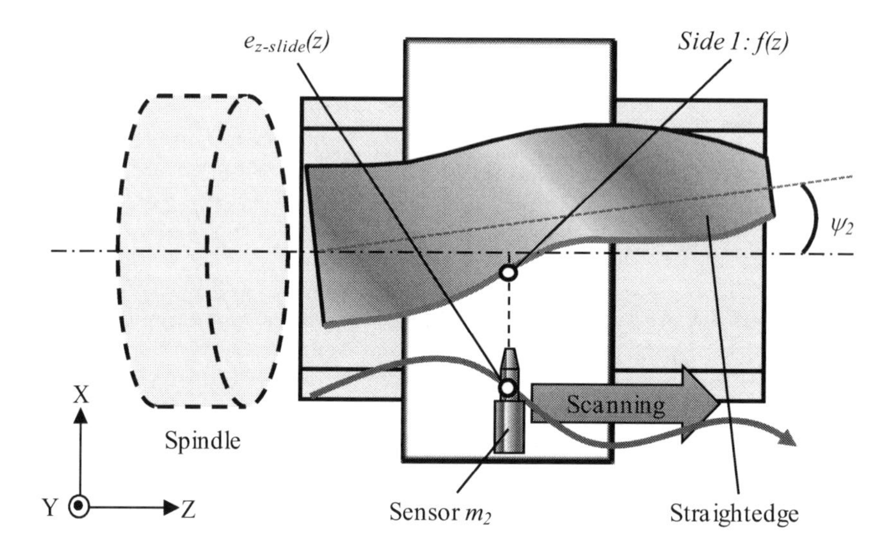

(b) The second scan after reversal

Figure 6.19. The conventional reversal method for separation of the surface form error from the slide error

After the first scan in Figure 6.19 (a), both the straightedge and the displacement sensor are reversed about the *Z*-axis to conduct the second scan as shown in Figure 6.19 (b). The sensor output in the second scan becomes:

$$m_2(z) = -e_{z-slide}(z) + f(x) - z\psi_2, \tag{6.46}$$

where ψ_2 is the mount error associated with the reversal operation, which is different from ψ_1. The following calculation is carried out to remove $f(x)$:

$$e_{z-slide}(z) = e_{z-slide\_s}(z) - z\varepsilon_{z-slide}$$

$$= \Delta m_{12}(z) - \frac{z(\psi_1 + \psi_2)}{2}, \tag{6.47}$$

where

$$\Delta m_{12}(z) = \frac{m_1(z) - m_2(z)}{2}. \tag{6.48}$$

This method is effective for the measurement of the out-of-straightness component $e_{z\text{-slide}\_s}(z)$, which can be obtained by removing the linear component of $\Delta m_{12}(z)$. However, because the mount errors ψ_1 and ψ_2 are unknown, the out-of-parallelism component $\varepsilon_{z\text{-slide}}$ cannot be evaluated. This is one of the shortcomings of the conventional reversal method in addition to the time-consuming reversal operation.

To overcome the shortcomings of the conventional reversal method, two error separation methods of using a rotating cylinder are presented in the following, which are referred to as the rotating-reversal method and the error-replication method, respectively. Figure 6.20 shows a schematic of the rotating-reversal method. In this method, a cylinder is mounted on the machine spindle and two sensors (m_1, m_2) are mounted on the *Z*-slide. The two sensors are moved by the *Z*-slide to simultaneously scan the cylinder, which is rotated by the spindle. The outputs of the two sensors sampled at the position of (z, θ), where z is the moving distance of the *Z*-slide and θ is the rotation angle of the spindle, respectively, can be expressed by:

$$m_1(z,\theta) = e_{z-slide}(z) + g(z,\theta) + e_{spindle}(z,\theta) - z\gamma, \tag{6.49}$$

$$m_2(z,\theta) = -e_{z-slide}(z) + g(z,\theta+\pi) - e_{spindle}(z,\theta) + z\gamma. \tag{6.50}$$

Here $g(z, \theta)$ is the surface form error of the cylinder, which includes an out-of-straightness component and a linear taper component. γ is the mount error of the cylinder axis with respect to the spindle axis.

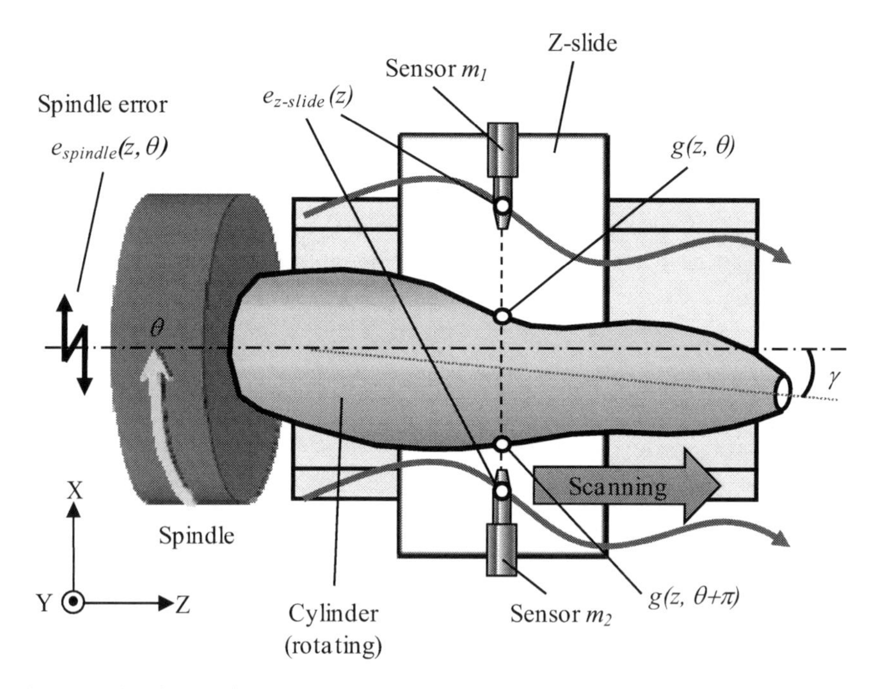

Figure 6.20. The rotating-reversal method of using a rotating cylinder for slide error measurement

Similarly, the outputs of the two sensors sampled at the position of (z, $\theta+\pi$) can be expressed by:

$$m_1(z,\theta+\pi) = e_{z-slide}(z) + g(z,\theta+\pi) + e_{spindle}(z,\theta+\pi) + z\gamma \ , \tag{6.51}$$

$$m_2(z,\theta+\pi) = -e_{z-slide}(z) + g(z,\theta) - e_{spindle}(z,\theta+\pi) - z\gamma \ . \tag{6.52}$$

The following operation based on Equations 6.47 and 6.52 provides:

$$\begin{aligned} e_{z-slide}(z) &= e_{z-slide\_s}(z) - z\varepsilon_{z-slide} \\ &= \Delta m_{12}(z,\theta) - \Delta e_{spindle}(z,\theta) \ , \end{aligned} \tag{6.53}$$

where

$$\Delta m_{12}(z,\theta) = \frac{m_1(z,\theta) - m_2(z,\theta+\pi)}{2} \ , \tag{6.54}$$

$$\Delta e_{spindle}(z,\theta)=\frac{e_{spindle}(z,\theta)+e_{spindle}(z,\theta+\pi)}{2}. \tag{6.55}$$

It can be seen that the surface form error of the cylinder and the mount error are successfully removed in Equation 6.53.

As shown in Equation 6.56, the spindle error $e_{spindle}(z, \theta)$ at the position z can be expressed as the superposition of infinite series of periodic functions [22]:

$$e_{spindle}(z,\theta)=\sum_{k=1}^{\infty} c_k \cos(k\theta-\phi_k), \tag{6.56}$$

where c_k and f_k are the amplitude and phase angle of the kth harmonic component. As can be seen from Equation 6.56, the average of the spindle error $e_{spindle}(z, \theta)$ over one rotation is zero. An averaging operation shown in the next equation can thus remove the residual spindle error $\Delta e_{spindle}(z, \theta)$.

$$\begin{aligned} e_{z-slide}(z) &= e_{z-slide\_s}(z)-z\varepsilon_{z-slide} \\ &= \frac{1}{M}\sum_{\theta=0}^{2\pi}\Delta m_{12}(z,\theta)-\frac{1}{M}\sum_{\theta=0}^{2\pi}\Delta e_{spindle}(z,\theta) \\ &= \overline{\Delta m_{12}(z,\theta)}, \end{aligned} \tag{6.57}$$

where M is the averaging number and

$$\overline{\Delta m_{12}(z,\theta)}=\frac{1}{M}\sum_{\theta=0}^{2\pi}\Delta m_{12}(z,\theta), \tag{6.58}$$

$$\frac{1}{M}\sum_{\theta=0}^{2\pi}\Delta e_{spindle}(z,\theta)=0. \tag{6.59}$$

As a result, both $\varepsilon_{z\text{-slide}}$ and $e_{z\text{-slide}\_s}(z)$ of the slide error can be evaluated by a linear fitting of $\Delta m_{12}(z, \theta)$, without the influence of surface form error of the cylinder, the spindle error and the cylinder mounting error.

Figure 6.21 shows a schematic of the error-replication method for the slide error measurement by using one displacement sensor and a rotating cylinder workpiece. The displacement sensor is mounted on the opposite position of the cylinder workpiece with respect to the cutting tool. In this method, the cylinder workpiece is first turned (self-cut) by the diamond turning machine and the measurement is carried out without removing the workpiece from the spindle. Because the surface form of the cylinder workpiece is a transfer of the tool motion, which is a combination of the slide motion and the spindle motion, the surface profile form error at the cutting position (z, θ) can be denoted as:

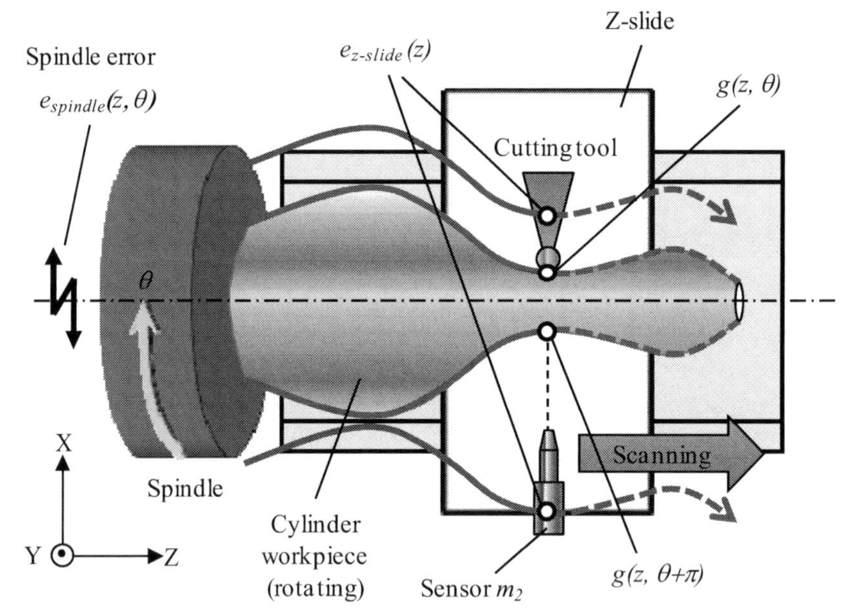

Figure 6.21. The error-replication method of using a cylinder workpiece turned by the precision lathe

$$g(z,\theta) = -e_{z-slide}(z) - e_{spindle}(z,\theta) . \tag{6.60}$$

The sensor output sampled at the position (*z*, θ) can be expressed by substituting Equation 6.50 into Equation 6.60:

$$m_2(z,\theta) = -e_{z-slide}(z) + g(z,\theta+\pi) - e_{spindle}(z,\theta)$$

$$= -2e_{z-slide}(z) - e_{spindle}(z,\theta+\pi) - e_{spindle}(z,\theta) . \tag{6.61}$$

It is noted that the mount error of the cylinder axis with respect to the spindle axis does not appear in the sensor output because the cylinder workpiece is self-cut on the machine.

Similar to the rotating-reversal method, both the out-of-straightness component and the out-of-parallelism component of the slide error can be evaluated based on the following result:

$$e_{z-slide}(z) = e_{z-slide\_s}(z) - z\varepsilon_{z-slide}$$

$$= -\frac{1}{M}\sum_{\theta=0}^{2\pi}\frac{m_2(z,\theta)}{2} - \frac{1}{M}\sum_{\theta=0}^{2\pi}\Delta e_{spindle}(z,\theta)$$

$$= -\frac{1}{M}\sum_{\theta=0}^{2\pi}\frac{m_2(z,\theta)}{2}, \tag{6.62}$$

where $\Delta e_{spindle}(z, \theta)$ is defined in Equation 6.55. Note that the error-replication method is only effective when the cylinder workpiece has not been removed from the spindle after the self-cut. In contrast, the rotating-reversal method is also effective for a remounted cylinder workpiece or any cylinders machined by other machines.

Figure 6.22 shows photographs of the experimental setups for the error-replication method and the rotating-reversal method, respectively. The experiment was carried out on the precision lathe shown in Figure 6.17. A round nose single-crystal diamond cutting tool with a nose radius of 2 mm and a capacitive displacement sensor (m_2) were mounted on the X-slide. An aluminum cylinder workpiece with a diameter of 50 mm and a length of 150 mm was mounted on the spindle. The cylinder workpiece was first turned (self-cut) over a length of 140 mm by the cutting tool. In the experiment of the error-replication method, the displacement sensor was covered by a waterproof cover during the turning. After the turning, the displacement sensor was moved by the Z-slide to scan the cylinder surface that was rotated by the spindle. Figure 6.23 (a) shows the measured slide error $e_{z\text{-slide}}(z)$ based on Equation 6.62. The slide error was measured to be approximately 620 nm, most of which was caused by the out-of-parallelism component. After the measurement by the error-replication method, the cutting tool in Figure 6.22 (a) was replaced with the displacement sensor m_1 to carry out the experiment by the rotating-reversal method shown in Figure 6.22 (b), without removing the cylinder workpiece from the spindle. Figure 6.23 (b) shows the measured slide error $e_{z\text{-slide}}(z)$ based on Equation 6.53. The slide error was measured to be approximately 630 nm. It can be seen that the result was consistent to that by the error-replication method with a difference of approximately 10 nm, which was on the order of the repeatability of the slide motion.

Figure 6.24 shows a result by the conventional reversal method measured in the experimental setup shown in Figure 6.22 (b). The reversal operation of the cylinder was carried out by rotating the spindle 180°. The slide error was measured to be approximately 700 nm, which had a 70 nm difference from that by the rotating-reversal method. Figure 6.25 shows the results of the out-of-straightness component $e_{z\text{-slide}\_s}(z)$ by the three methods, which were consistent with each other. This verified that the conventional reversal method is only effective for measurement of the out-of-straightness component of the slide error and the proposed error-replication method and the rotating-reversal method are effective for measurement of both the out-of-straightness component and the out-of-parallelism component.

Figure 6.26 shows the experimental result carried out by the rotating-reversal method after the cylinder workpiece was remounted. The slide error was measured to be 620 nm, which was almost the same as that shown in Figure 6.23 (b). This indicates that the rotating-reversal method is effective for a re-mounted workpiece.

The reversal method is employed to measure the out-of-straightness component of the X-slide of the precision lathe, which is difficult to use the error-replication method or the rotating-reversal method. Figure 6.27 shows the experimental setup.

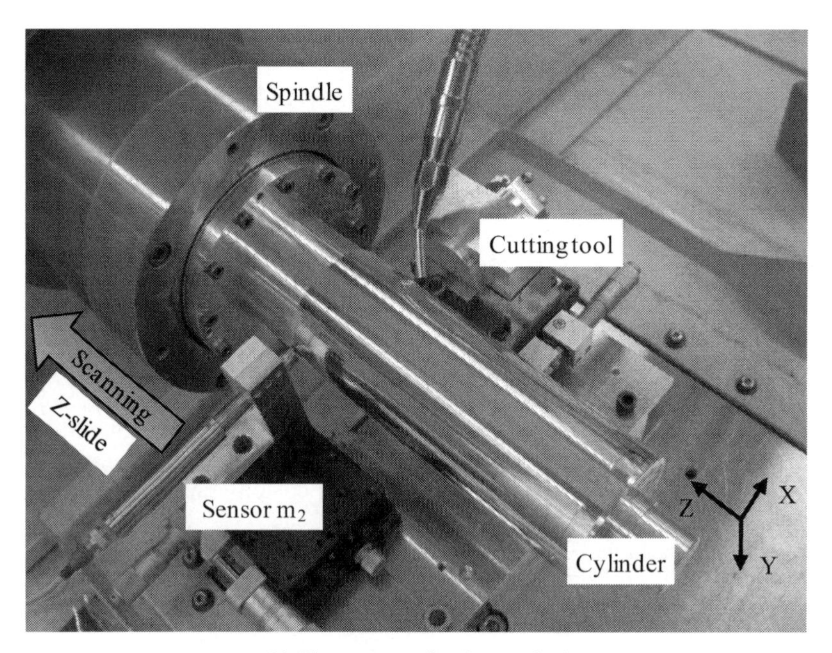

(a) The error-replication method

(b) The rotating-reversal method

Figure 6.22. Photographs of experimental setups for slide error measurement

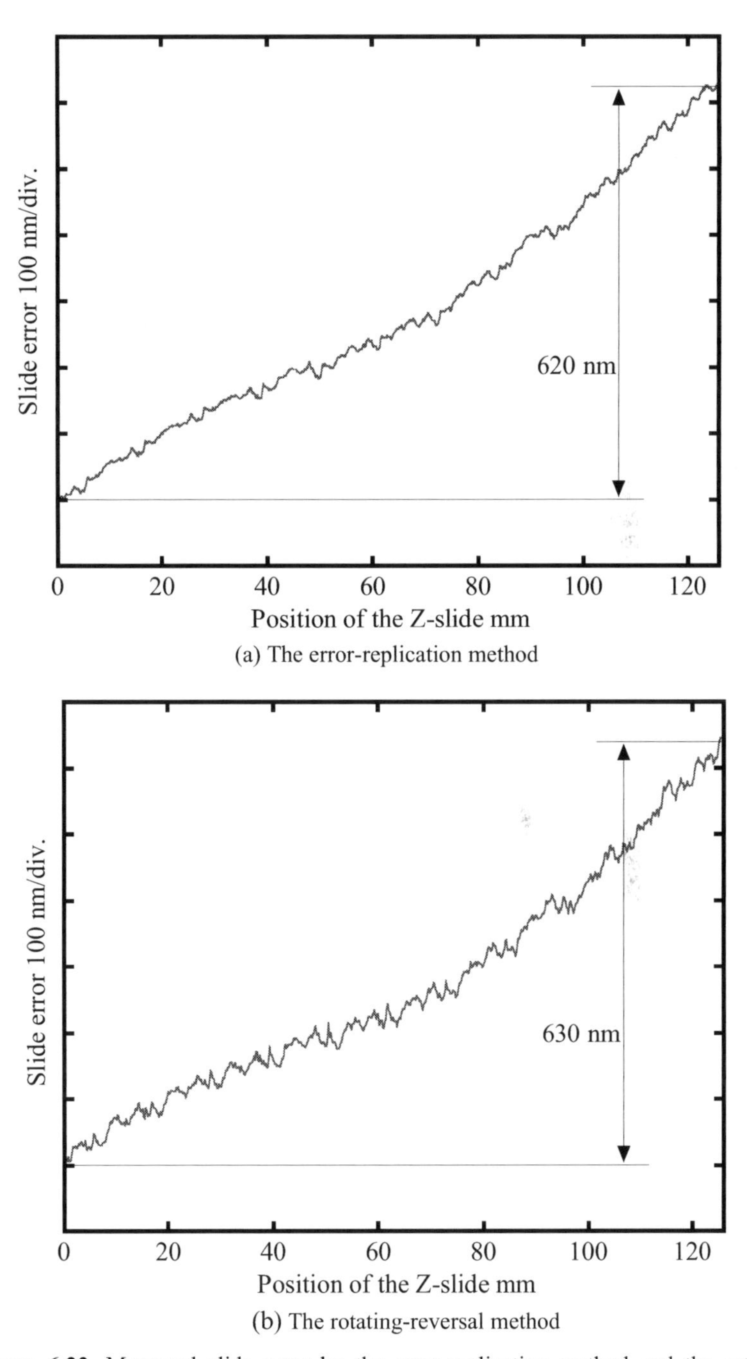

(a) The error-replication method

(b) The rotating-reversal method

Figure 6.23. Measured slide error by the error-replication method and the rotating-reversal method

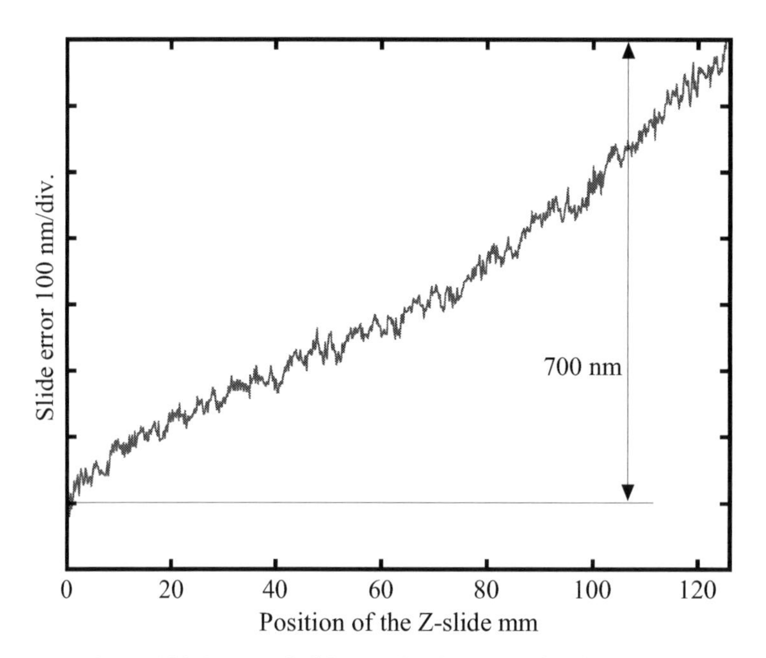

Figure 6.24. Measured slide error by the conventional reversal method

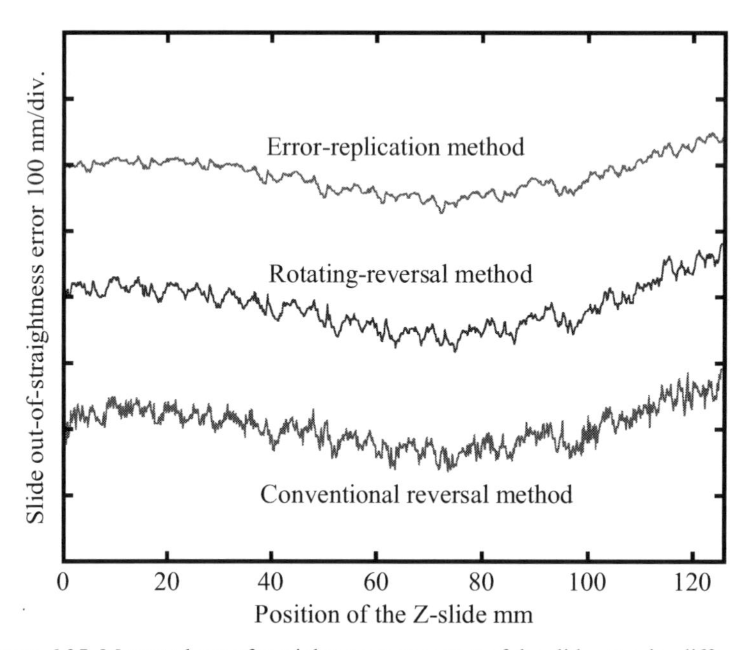

Figure 6.25. Measured out-of-straightness component of the slide error by different error separation methods

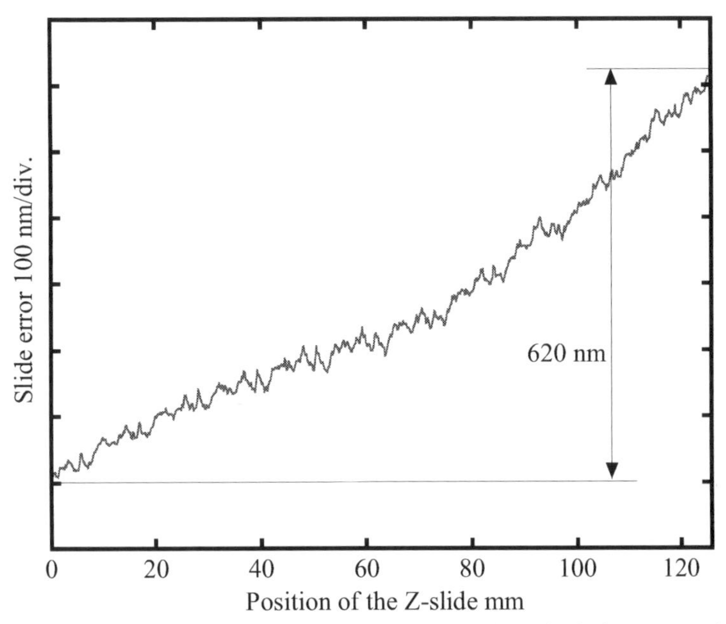

Figure 6.26. Measured slide error by the rotating-reversal method after remounting the workpiece

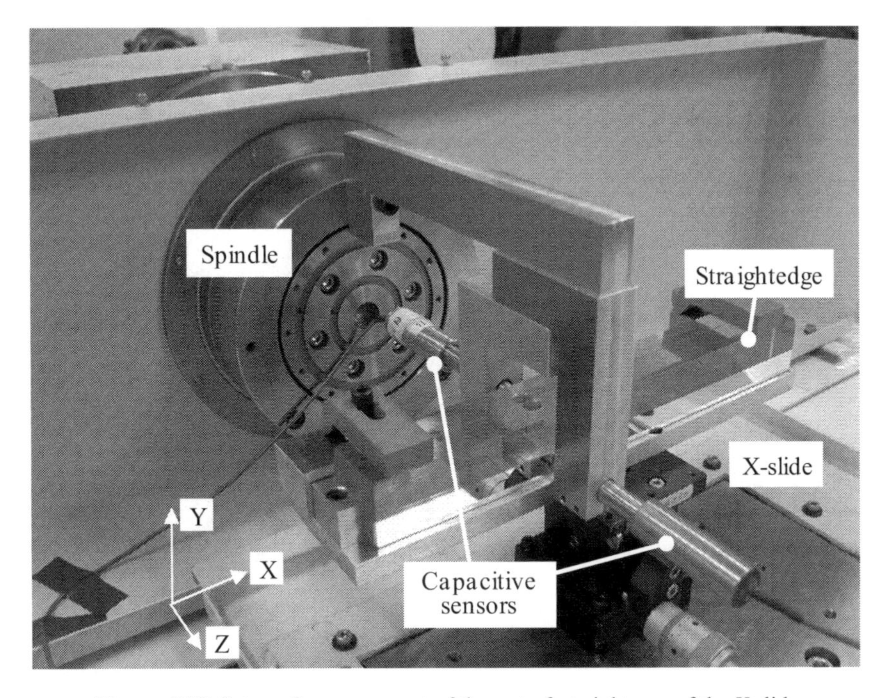

Figure 6.27. Setup of measurement of the out-of-straightness of the *X*-slide

As shown in Figure 6.27, two capacitive displacement sensors were employed. The material of the straightedge was Zerodur with an aluminum coating. The Zerodur straightedge was mounted on the *X*-slide and the capacitive displacement sensors were kept stationary on the spindle. The measurement started from $x = 80$ mm and ended at $x = 0$ mm. The position $x = 0$ mm corresponded to the position of the spindle center. The movement speed of the *X*-slide was 24 mm/min. Figure 6.28 shows the out-of-straightness component of the *X*-slide and the surface profile of the straightedge. As can be seen in Figure 6.28 (a), the *X*-slide had an out-of-straightness component of approximately 60 nm over the 80-mm travel. It is mainly composed of a parabolic component and a periodic component with a pitch of 11 mm. The peak-to-valley (PV) of the periodic component with the 11-mm pitch was approximately 20 nm. As can be seen in Figure 6.29, the surface profile of the straightedge was approximately 15 nm.

6.4 Summary

A scanning multi-sensor system consisting of two sensor units of the three-displacement sensor method has been presented for precision nanometrology of straightness profiles of workpieces. No accurate reference flat surfaces or auxiliary artifacts are necessary for the zero-adjustment of sensors in the system. The zero-difference between sensors in each sensor unit, which greatly influences the profile evaluation accuracy, can be obtained from the sensor outputs of scanning the workpiece surface before and after a rotation of the cylinder. Influences of positioning errors of sampling and random errors in the outputs of the sensors are reduced from an averaging operation in the calculation of the zero-difference. An improved zero-adjustment method, which can evaluate the variation of the zero-difference, has also been presented. The feasibility of the three-displacement sensor method with self zero-adjustment has been verified by experiments. Good correspondence with the two-displacement sensor method with angle sensor compensation and the reversal method has also been confirmed.

Two measurement methods, which are referred to as the error-replication method and the rotating-reversal method, respectively, have been presented for measurement of the *Z*-slide error of a precision lathe. A rotating cylinder workpiece, which has been turned (self-cut) on the machine, is used as the specimen for the measurement. The influence of the spindle error and the surface form error of the cylinder workpiece can be removed from the measurement result of the slide error by an averaging operation of the sensor outputs over one rotation. In addition to the out-of-straightness component of the slide error, the proposed methods can measure the out-of-parallelism of the slide axis with respect to the spindle axis, which is a main error factor influencing the machining accuracy of cylinders. The error-replication method is simpler than the rotating-reversal method. On the other hand, the rotating-reversal method, which is also effective for a re-mounted cylinder workpiece or cylinders turned by other machines, is more flexible than the error-replication method. The reversal method has also been applied to the measurement of the out-of-straightness component of the *X*-slide error of the precision lathe.

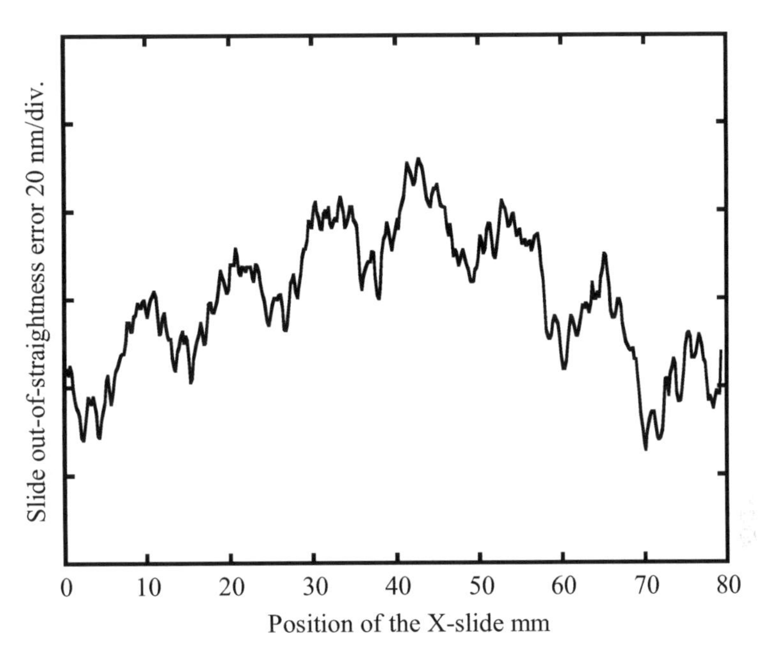

Figure 6.28. Measured out-of-straightness error of the *X*-slide

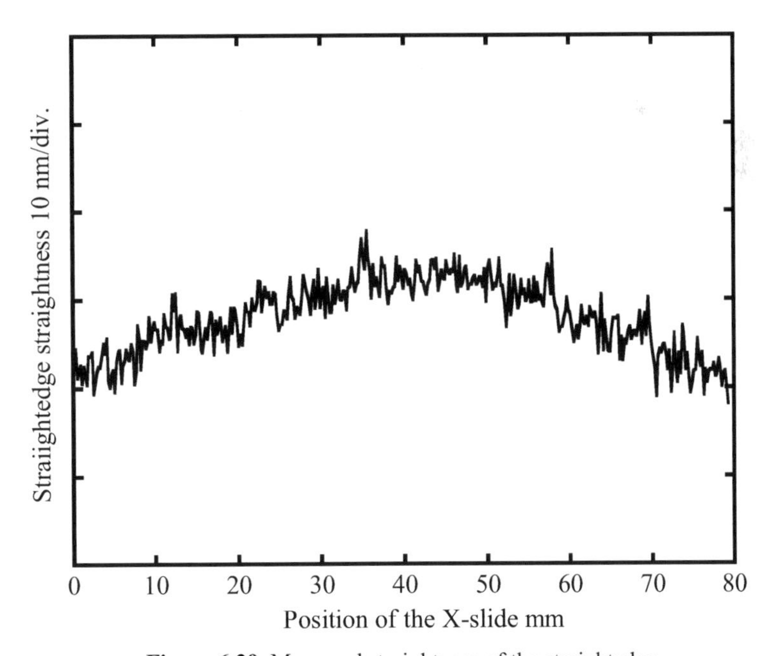

Figure 6.29. Measured straightness of the straightedge

References

[1] Toshiba Machine Co., Ltd. (2010) http://www.toshiba-machine.co.jp. Accessed 1 Jan 2010
[2] Whitehouse DJ (1978) Measuring instrument. US Patent 4,084,324, 18 April 1978
[3] Tanaka H, Sato H (1986) Extensive analysis and development of straightness measurement by sequential two-point method. ASME Trans 108:176–182
[4] Gao W, Kiyono S (1997) On-machine profile measurement of machined surface using the combined three-probe method. JSME Int J 40(2):253–259
[5] Yamaguchi J (1993) Measurement of straight motion accuracy using the improved sequential three-point method. J JSPE 59(5):773–778 (in Japanese)
[6] Gao W, Yokoyama J, Kojima H, Kiyono S (2002) Precision measurement of cylinder straightness using a scanning multi-probe system. Precis Eng 26(3):279–288
[7] Gao W, Lee JC, Arai Y, Park CH (2009) An improved three-probe method for precision measurement of straightness. Technisches Messen 76(5):259–265
[8] Kiyono S, Gao W (1994) Profile measurement of machined surface with a new differential surface with a new differential method. Precis Eng 16(3):212–218
[9] Gao W, Kiyono S (1996) High accuracy profile measurement of a machined surface by the combined method. Measurement 19(1):55–64
[10] Estler WT (1985) Calibration and use of optical straightedges in the metrology of precision machines. Opt Eng 24(3):372–379
[11] Kim SD, Chang IC, Kim SW (2002) Microscopic topographical analysis of tool vibration effects on diamond turned optical surfaces. Precis Eng 26(2):168–174
[12] Taniguchi N (1994) The state of the art of nanotechnology for processing of ultra-precision and ultra-fine products. Precis Eng 16(1):5–24
[13] Fawcett SC, Engelhaupt D (1995) Development of Wolter I X-ray optics by diamond turning and electrochemical replication. Precis Eng 17(4):290–297
[14] Fan KC, Chao YH (1991) In-process dimensional control of the workpiece during turning. Precis Eng 13(1):27–32
[15] Gao W, Tano M, Sato S, Kiyono S (2006) On-machine measurement of a cylindrical surface with sinusoidal micro-structures by an optical slope sensor. Precis Eng 30(3):274–279
[16] Hwang J, Park CH, Gao W, Kim SW (2007) A three-probe system for measuring the parallelism and straightness of a pair of rails for ultra-precision guideways. Int J Mach Tools Manuf 4:1053–1058
[17] Campbel A (1995) Measurement of lathe Z-slide straightness and parallelism using a flat land. Precis Eng 17(3):207–210
[18] Gao W, Tano M, *et al* (2007) Measurement and compensation of error motions of a diamond turning machine. Precis Eng 31(3):310–316
[19] Irick SC, McKinney WR, Lunt DLJ, Takacs PZ (1992) Using a straightness reference in obtaining more accurate surface profiles from a long trace profiler. Rev Sci Instrum 63(1):1436–1438
[20] Gao W, Lee JC, Arai Y, Noh YJ, Hwang J, Park CH (2010) Measurement of slide error of an ultra-precision diamond turning machine by using a rotating cylinder workpiece. Int J Mach Tools Manuf, 50(4), 404-410.
[21] Jywe W, Chen CJ (2007) A new 2D error separation technique for performance tests of CNC machine tools. Precis Eng 31(4):369–375
[22] Hii KF, Vallance RR, Grejda RD, Marsh ER (2004) Error motion of a kinematic spindle. Precis Eng 28(2):204–217

7

Scanning Micro-stylus System for Measurement of Micro-aspherics

7.1 Introduction

Aspheric micro-lenses are key components in medical endoscopes for diagnosis and surgery of internal organs [1, 2]. The diameter of the micro-lens is less than 1 mm for fitting into the thin endoscope tube. The aspheric surface of the lens makes it possible to improve the imaging performance of the endoscope while reducing the number of optical elements. To obtain clear and high-quality endoscopic images, it is necessary to employ micro-lenses with high surface form accuracy and good surface finish.

Most of the aspheric micro-lenses are manufactured by the glass molding technology [3, 4]. The form accuracy of the lens is mainly determined by that of the master mold, which is generally machined by a multi-axis CNC grinding machine [5–7]. The state-of-the-art micro-lens molds are required to be machined with 100 nm form accuracy and 10 nm surface finish. The latter is relatively easily achieved by using the post polishing process. However, the 100 nm form accuracy is extremely difficult because the grinding process is influenced by a lot of factors such as the grinding tool, grind fluid, machine error motions, grinding parameters, etc. In addition, the aspheric surface form of the micro-lens is complicated and with large deviations from the spherical profile, which makes it more difficult to machine the master mold with the necessary accuracy.

On the other hand, accurate profile measurement of the aspheric surface is another challenge for realizing the precision micro-lens and lens mold. The measurement result can be used not only in quality control of the machined aspheric surface but also in compensation grinding of the aspheric surface for higher machining accuracy. Although the optical method is ideal from the viewpoint of non-contact and high-speed feature [8–11], the method using contact-stylus sensor is more realistic and more reliable because of the large profile variation and steep surface slope of the aspheric micro-lens [12, 13]. In a measurement system using a contact-stylus sensor, a relative scanning motion is generated between the aspheric surface and the stylus by scanning stages.

This chapter presents an accurate scanning micro-stylus system for aspheric micro-lenses. A sensor unit combining a ring artifact and two capacitive-type displacement sensors is constructed for compensation of the scanning error motions. Micro-ball styluses are also described in this chapter.

7.2 Compensation of Scanning Error Motion

Figure 7.1 shows an overview of the scanning micro-stylus measurement system for aspheric micro-lenses and micro-lens molds. The micro-aspheric specimen is chucked on the spindle and the micro-stylus probe is mounted on the *X*-slide. The scanning motion for the profile measurement is generated by the rotational motion of the spindle about the *Z*-axis and the linear motion of the slide along the *X*-axis. The sensor unit for separating the scanning error motion is composed of a ring artifact and two displacement sensors (compensation sensors 1 and 2). The ring artifact is mounted around the micro-aspheric specimen on the spindle so that the ring artifact and the specimen are subjected to the same spindle error motions. Compensation sensors 1 and 2 are symmetrically set on the two sides of the micro-stylus probe on the slide to sense the same error motions of the slide and the spindle. The intervals between the sensors and the stylus probe are set to be d. Figures 7.2 and 7.3 show the measurement parameters and the motions associated with the scanning, respectively. The surface profile of the aspheric sample and that of the ring artifact are let to be $g(x, \theta)$ and $f(x, \theta)$, respectively, where θ is the rotation angle of the spindle.

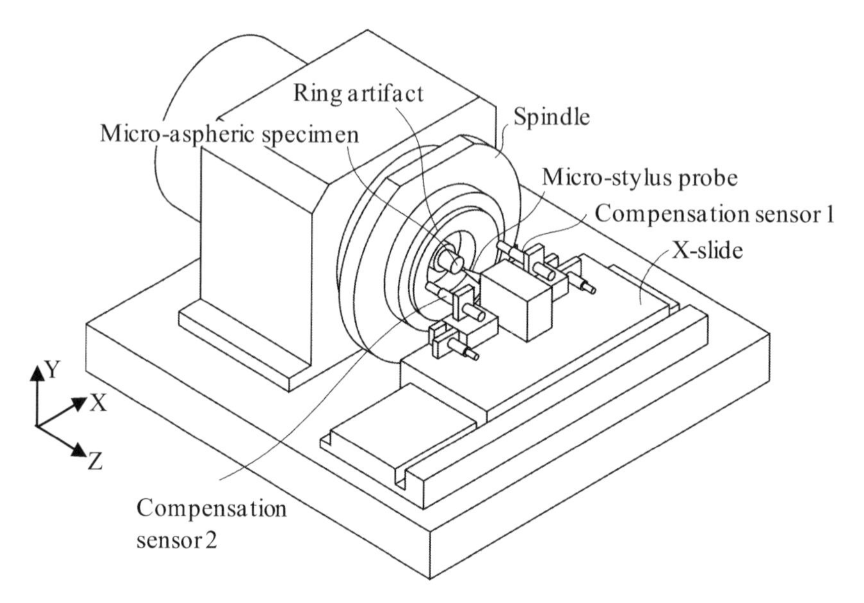

Figure 7.1. Overview of the scanning micro-stylus system for surface profile measurement of aspheric micro-lens

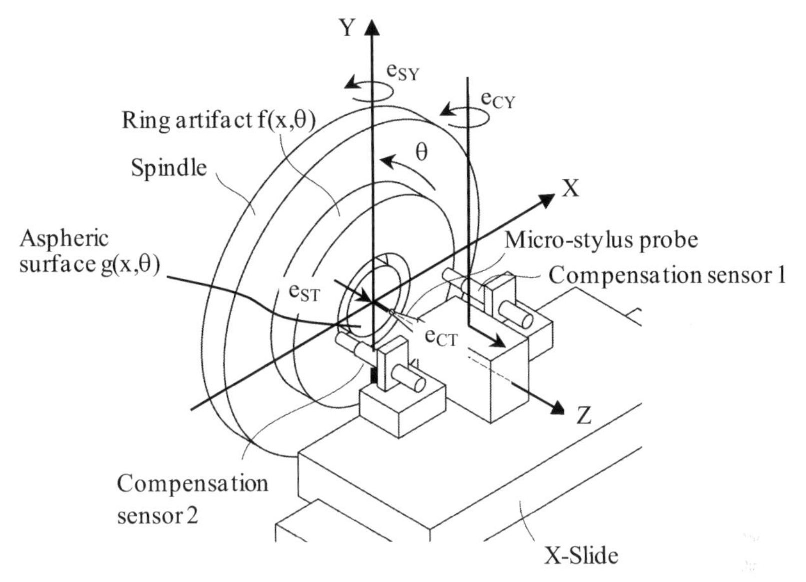

Figure 7.2. The measurement parameters

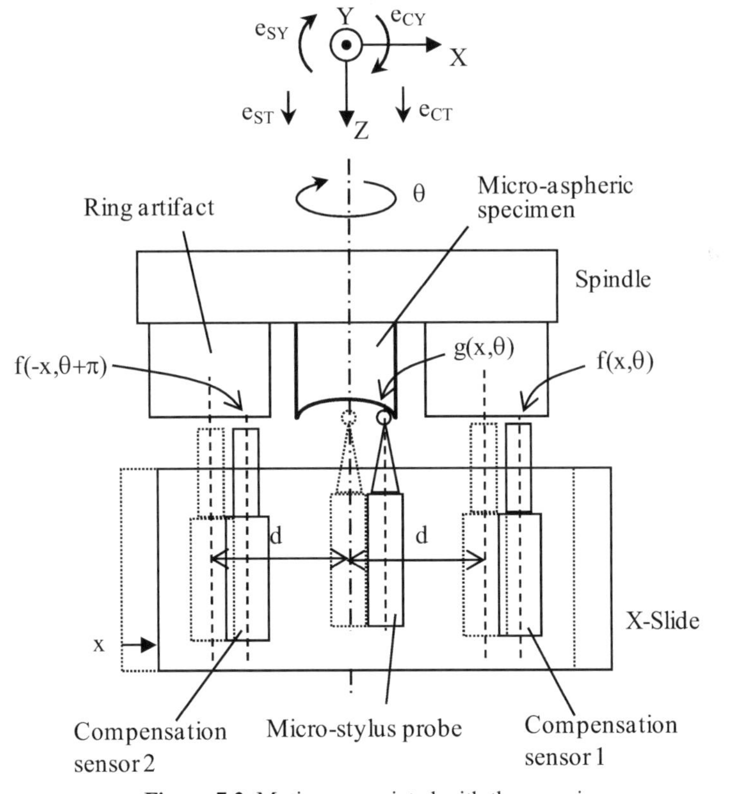

Figure 7.3. Motions associated with the scanning

Meanwhile, error motions of the spindle and the slide occur when the sensors are moved to scan the rotating specimen. $e_{ST}(x, \theta)$ is the axial error motion of the spindle, which corresponds to the error motion along the spindle rotation axis (*Z*-axis). $e_{CT}(x, \theta)$ is the *Z*-directional straightness error of the slide. $e_{SY}(x, \theta)$ is the tilt error motion of the spindle about the *Y*-axis. $e_{CY}(x, \theta)$ is the yaw error motion of the slide about the *Y*-axis. As can be seen in Figure 7.3, when the three sensors are moved by the slide to a position x and the spindle rotates the aspheric surface and the ring artifact to a position θ, the output of the micro-stylus probe to measure the profile of the aspheric surface at the position of (x, θ) can be expressed as follows:

$$m_m(x,\theta) = g(x,\theta) + e_{ST}(x,\theta) + e_{CT}(x,\theta) - x(e_{SY}(x,\theta) + e_{CY}(x,\theta)). \quad (7.1)$$

The outputs of the capacitive sensors (sensors 1 and 2) to detect the ring artifact can be expressed by:

$$\begin{aligned} m_1(x,\theta) = f(x,\theta) + e_{ST}(x,\theta) + e_{CT}(x,\theta) \\ -(d+x)(e_{SY}(x,\theta) + e_{CY}(x,\theta)) \end{aligned}, \quad (7.2)$$

$$\begin{aligned} m_2(x,\theta) = f(-x,\theta+\pi) + e_{ST}(x,\theta) + e_{CT}(x,\theta) \\ +(d-x)(e_{SY}(x,\theta) + e_{CY}(x,\theta)) \end{aligned}. \quad (7.3)$$

Taking the sum of Equations 7.2 and 7.3 gives:

$$\begin{aligned} m_1(x,\theta) + m_2(x,\theta) = f(x,\theta) + f(-x,\theta+\pi) + 2e_{ST}(x,\theta) \\ + e_{CT}(x,\theta) - 2x(e_{SY}(x,\theta) + e_{CY}(x,\theta)) \end{aligned}. \quad (7.4)$$

The profile of the aspheric surface $g(x, \theta)$ can thus be calculated from Equations 7.1 and 7.2 as follows:

$$g(x,\theta) = m_m(x,\theta) - \frac{m_1(x,\theta) + m_2(x,\theta)}{2} - \frac{f(x,\theta) + f(-x,\theta+\pi)}{2}. \quad (7.5)$$

It can be seen that the scanning error motions are separated from the aspheric surface profile in Equation 7.5.

On the other hand, however, it is necessary to know the surface profile of the ring artifact $f(x, \theta)$ for getting $g(x, \theta)$ accurately. The reversal technique [14] shown in Figure 7.4 is employed for this purpose. Two measurements are carried out before and after the ring artifact is turned 180° with respect to the spindle, which is called the reversal operation. In each of the measurements, the slide is kept stationary and only the spindle rotates for sensors to scan the aspheric sample and the ring artifact simultaneously. The micro-stylus probe is aligned to the axis of the spindle and the center of the aspheric sample so that the probe is sensitive only to the scanning error motions but not the aspheric surface profile. The sensor interval

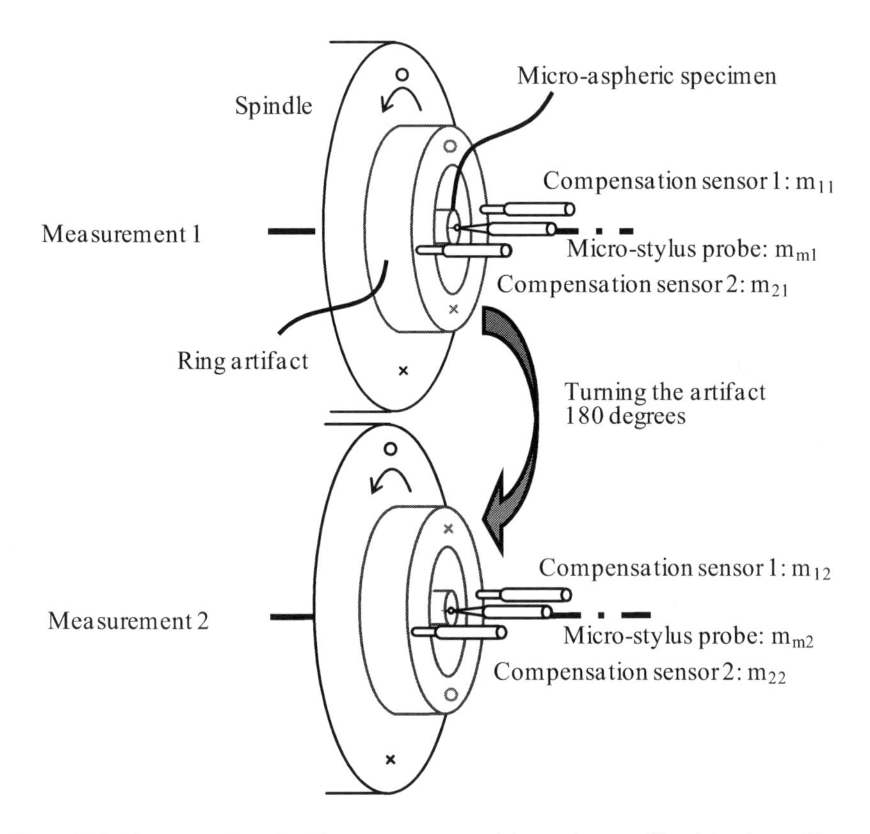

Figure 7.4. The reversal method for measurement of the surface profile of the ring artifact

is set to be $d + x$ as shown in Figure 7.5 so that sensors 1 and 2 can detect $f(x, \theta)$ and $f(x, \theta+\pi)$. Assume that the scanning error motions of the spindle are repeatable in the two measurements. The outputs of the sensors can be expressed by:

Measurement 1 (before reversal):

$$m_{m1}(\theta) = e_{ST}(\theta) + e_{CT}(\theta) \ , \tag{7.6}$$

$$m_{11}(\theta) = f(x,\theta) + e_{ST}(\theta) + e_{CT}(\theta) - (d + x)(e_{SY}(\theta) + e_{CY}(\theta)) \ , \tag{7.7}$$

$$m_{21}(\theta) = f(x,\theta + \pi) + e_{ST}(\theta) + e_{CT}(\theta) + (d + x)(e_{SY}(\theta) + e_{CY}(\theta)) \ . \tag{7.8}$$

Measurement 2 (after reversal):

$$m_{m2}(\theta) = e_{ST}(\theta) + e_{CT}(\theta) = m_{m1}(\theta) \ , \tag{7.9}$$

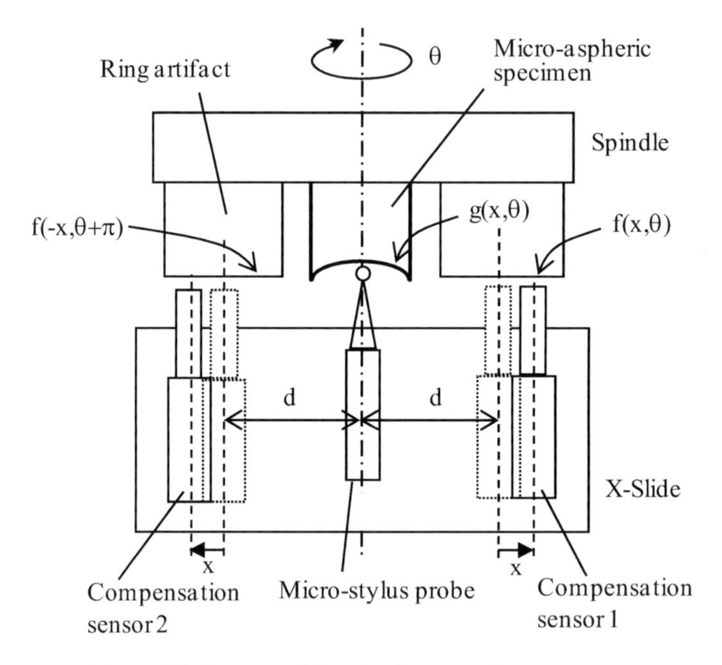

Figure 7.5. Sensor positions for the reversal measurement

$$m_{12}(\theta) = f(x, \theta + \pi) + e_{ST}(\theta) + e_{CT}(\theta) - (d + x)(e_{SY}(\theta) + e_{CY}(\theta)), \quad (7.10)$$

$$m_{22}(\theta) = f(x, \theta) + e_{ST}(\theta) + e_{CT}(\theta) + (d + x)(e_{SY}(\theta) + e_{CY}(\theta)). \quad (7.11)$$

The term $f(x, \theta)$ can thus be calculated by:

$$f(x, \theta) = \frac{m_{11}(\theta) + m_{22}(\theta)}{2} - m_{m1}(\theta) , \quad (7.12)$$

or by

$$f(x, \theta) = \frac{m_{21}(\theta - \pi) + m_{12}(\theta - \pi)}{2} - m_{m1}(\theta - \pi) . \quad (7.13)$$

The combination of the spindle tilt error motion and the side yaw motion can also be obtained as follows:

$$e_{SY}(\theta) + e_{CY}(\theta) = \frac{m_{22}(\theta) - m_{11}(\theta)}{2(d + x)} , \quad (7.14)$$

$$e_{SY}(\theta) + e_{CY}(\theta) = \frac{m_{21}(\theta) - m_{12}(\theta)}{2(d + x)} . \quad (7.15)$$

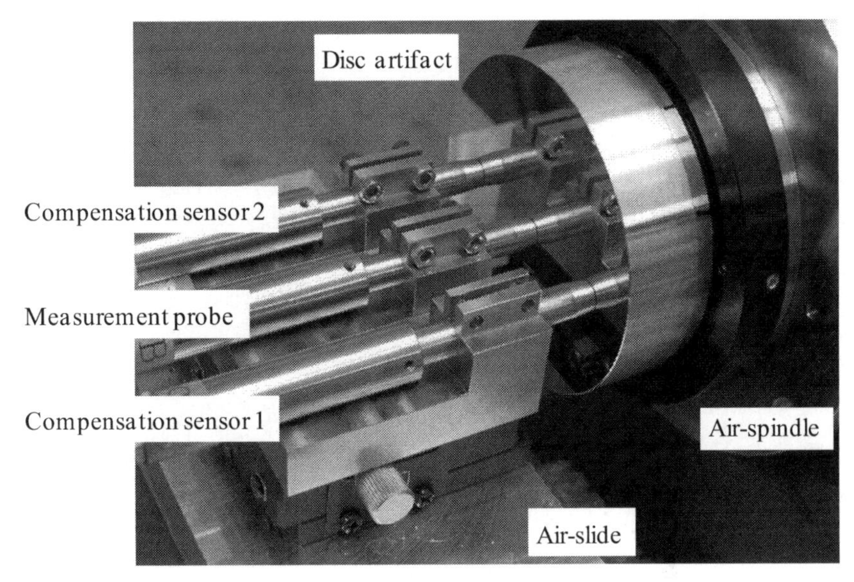

Figure 7.6. Sensor positions for the reversal measurement

It should be pointed out that the measurement accuracy of $f(x, \theta)$ is influenced by non-repeatability of the spindle error motions, which is a main drawback of the reversal technique.

Experiments were carried out to verify the effect of compensating the scanning errors. Figure 7.6 shows a photograph of the experimental setup. An air-bearing spindle (air-spindle) and an air-bearing slide (air-slide) with good motion repeatability were employed. The spindle had a rotary encoder output of 4096 pulses/revolution, which was used as the external trigger signal data acquisition of the sensor outputs. The slide had a linear encoder with a resolution of 20 nm. An aluminum disk artifact was employed as the measurement object instead of the ring artifact and the micro-aspheric specimen. Three capacitive sensors were set on the slide, and the aluminum disk artifact was vacuum chucked on the spindle. The sensor in the center was employed instead of the micro-stylus probe and those in the two sides were used as the compensation sensors. The sensor interval was 25 mm. The sensing electrode diameter of the capacitive sensor was 1.7 mm. The sensor measurement range was 50 μm corresponding to a voltage output range of 20 V. The voltage outputs of the three sensors were simultaneously taken into a data logger with 16-bit analog-to-digital converters. The disc artifact had a diameter of 80 mm and a thickness of 24 mm. Both the top surface and the side surface of the disc artifact were cut on a diamond turning machine. The top surface was used as the target for the capacitive sensors and the side surface was used for the alignment of the disc artifact on the spindle.

Figure 7.7 shows the sensor outputs acquired when both the spindle and the slide were kept stationary. The sampling rate was 10 kHz. The sensor outputs had similar vibration components with an amplitude of approximately 8 nm and a frequency of approximately 250 Hz. The vibration errors were reduced to approximately 2 nm in the differential output of sensors shown in Equation 7.5.

Figure 7.8 shows the experimental results of investigating the repeatability of the spindle error motions. In the experiment, the spindle was rotated at a speed of 15 rpm and the slide was kept stationary. The outputs of the measurement probe over 100 spindle revolutions were acquired consecutively by using the spindle rotary encoder output as the external trigger signal for the data logger. The differences between outputs in different revolutions, which represent the repeatability of spindle error motion, are plotted in Figure 7.8. It can be seen that the repeatability was better than 10 nm. In the following experiment, the repeatability was further improved by averaging the sensor outputs over 100 revolutions.

Figure 7.9 shows the experiment procedure for the surface profile measurement of the disc artifact. The experiment included four steps. In each of the steps, the disc artifact was rotated by the spindle so that it could be scanned by the three sensors. Then the artifact was turned 90° with respect to the spindle for the next step measurement. The surface profile of the disc artifact could be calculated from the sensor outputs in steps 1 and 3 based on Equation 7.12. The sensor outputs in steps 2 and 4 could also be employed for the calculation of the same surface profile. The correspondence of the two calculation results could thus be used to confirm the feasibility of the reversal method. Figure 7.10 shows the outputs of sensors 1 and 2 in each of the steps. The component of one undulation per revolution caused by the artifact inclination was removed from the output. The range of the sensor output was approximately 40 nm, which was a combination of the axial error motion, the tilt error motion and the artifact surface profile.

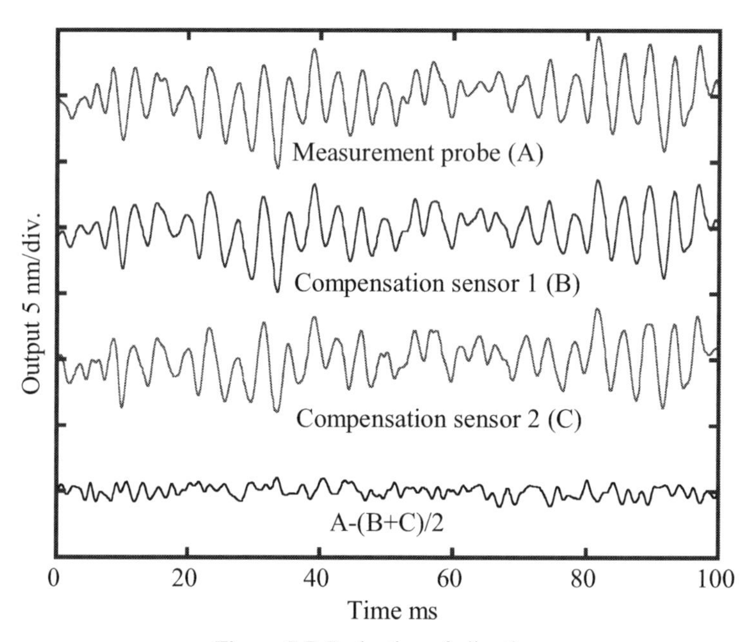

Figure 7.7. Reduction of vibration

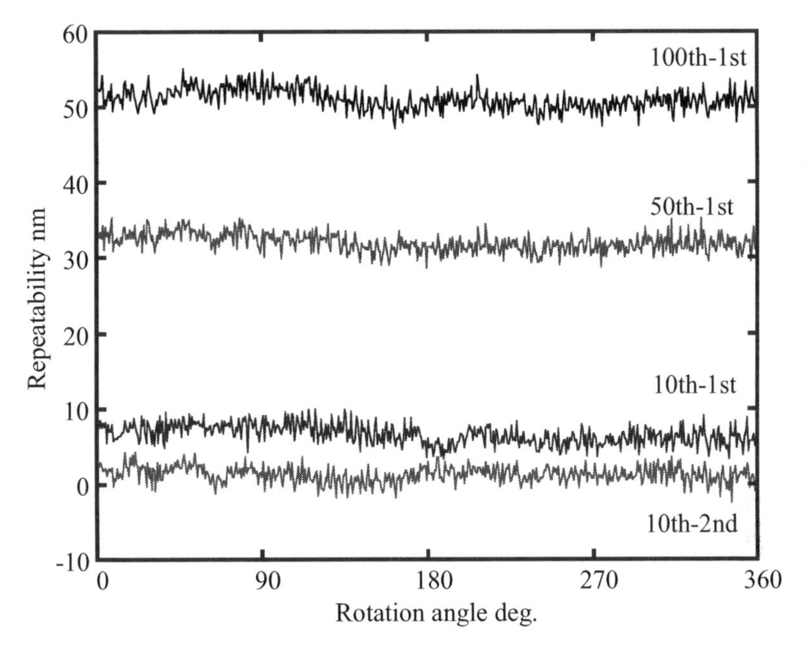

Figure 7.8. Repeatability of the scanning motion

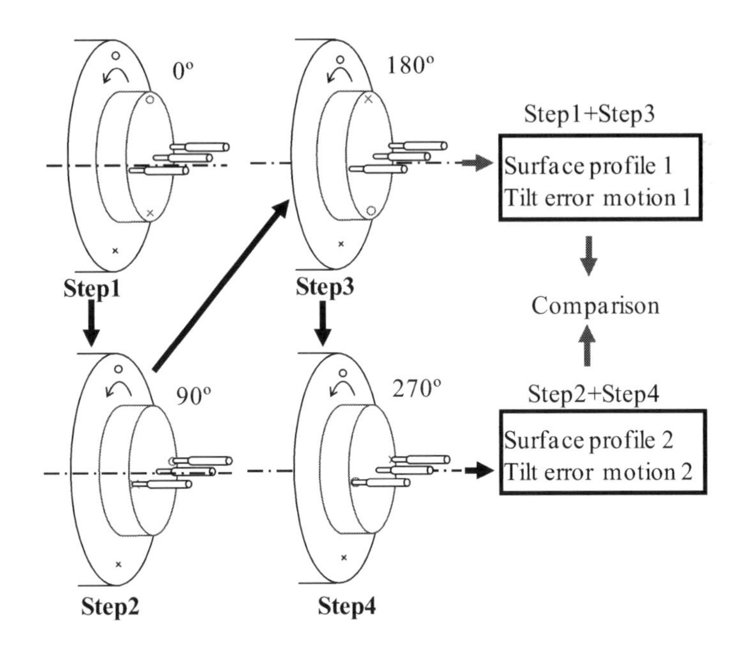

Figure 7.9. Procedure of experiment

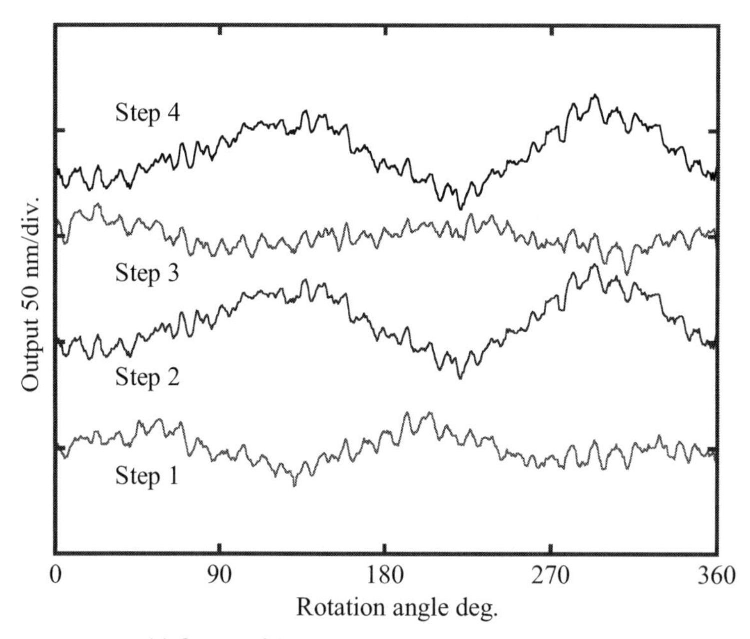

(a) Output of the compensation sensor 1 at each step

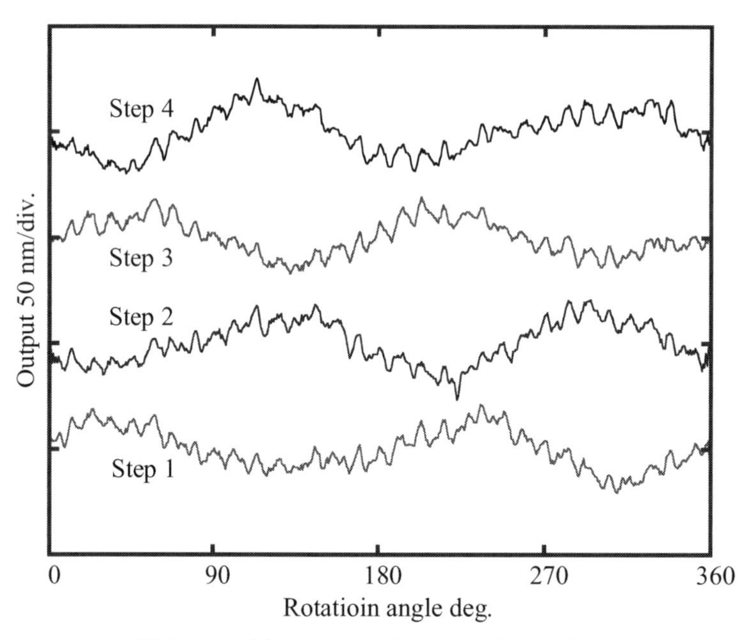

(b) Output of the compensation sensor 2 at each step

Figure 7.10. Output data of sensors in the reversal measurement of the disc artifact

Figure 7.11 shows the outputs of the measurement probe. Because the measurement probe was pointing to the center of the artifact as well as that of the spindle, only the spindle axial error motion was included in the output. The axial error motion was approximately 10 nm. It was repeatable on the nanometer level in the measurements of the four steps. The average of the axial error motions in the measurements was employed in the calculation of the surface profile of the disc artifact.

Figure 7.12 shows the surface profiles of the disc artifact. Figures 7.12 (a) and 7.12 (b) show the profiles with and without compensation of the axial error motion, respectively. As can be seen in Figure 7.12 (b), the 10 nm vibration component caused by the axial error motion was successfully removed from the surface profile. The peak-to-valley value of the surface profile was measured to be approximately 20 nm. Surface profile 1 calculated from the data of steps 1, 3 and surface profile 2 from the data of steps 2, 4 also corresponded with each other.

Figure 7.13 shows the corresponding tilt error motions calculated by Equation 7.14. The tilt error motion was approximately 0.5 μrad. Compared with the spindle axial error motion shown in Figure 7.11, the spindle tilt error motion had higher-frequency components.

To further confirm the reliability of the result by the reversal method, the sample was measured by a commercial interferometer [15]. Figures 7.14 and 7.15 show the results in 2D and 3D expressions, respectively. As can be seen in the figures, the results by the two different methods were coincident with each other in nanometer level. High-frequency components were observed in the result by the interferometer, which were caused by external vibrations associated with the measurement of the interferometer.

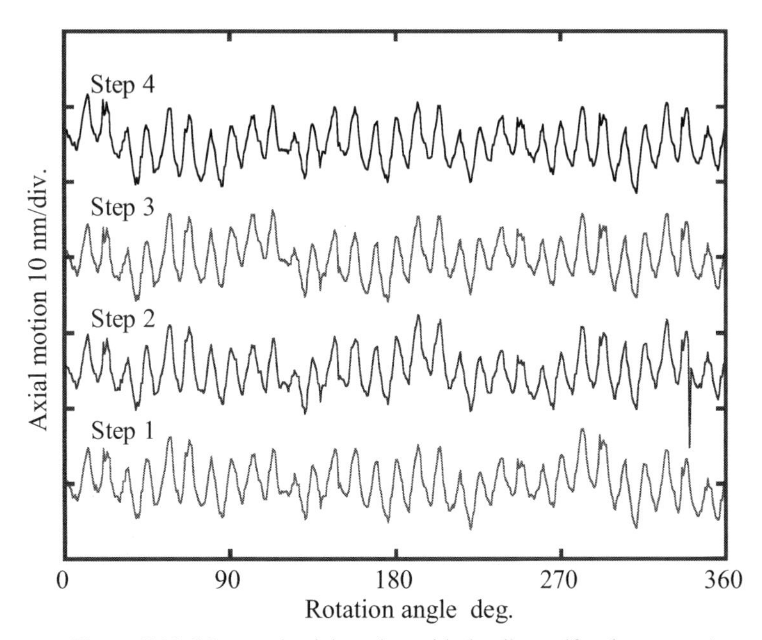

Figure 7.11. Measured axial motion with the disc artifact by sensor 1

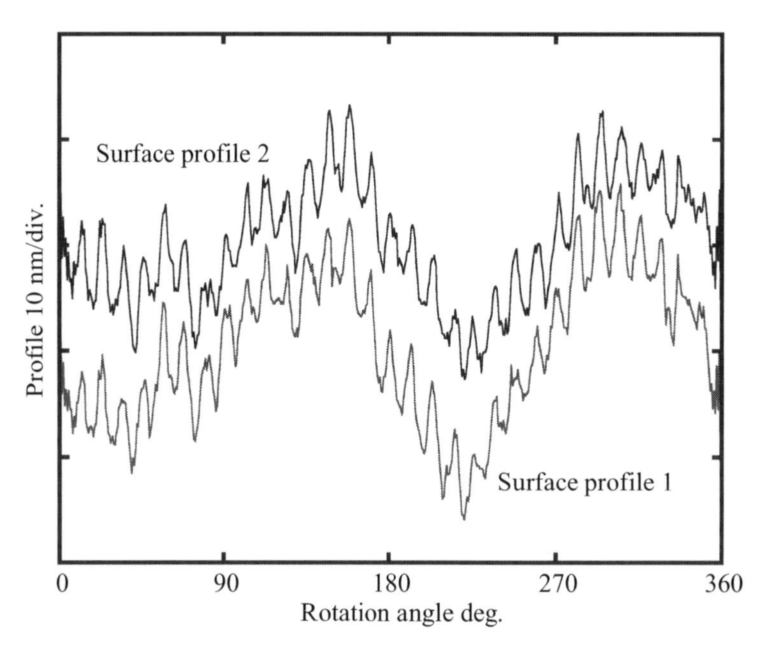

(a) Measured surface profiles without compensation of axial error motion

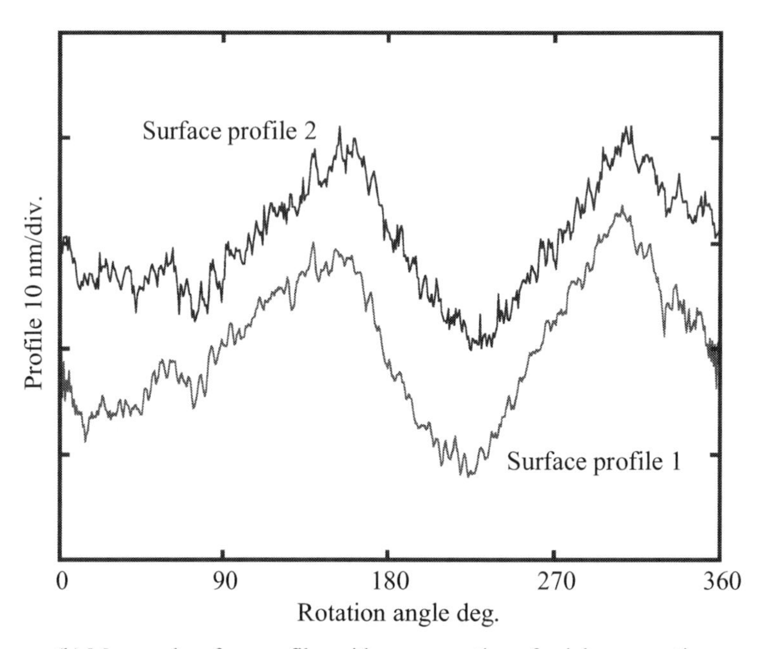

(b) Measured surface profiles with compensation of axial error motion

Figure 7.12. Comparison with the result by an interferometer

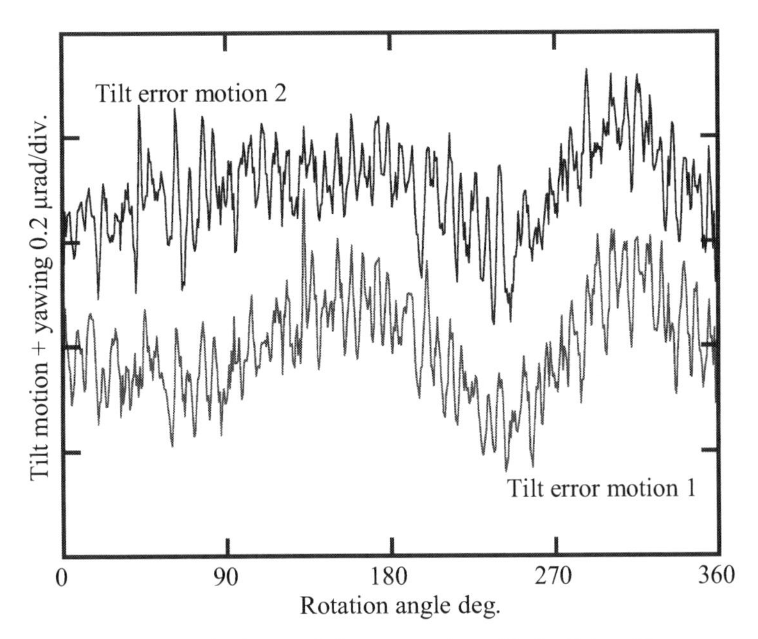

Figure 7.13. Measured tilt error motion with the disk artifact

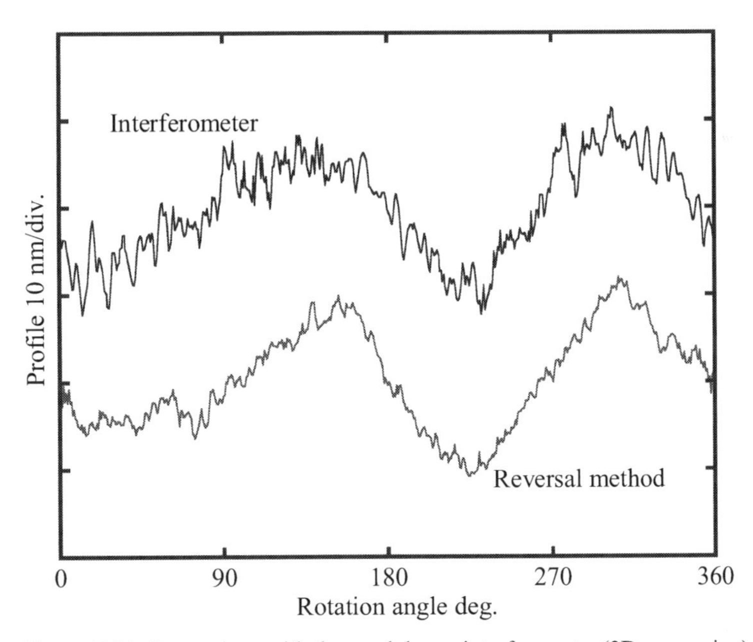

Figure 7.14. Comparison with the result by an interferometer (2D expression)

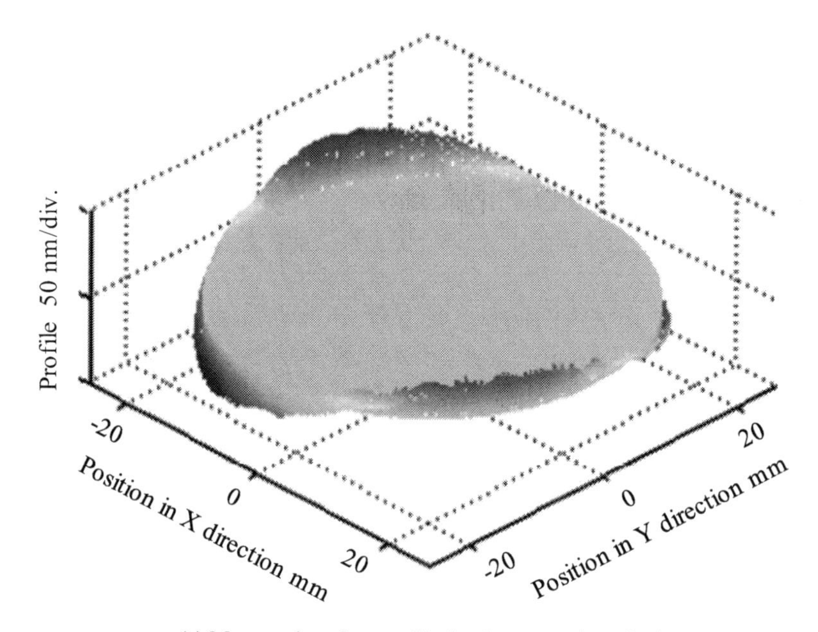

(a) Measured surface profile by the reversal method

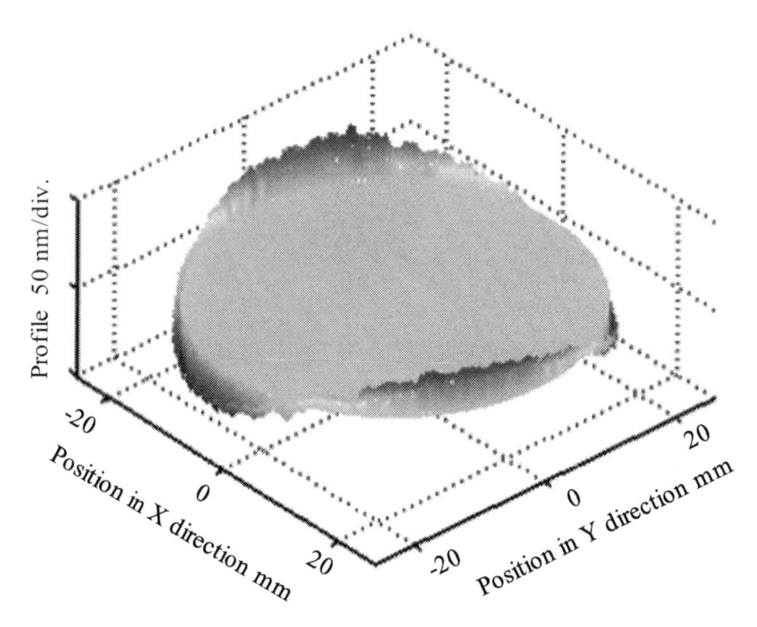

(b) Measured surface profile by the interferometer

Figure 7.15. Comparison with the result by an interferometer (3D expression)

7.3 Micro-stylus Probe

The tip ball and the stylus shaft are the fundamental elements of a tactile probe. The radius of the tip ball is required to be small and the aspect ratio (the ratio of length to diameter) of the stylus shaft is required to be large for the micro-stylus probe used in measurement of micro-aspheric surfaces. In this section, a micro-ball glass tube stylus, which is composed of a 100 μm diameter tip ball, a long and thin glass tube shaft, is presented. A micro-tapping stylus probe, which can reduce the influence of the friction force between the tip ball and the measurement specimen, is also described.

7.3.1 Micro-ball Glass Tube Stylus

Glass tubes can be stretched and thinned by heating and the ratio between the inner diameter and the outer diameter does not change during the heating process. The shaft of the micro-ball glass tube stylus is fabricated based on this characteristic. A micro-ball is bonded at the tip of the shaft. The glass tube is stretched by a device called a puller shown in Figure 7.16. The tube is heated by the filament and pulled to be separated into two parts by the weight attached at the bottom end of the glass tube at the same time. The heating temperature can be adjusted by changing the current of the filament. The end of the stretched glass tube is cone shaped and the cone angle is an important factor to determine the stiffness of the shaft. The larger the cone angle, the stiffer the shaft as shown in Figure 7.17. Although it is necessary to reduce the heating temperature to obtain a larger cone angle as shown in Figure 7.18, it is difficult to separate the two stretched parts for glass tubes with the necessary wall thickness. A multi-step heating-and-stretching method is thus employed to make large cone angles for thick-wall glass tubes as shown in Figure 7.19. Figure 7.20 shows the procedure for fabrication of the micro-stylus.

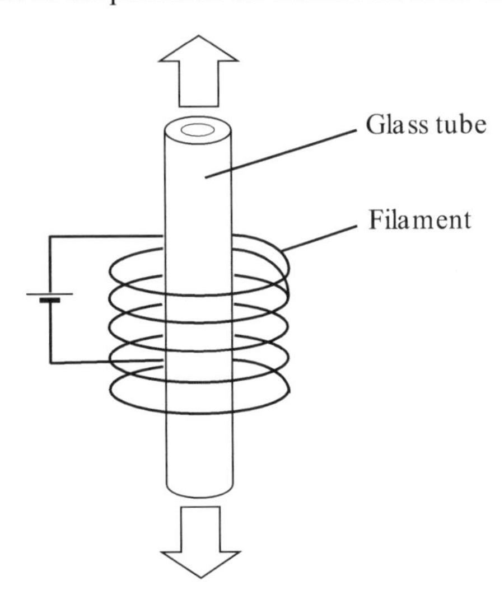

Figure 7.16. Function of the puller for stretching the glass tube

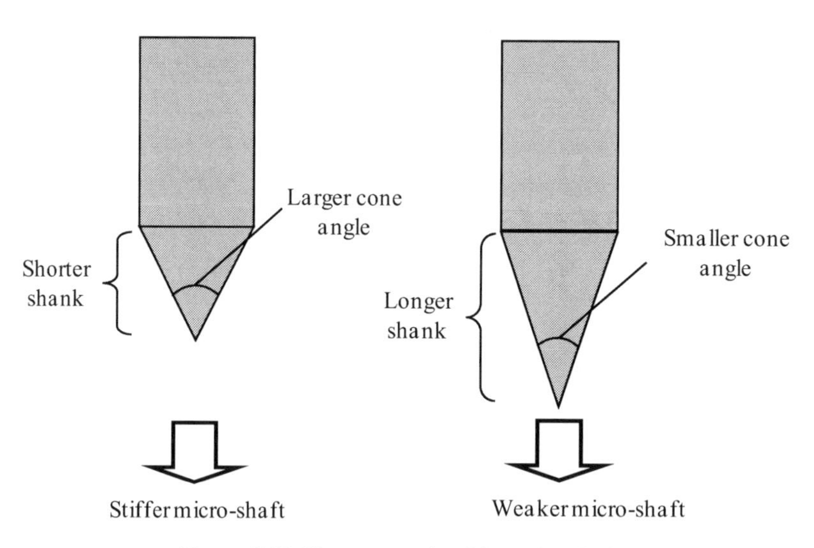

Figure 7.17. The cone angle of the stylus shaft

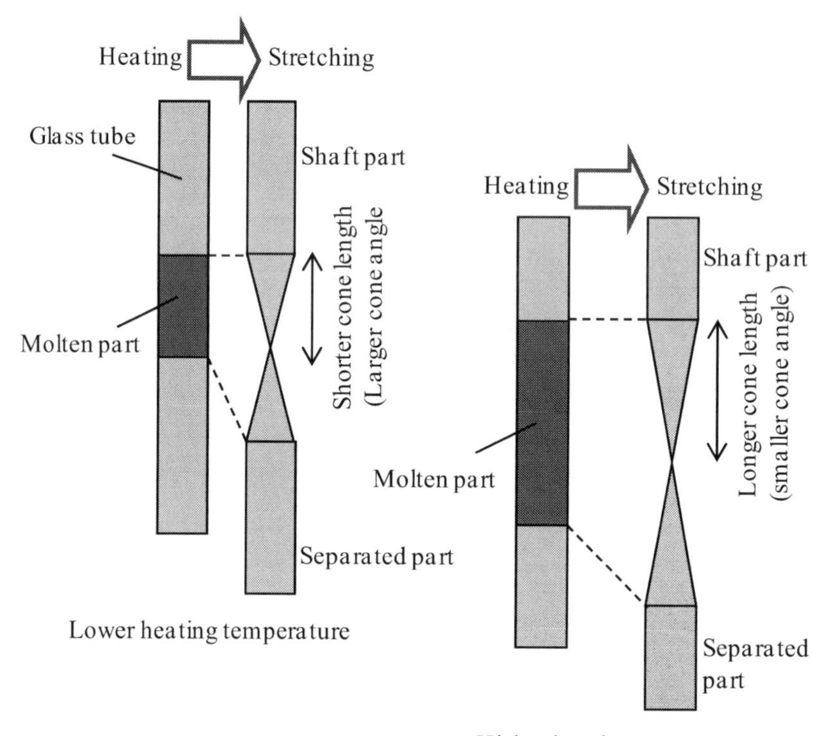

Figure 7.18. Influence of heating temperature on the cone angle of the stylus shaft

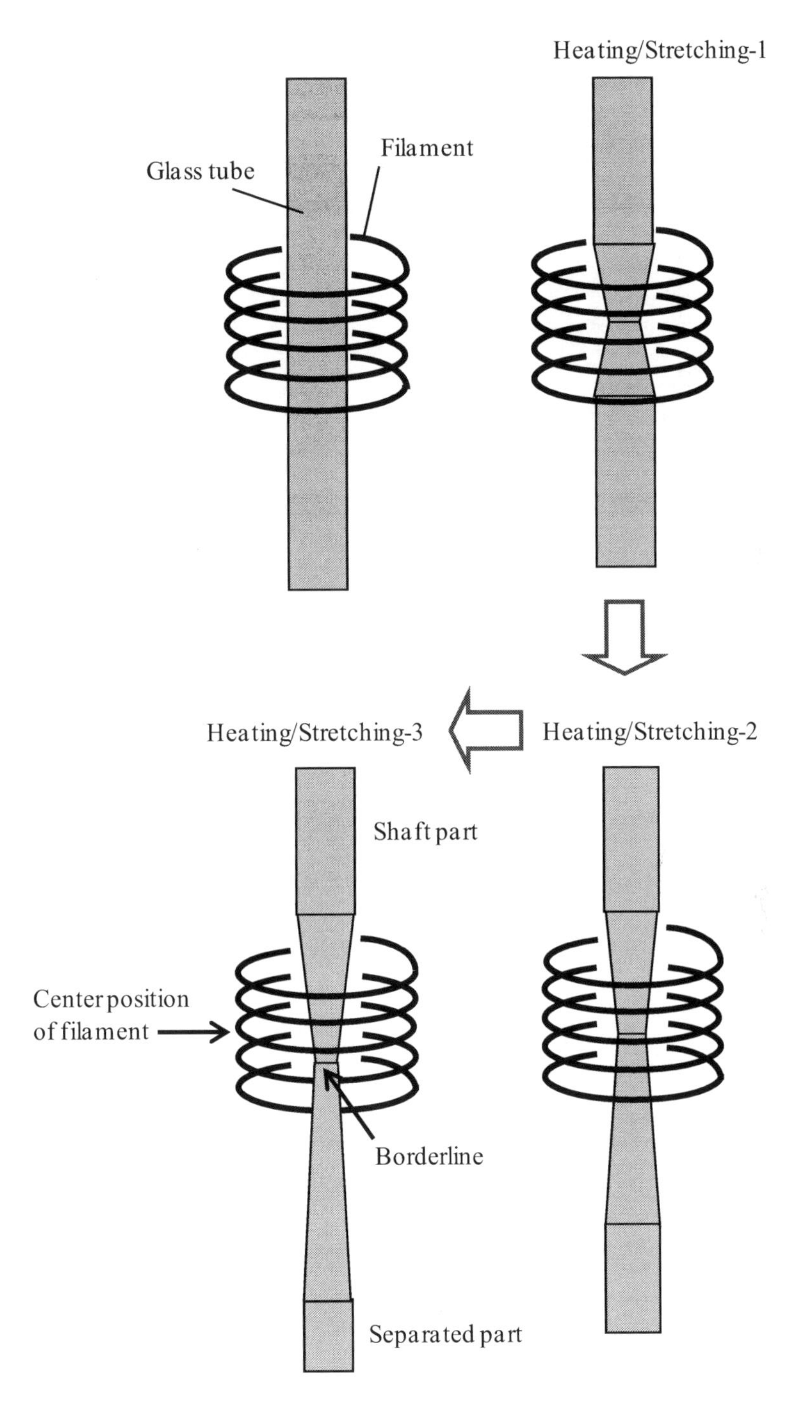

Figure 7.19. Schematic of the multi-step heating-and-stretching method for fabrication of larger cone angle shaft

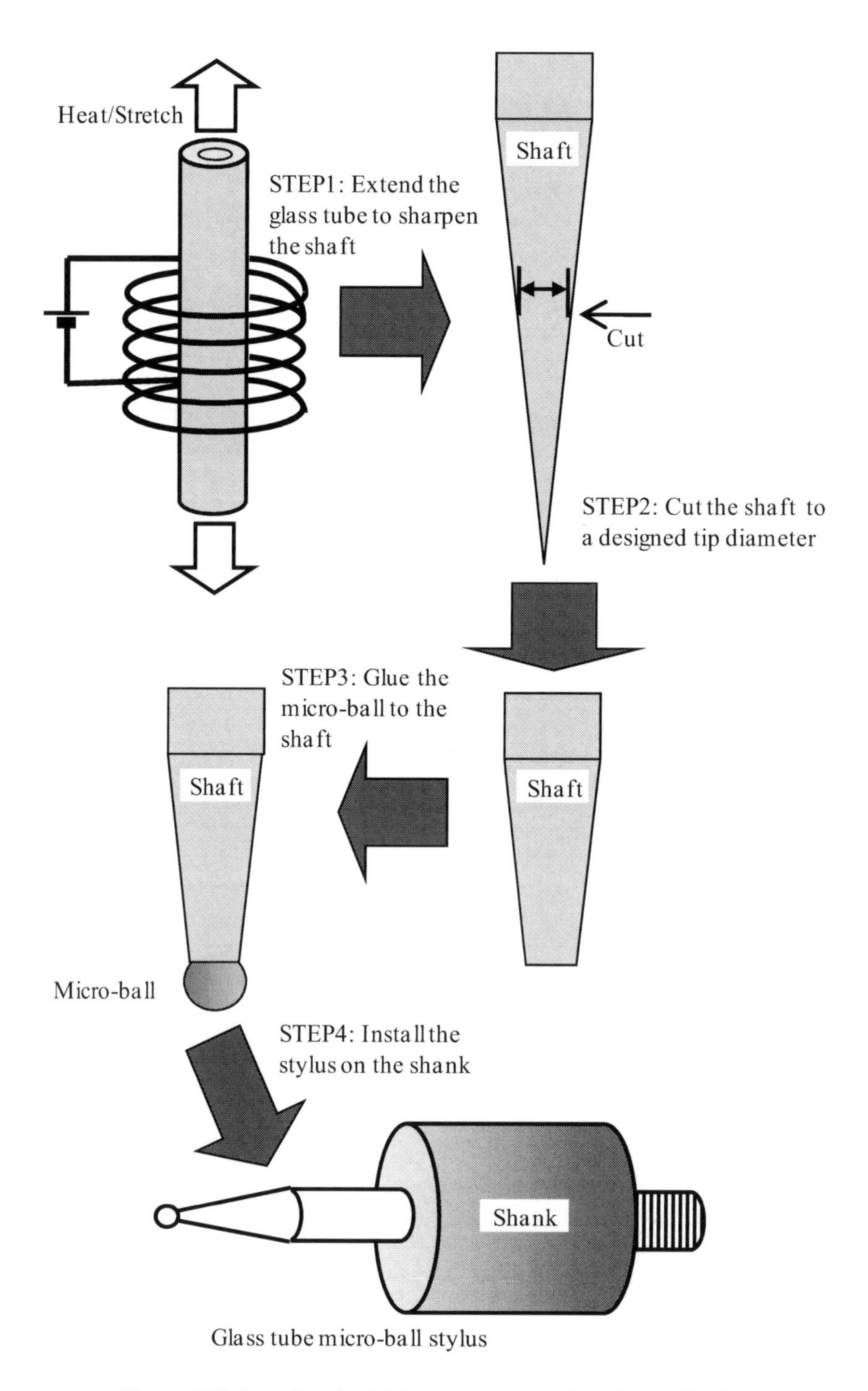

Figure 7.20. Procedure for fabricating the glass tube micro-ball stylus

A glass tube with an outer diameter of 1 mm and an inner diameter of 0.6 mm is used to make the glass tube stylus based on the multi-step heating-and-stretching method shown in Figure 7.19. The temperature of the puller heater is set to be as low as possible and the glass is stretched in such a way that the length of the cone is short (about 0.5 mm) at each step till the stretched parts (the shaft part and the separated part) are separated from each other. The multi-step heating-and-stretching method is effective to make the cone angle larger and the cone length shorter because the center position of the filament rises with respect to the borderline between the shaft part and the separated part of the glass tube step-by-step, as shown in Figure 7.19. A microscope image of the stretched glass tube is shown in Figure 7.21. The cone angle is approximately 20°. Figure 7.22 shows a photograph of the glass tube after its tip is cut at the position with a tip width of approximately 100 μm by using a laser cutting machine.

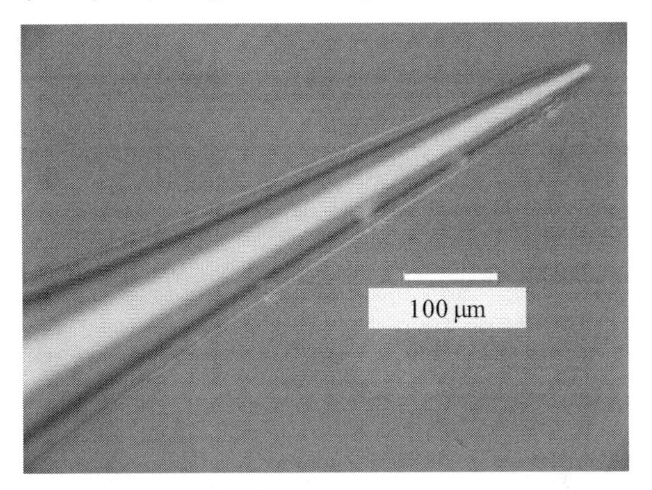

Figure 7.21. The fabricated glass tube shaft

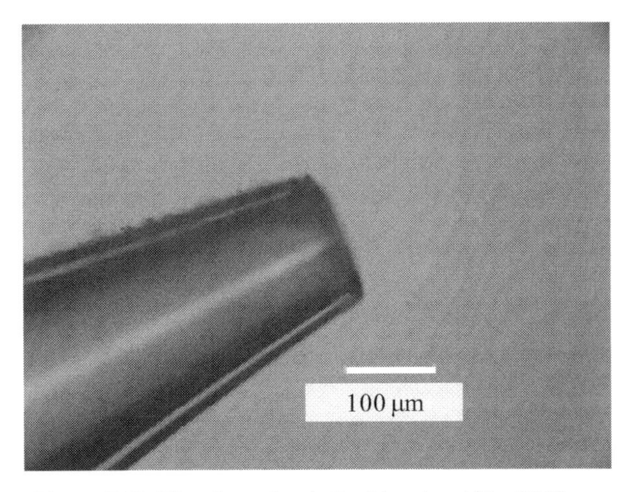

Figure 7.22. The glass tube shaft with a tip width of 100 μm

The next step in the fabrication of the micro-ball glass tube stylus is to glue a micro-ball to the tip of the shaft. There are two important points in this process. One is to align the axis of the shaft with the center of the ball. The other is the strength of the adhesive between the tip ball and the shaft. A low cost is also desired for the fabrication of the stylus of the tactile probe. Precision glass micro-balls with diameters of less than 100 μm were used for the stylus [16, 17]. The glass ball, which is used as spacers in crystalline liquid panels, has a low cost and high form accuracy.

A photograph of the system for gluing the micro-ball is shown in Figure 7.23. Two *XYZ*-stages and two CCD cameras are employed to glue the micro-ball on the tip of the stylus shaft. The stylus shaft is attached to one of the *XYZ*-stages. The glass micro-ball is put on the other *XYZ*-stage. The position of the stylus tip relative to the micro-ball can be adjusted by using these *XYZ*-stages. The micro-ball and the stylus tip are also monitored by the two CCD cameras from the *X*- and *Y*-directions so that the micro-ball can be glued at the center of the stylus tip.

To glue the micro-ball, the adhesive, which is the heat-hardening-type, is first attached to the tip of the stylus shaft. When the tip of the micro-shaft gets close to the adhesive bond, the hole of the shaft around the tip is filled with the adhesive bond by the capillary phenomenon. Then the tip of the stylus shaft comes in contact with the micro-ball by using the *XYZ*-stage. When the micro-ball touches the adhesive at the tip of the stylus shaft, the micro-ball is attached to the tip of the shaft by the effect of surface tension. The micro-ball can then be glued on the tip of the stylus by heating the adhesive to be hardened. Microscope images of the micro-styluses with glass micro-balls of 100, 50 and 30 μm diameters are shown in Figure 7.24.

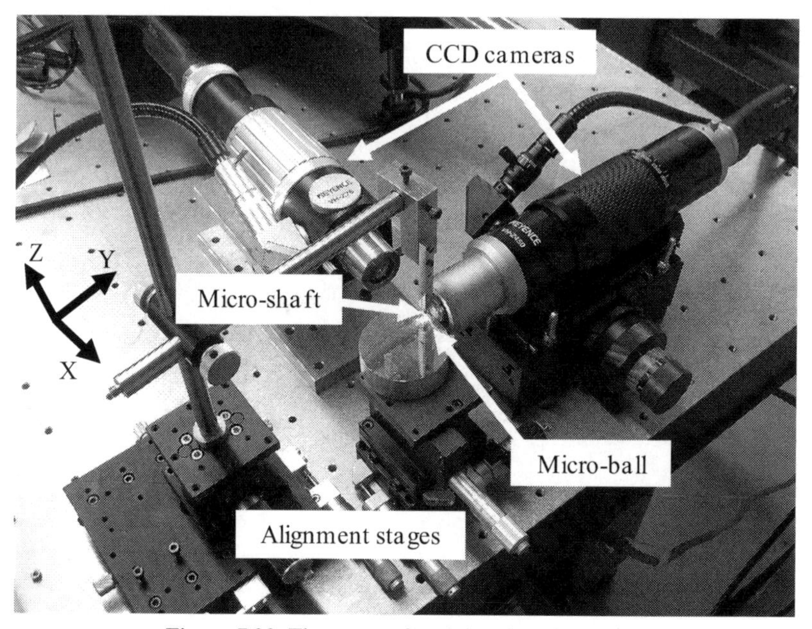

Figure 7.23. The system for gluing the micro-ball

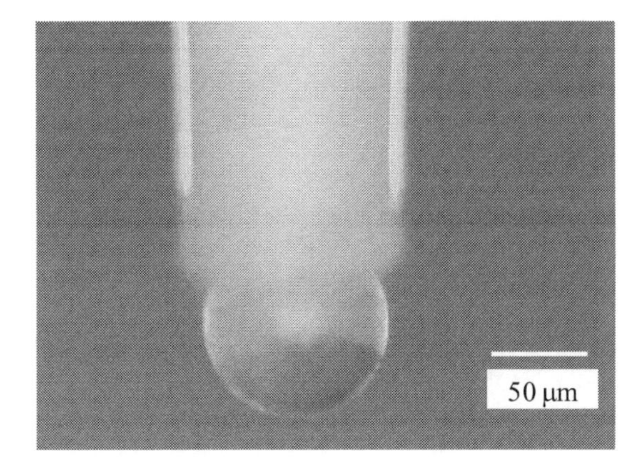

(a) Ball diameter 100 μm

(b) Ball diameter 50 μm

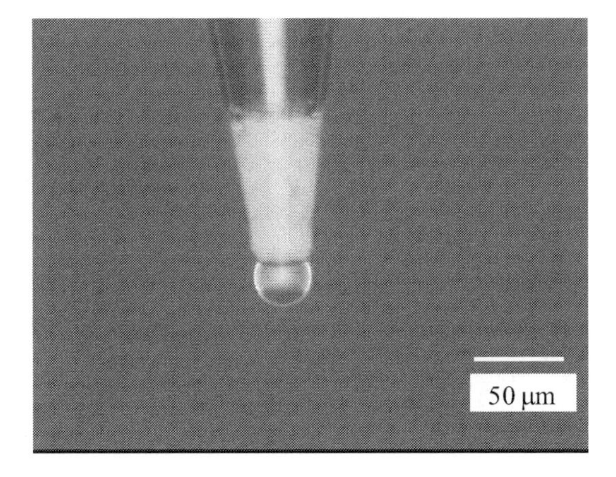

(c) Ball diameter 30 μm

Figure 7.24. Micro-balls glued on the glass tube

The strength of the micro-ball stylus was tested by using a strain gauge-type force sensor with a 100 mN range. The stylus was inclined to 45° and pushed against the force sensor by moving a manual stage so that an increasing load could be applied. Figure 7.25 shows the result of the stylus with the 100 μm diameter micro-ball. The output of the force sensor, which represents the strength of the micro-ball stylus, is shown in the figure. It can be seen that the micro-ball stylus can tolerate a 100 mN load, which was limited by the range of the force sensor. Similar results have also been obtained for the 50 and 30 μm diameter balls. Figure 7.26 shows a picture of a fabricated glass tube micro-ball stylus mounted on a displacement sensor of low measuring force [18] with a measuring axis supported by air-bearing. The cone angle of the glass tube shaft is approximately 10° and the length is 3 mm.

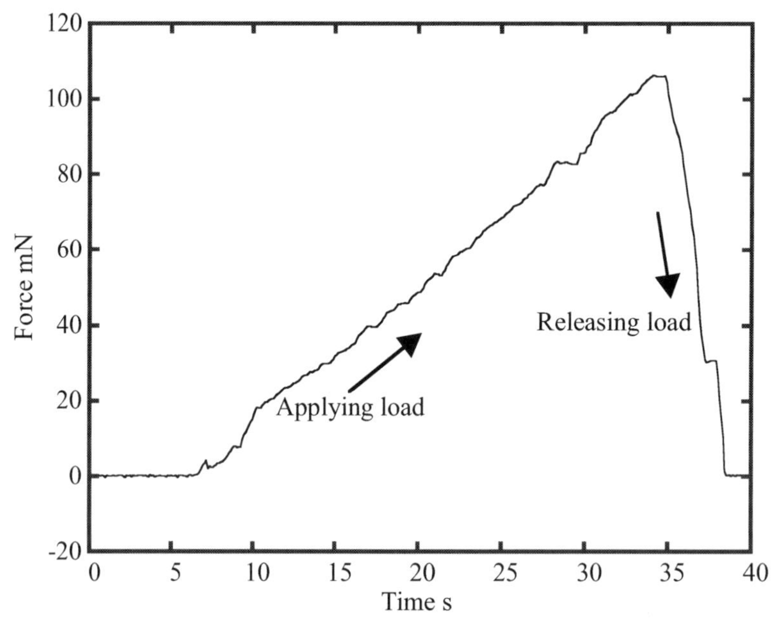

Figure 7.25. Result of testing the strength of the glued micro-ball

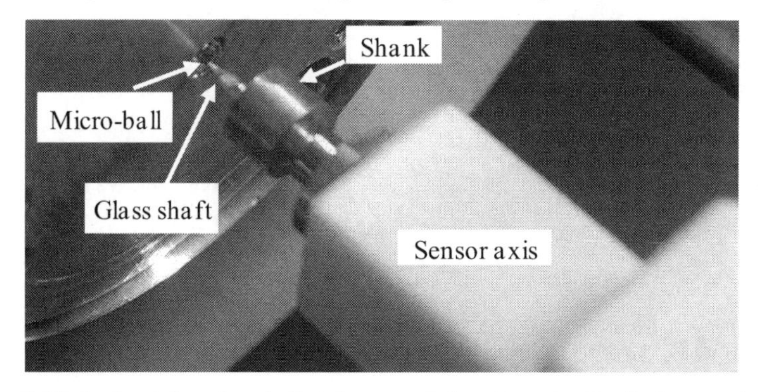

Figure 7.26. A glass tube micro-ball stylus mounted on the axis of a displacement sensor

7.3.2 Tapping-type Micro-stylus

The stick-slip phenomenon is a critical problem for a tactile probing system. A tapping-type micro-stylus, which can reduce the influence of the stick-slip phenomenon, is presented in this section.

Figure 7.27 (a) shows a schematic of the tapping-type micro-stylus mounted on the displacement sensor of low measuring force with its axis supported by air-bearing [18]. A piezoelectric actuator (PZT) is added to the micro-ball stylus shown in Figure 7.26 to tap the micro-ball on the specimen surface. The important point is to find a condition that the oscillation of the PZT is transmitted only to the micro-stylus but not to the axis of the displacement sensor.

Figure 7.27 (b) shows a mechanical model of the system in Figure 7.27 (a). m_g and m_h represent the masses of the micro-stylus and the axis of the displacement sensor, respectively. The center of the PZT is treated as the center of oscillation.

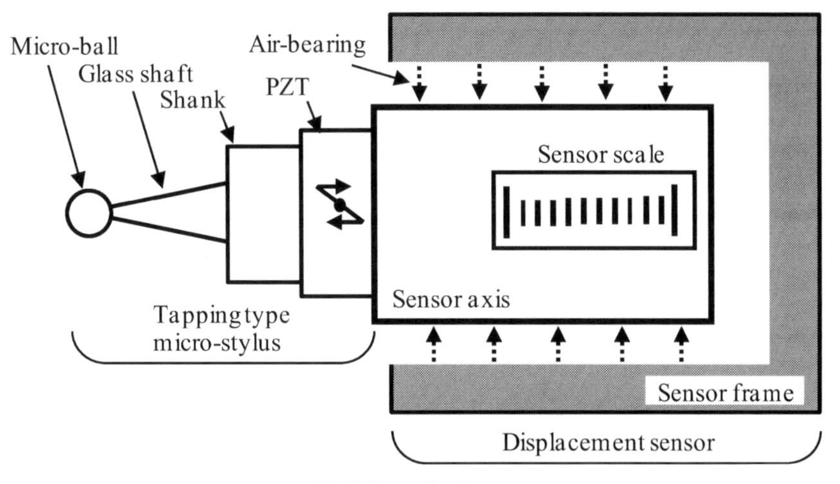

(a) Schematic view

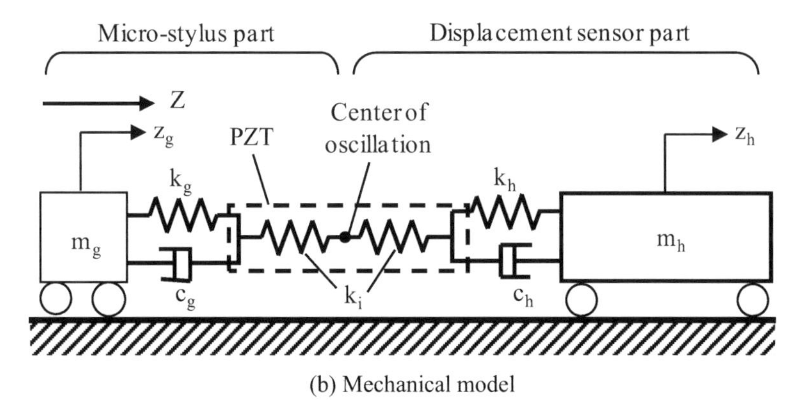

(b) Mechanical model

Figure 7.27. Schematic of the tapping-type micro-stylus mounted on a low measuring force displacement sensor with its axis supported by air-bearing

The mass of the PZT is divided into two equal parts. One belongs to m_g and the other to m_h. z_g and z_h represent the displacements of m_g and m_h, respectively. k_g and c_g are the spring constant and the viscosity of the junction part between m_g and the PZT, respectively. k_h and c_h are the spring constant and the viscosity of the junction part between m_h and the PZT, respectively. k_i is the spring constant of the PZT. Assume that the oscillation of the PZT is expressed by $a\sin\omega t$, where a and ω are the amplitude and the angular frequency, respectively. z_g can be expressed as:

$$\begin{cases} z_g = a_g \sin(\omega t) \\ a_g = \dfrac{a\sqrt{1+\left(2\varsigma_g \dfrac{\omega}{p_g}\right)^2}}{\sqrt{\left(1-\left(\dfrac{\omega^2}{p_g^2}+\dfrac{\omega^2}{p_{ig}^2}\right)\right)^2+\left(2\varsigma_g \dfrac{\omega}{p_g}\left(1-\dfrac{\omega^2}{p_{ig}^2}\right)\right)^2}} \end{cases}, \tag{7.16}$$

where

$$\varsigma_g = \frac{c_g}{2\sqrt{m_g k_g}}, \quad p_g = \sqrt{\frac{k_g}{m_g}} \text{ and } p_{ig} = \sqrt{\frac{k_i}{m_g}}. \tag{7.17}$$

The term z_h can be similarly obtained as:

$$\begin{cases} z_h = a_h \sin(\omega t) \\ a_h = \dfrac{a\sqrt{1+\left(2\varsigma_h \dfrac{\omega}{p_h}\right)^2}}{\sqrt{\left(1-\left(\dfrac{\omega^2}{p_h^2}+\dfrac{\omega^2}{p_{ih}^2}\right)\right)^2+\left(2\varsigma_h \dfrac{\omega}{p_h}\left(1-\dfrac{\omega^2}{p_{ih}^2}\right)\right)^2}} \end{cases}, \tag{7.18}$$

where

$$\varsigma_h = \frac{c_h}{2\sqrt{m_h k_h}}, \quad p_h = \sqrt{\frac{k_h}{m_h}} \text{ and } p_{ih} = \sqrt{\frac{k_i}{m_h}}. \tag{7.19}$$

a_g/a and a_h/a with respect to the frequency of the PZT's oscillation are shown in Figures 7.28 (a) and 7.28 (b), respectively. m_g and m_h are decided to be 1 and 35 g by actual measurements. k_g and k_i are decided by an FEM analysis to be 3.0×10^6 N/m and 2.0×10^7 N/m. ζ_g, ζ_h and k_h are chosen to be 0.3, 0.3 and 1.0×10^5 N/m, respectively.

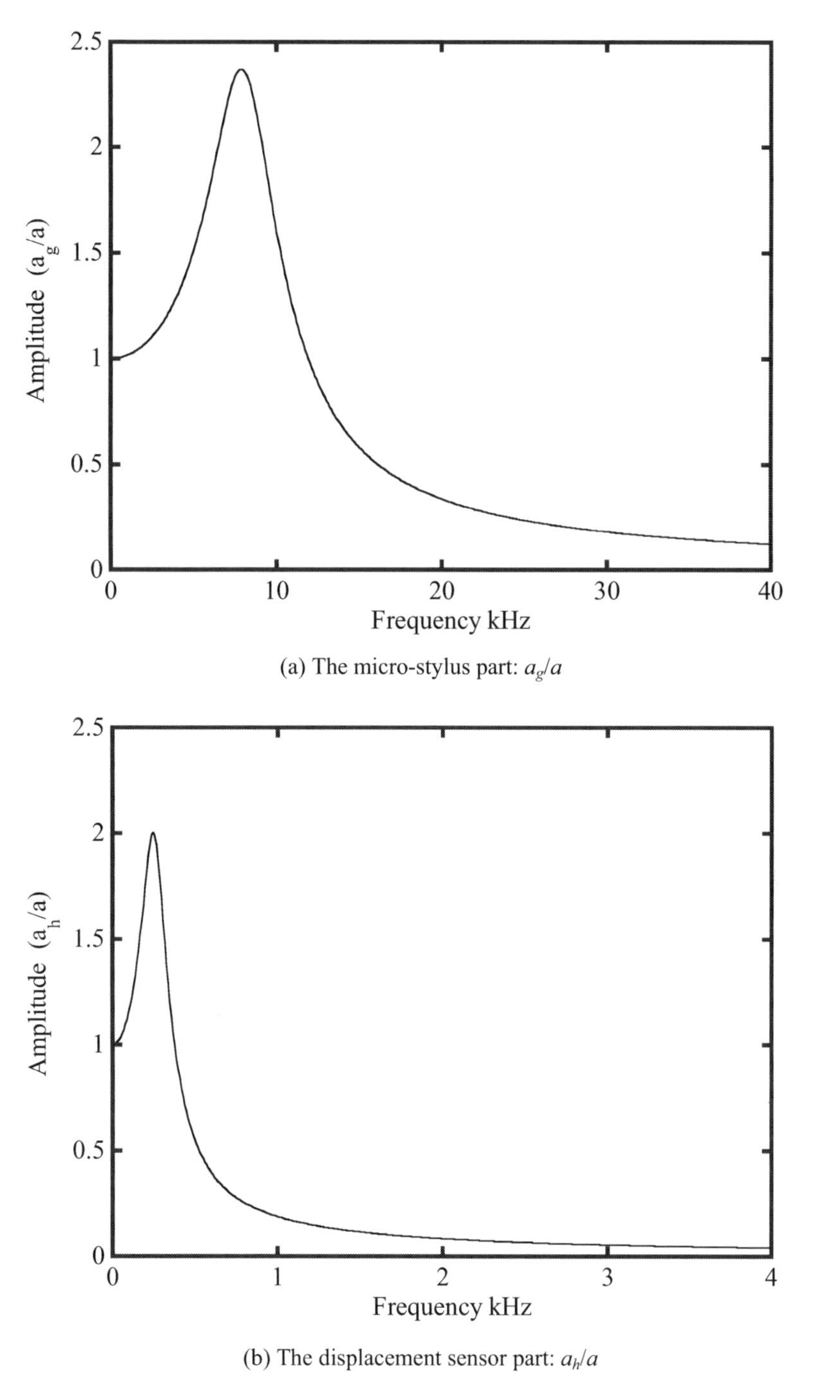

(a) The micro-stylus part: a_g/a

(b) The displacement sensor part: a_h/a

Figure 7.28. Calculated frequency responses of the micro-stylus part and the displacement sensor part

As can be seen from the figure, the amplitude a_h of the sensor part increases in the range from DC to the resonance frequency, which is approximately 250 Hz. Then a_h decreases from the resonance frequency to a value of 0.1a at around 2 kHz. On the other hand, the amplitude a_g of the micro-stylus part keeps increasing till the resonance frequency at around 8 kHz. This means if the oscillation frequency of the PZT is set to be between 2 and 8 kHz, the oscillation of the PZT will be transmitted to the micro-stylus so that the micro-ball can tap the surface of the specimen to reduce the influence of the stick-slip phenomenon. On the other hand, only one tenth of the amplitude of the PZT oscillation is transmitted to the axis of the displacement sensor. If the amplitude of the PZT oscillation is chosen to be less than 10 nm, the oscillation of the sensor axis will be less than 1 nm, which almost does not influence the output of the sensor.

Figure 7.29 show a picture of the fabricated tapping-type micro-ball stylus. The stylus is composed of a micro-ball, a glass shaft, a shank, a PZT and a connector for connection with the axis of the displacement sensor. The glass shaft is attached to the shank by a small screw so that it can be easily replaced. The PZT is sandwiched between the shank and the connector. The PZT, the shank and the connector are bonded by using an adhesive bond. The electric wires of the PZT are thin enough that they do not affect the measuring force of the displacement sensor. The specifications of the PZT are listed as follows:

- Dimension (diameter × height) = 5 mm × 2 mm
- Piezoelectric charge constant (d_{33}) = 600 × 10^{-12} m/V
- Young's modulus (Y33) = 5.1 × 10^{10} N/m^2
- Poisson's ratio = 0.34

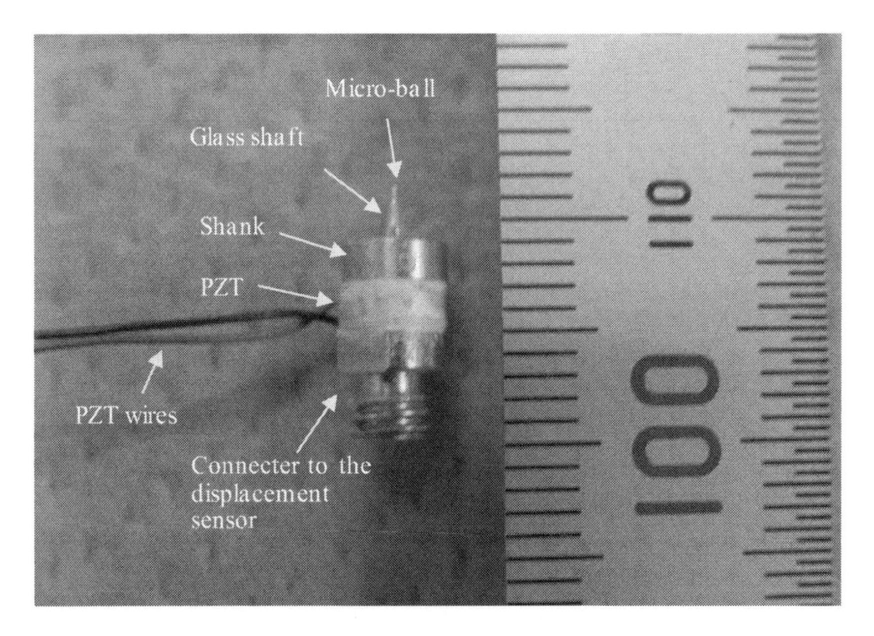

Figure 7.29. The fabricated tapping-type micro-stylus

7.4 Small-size Measuring Instrument for Micro-aspherics

The schematic of a small-size measuring instrument designed for micro-aspheric surfaces is shown in Figure 7.30. The footprint of the machine is 400 mm × 275 mm and the weight of the machine is 57 kg. The micro-aspheric specimen is mounted on the spindle and the sensor unit is mounted on the slide. The sensor consists of a stylus probe for measurement of the micro-aspheric specimen and two capacitive sensors for compensation of the scanning error motions [19].

Figure 7.31 shows a picture of the scanning stages. The air-bearing spindle is

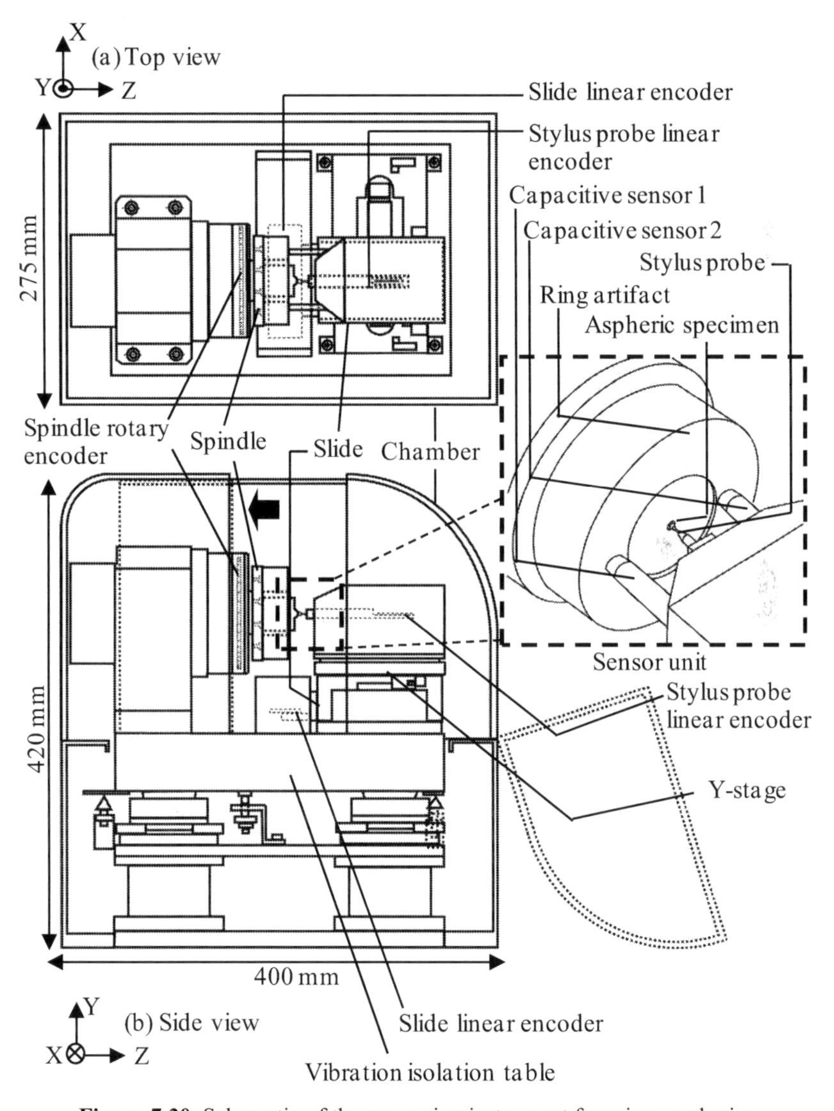

Figure 7.30. Schematic of the measuring instrument for micro-aspherics

equipped with a rotary encoder with an angular resolution of 0.0038 arcsec. The air-bearing slide, which has a moving stroke of 70 mm, has a linear encoder with a resolution of 0.28 nm. The scale of the slide linear encoder is set right below the micro-aspheric specimen to avoid Abbe errors. The main parts of the spindle and the slide are made of ceramics for high stiffness. The scanning stages are mounted on a vibration isolation table and the entire instrument is covered by a chamber to reduce the influence of vibrations and temperature variations.

Figure 7.32 schematizes the data flow of the measuring instrument. Commands for stage positions are sent from a PC to the stage controller. The positions of the spindle and the slide from the encoders are input to the controller as feedback signals for controlling the stages. The output from the linear encoder of the stylus probe sensor, which corresponds to the height information of the specimen surface, is also input to the PC as 32-bit parallel data. The analog outputs of the capacitive sensors are input to the PC through data logger. Figure 7.33 shows a close-up view of the sensor unit, the micro-aspheric specimen and the ring artifact. Figure 7.34 shows the stability data of the sensors. It can be seen that the thermal drift in each sensor is reduced to approximately 10 nm in the differential output. A measurement result of surface profile of the micro-aspheric specimen is shown in Figure 7.35. Figure 7.36 shows an overview of the instrument.

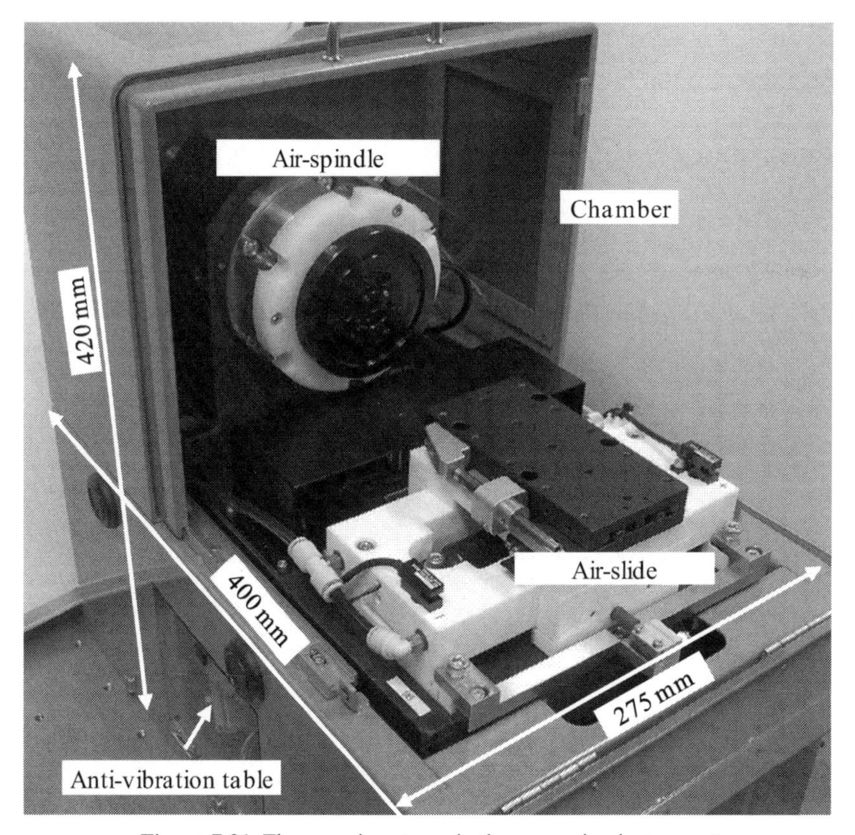

Figure 7.31. The scanning stages in the measuring instrument

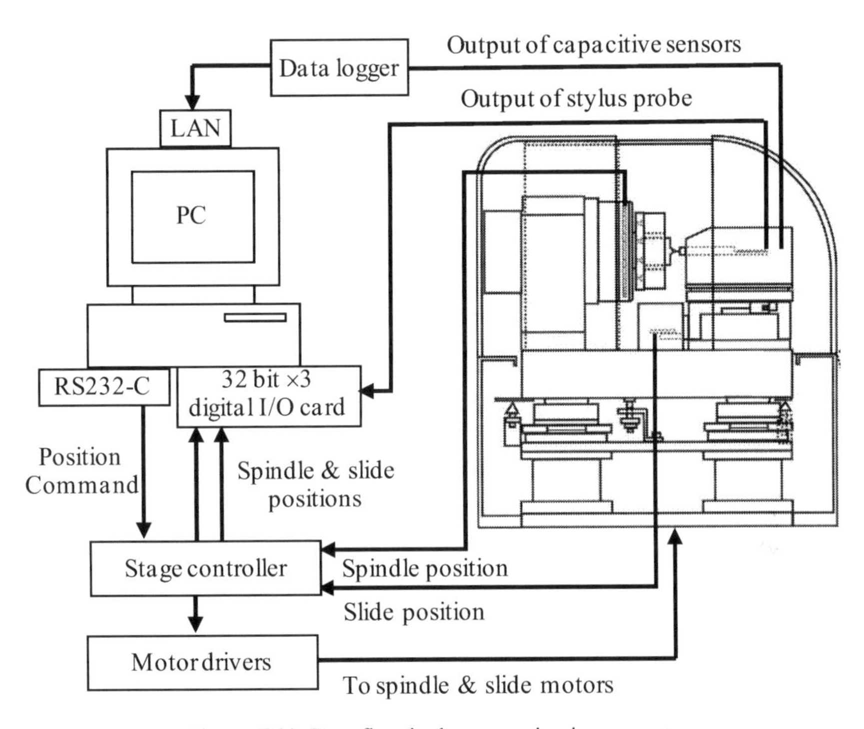

Figure 7.32. Data flow in the measuring instrument

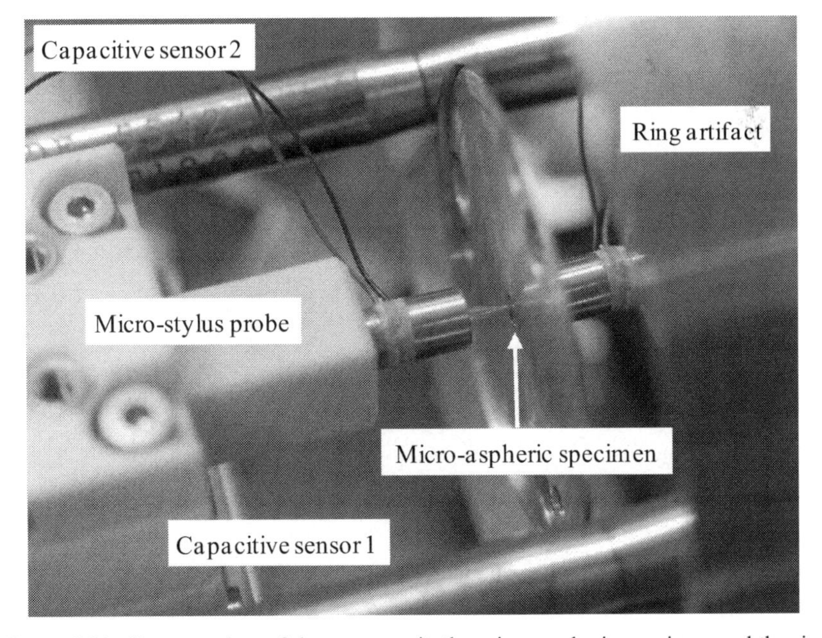

Figure 7.33. Close-up view of the sensor unit, the micro-aspheric specimen and the ring artifact in the measuring instrument

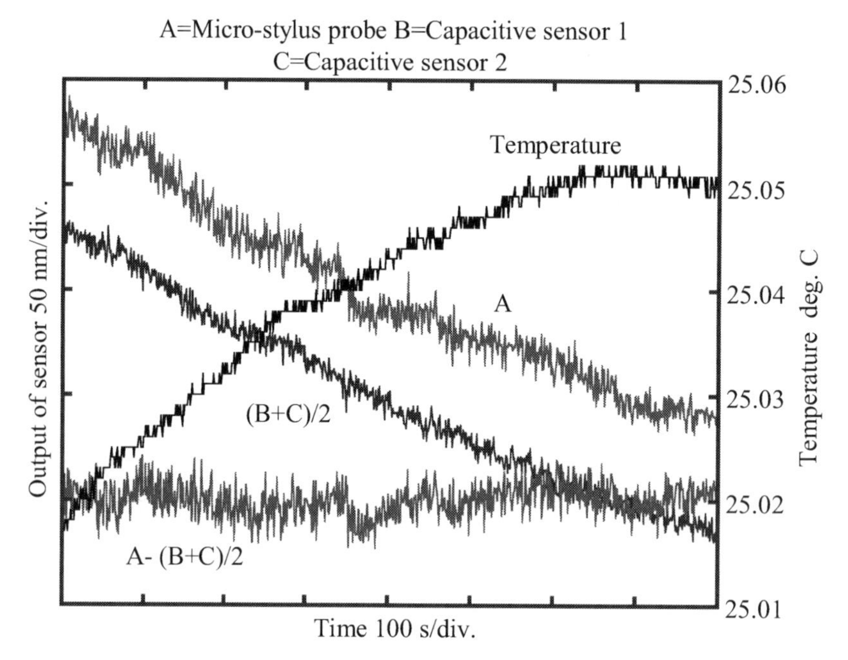

Figure 7.34. Reduction of thermal drift

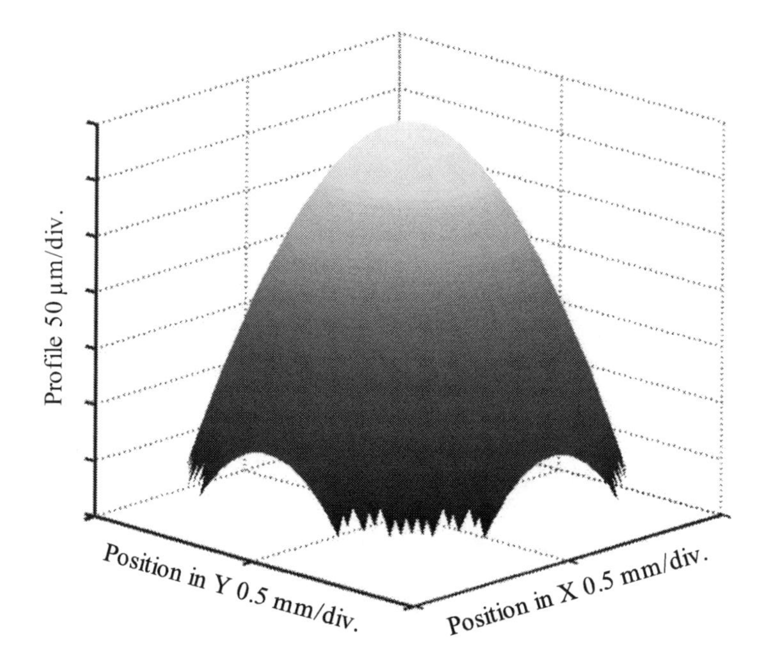

Figure 7.35. Measured surface profile of a micro-aspheric specimen

(a) Inside view

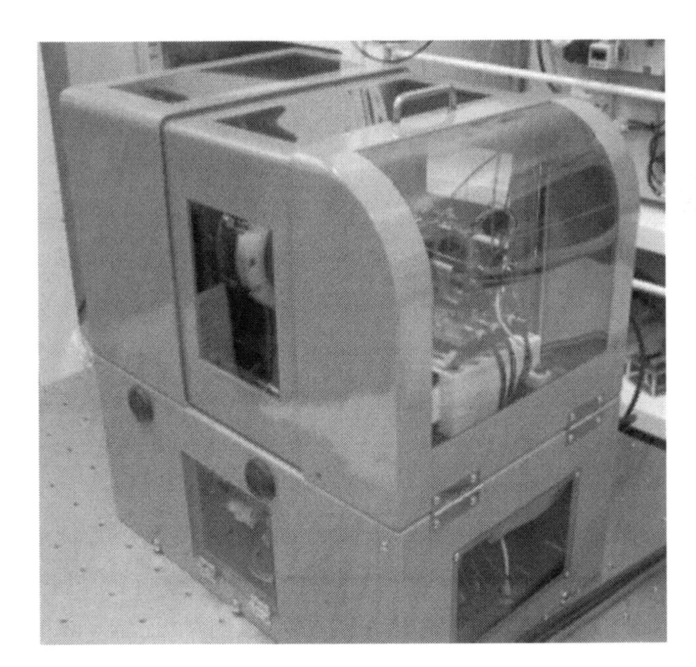

(b) Outside view

Figure 7.36. Overview of the measuring instrument for micro-aspherics

7.5 Summary

A scanning micro-stylus system has been constructed for accurate surface profile measurement of micro-aspherics. The scanning motion is generated by an air-bearing spindle and an air-bearing linear slide. The micro-aspheric specimen is mounted on the center of the spindle and a micro-stylus probe is mounted on the slide. A unit combining a ring artifact and two capacitive sensors is employed for separation of the scanning error motions. The ring artifact is mounted around the aspheric sample and the capacitive sensors are mounted on the two sides of the contact-stylus. Because the axial error motion component and the tilt error component are sensed by the three sensors simultaneously, a differential operation of the sensor outputs can remove the error motion components from the output of the micro-stylus probe and result in an accurate surface profile of the micro-aspheric specimen.

A glass tube micro-ball stylus with a tip ball whose diameter is less than 100 μm has been developed. The stylus is composed of a glass shaft and a glass micro-ball. The glass shaft is sharpened by heating and stretching a glass tube. The glass micro-ball is glued on the end of the glass shaft. From the above, it has been verified that the micro-ball stylus can tolerate a large tangential force. A PZT actuator is integrated into the glass tube micro-ball stylus to construct a tapping mode micro-ball stylus, which can reduce the influence of the stick-slip phenomenon that occurs during scanning the stylus over the specimen surface.

References

[1] Eric JS, Mark F, Janet CB, Quinn YJS, Chris MB (2003) Microfabricated optical fiber with microlens that produces large field of-view, video rate, optical beam scanning for microendoscopy applications. Proc SPIE 4957:46–55

[2] Fankhauser F, Kwasniewska S (2004) Cyclodestructive procedures. II Optical fibers, endoscopy, physics: a review. Ophthalmologica 218(3):147–161

[3] Shan XC, Maeda R, Murakoshi Y (2003) Microhot embossing for replication of microstructures. Jpn J Appl Phys Part 1 42(6B):3859–3862

[4] Firestone GC, Yi AY (2005) Precision compression molding of glass microlenses and microlens arrays: an experimental study. Appl Opt 44(29):6115–6122

[5] Venkatesh VC (2003) Precision manufacture of spherical and aspheric surfaces on plastics, glass, silicon and germanium. Curr Sci 84(9):1211–1229

[6] Chen WK, Kuriyagawa T, Huang H, Yosihara N (2005) Machining of micro aspherical mould inserts. Precis Eng 29(3):315–323

[7] Milton G, Gharbia Y, Katupitiya J (2005) Mechanical fabrication of precision microlenses on optical fiber endfaces. Opt Eng 44(12):123402

[8] Lee, HJ, Kim SW (1999) Precision profile measurement of aspheric surfaces by improved Ronchi test. Opt Eng 38(6):1041–1047

[9] Hill M, Jung M, McBride JW (2002) Separation of form from orientation in 3D measurements of aspheric surfaces with no datum. Int J Mach Tools Manuf 42(4):457–466

[10] Tomlinson R, Coupland JM, Petzing J (2003) Synthetic aperture interferometry: in-process measurement of aspheric optics. Appl Opt 42(4):701–707

[11] Greivenkamp JE, Gappinger R (2004) Design of a non-null interferometer for aspheric wave fronts. Appl Opt 43(27):5143–5151

[12] Arai Y, Gao W, Shimizu H, Kiyono S, Kuriyagawa T (2004) On-machine measurement of aspherical surface profile. Nanotechnol Precis Eng 2(3):210–216

[13] Lee CO, Park K, Park BC, Lee YW (2005) An algorithm for stylus instruments to measure aspheric surfaces. Meas Sci Technol 16(5):215–1222

[14] Evans CJ, Hocken RJ, Estler WT (1996) Self-calibration: reversal, redundancy, error separation, and "absolute testing". Ann CIRP 45(2):617–634

[15] Zygo Corporation (2010) http://www.zygo.com. Accessed 1 Jan 2010

[16] Moritex Corporatoin (2010) http://www.moritex.co.jp. Accessed 1 Jan 2010

[17] Takaya Y, Imai K, Dejima S, Miyoshi T, Ikawa N (2005) Nano-position sensing using optically motion-controlled microprobe with PSD based on laser trapping technique. Ann CIRP 54(1):467–470

[18] Nagaike Y, Nakamura Y, Ito Y, Gao W, Kuriyagawa T (2006) Ultra-precision on-machine measurement system for aspheric optical elements. J CSME 27(5):535–540

[19] Gao W, Shibuya A, Yoshikawa Y, Kiyono S, Park CH (2006) Separation of scanning error motions for surface profile measurement of aspheric micro lens. Int J Manuf Res 1:267–282

8

Large Area Scanning Probe Microscope for Measurement of Micro-textured Surfaces

8.1 Introduction

Large area three-dimensional (3D) micro-structured surfaces can be found in holograms, diffractive optical elements (DOEs) and anti-reflective films, etc. [1]. A large number of the surfaces are composed of periodical micro-structures with a small structure width (in the *X*- and *Y*-directions) from several microns to several tens microns. Most of the surfaces are required to be fabricated accurately over an area lager than several millimeters.

The small structure width makes it difficult to use optical microscopes (OMs) including interference microscopes [2]. The scanning electron microscopes (SEMs) also cannot provide accurate three-dimensional profile data because the SEM image is basically a 2D projection of a 3D surface. The contact-type surface roughness measuring instrument cannot measure the micro-structures because the tip radius of probe is typically too large. Compared with the OM, SEM and surface roughness measuring instruments, scanning probe microscopes (SPM) such as atomic force microscopes (AFM) [3], have higher potential for precision nanometrology of 3D micro-structured surfaces [4]. However, conventional SPMs are qualitative imaging oriented [5, 6]. The metrological SPMs, which are being built in national standard institutes for establishment of traceability, are too expensive and complicated for practical use [7, 8]. Most of the SPMs also do not have a large enough measurement range in the *XYZ*-directions. Since the sample is typically moved by PZT scanners or scanning stages in a raster scanning pattern, many scans are needed to measure the large area micro-structured surface. As a result, the measurement is time-consuming.

This chapter describes measuring systems based on AFM to measure large area micro-structured surfaces accurately and rapidly. Methods using the capacitive displacement sensor and the linear encoder for measurement and compensation of the *Z*-directional error of the AFM probe unit are presented in this chapter, respectively. A spiral scanning system for fast *XY*-scanning is also discussed.

8.2 Capacitive Sensor-compensated Scanning Probe Microscope

Figure 8.1 shows the schematic of a typical scanning system for the AFM. The AFM image is obtained by the *X*-, *Y*- and *Z*-directional scans of the AFM cantilever probe. The *XYZ*-coordinates of the AFM probe tip, which indicates those of the measurement point on the target surface, are provided by positions of the scanning actuators/stages. Accurate position measurement of the scanning actuators/stages, which can be accomplished by using displacement sensors, is thus important for accurate AFM imaging.

In addition to this, error motions of scanning actuators/stages including the tilt

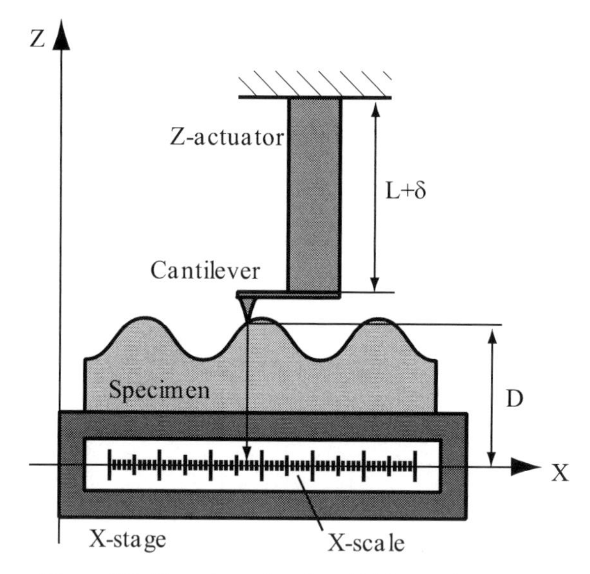

Figure 8.1. The schematic of an AFM scanning system

error motions are also important factors influencing the accuracy of the AFM with large imaging areas. Figure 8.2 shows errors of the AFM caused by the tilt error motions of the *Z*-actuator and *X*-stage. The tilt error motion ($\Delta\theta_{PZT}$) of the *Z*-actuator (typically a PZT actuator) will generate an error component ΔX_{PZT} in the measured *X*-coordinate, which can be expressed as:

$$\Delta X_{PZT} = (L + \delta)\sin\Delta\theta_{PZT} \ , \tag{8.1}$$

where *L* is the length of the *Z*-actuator and δ is the displacement of the *Z*-actuator.

The pitching error ($\Delta\theta_{stage}$) of the *X*-stage will also generate an error component ΔX_{stage} (Abbe error) in the *X*-coordinate as:

$$\Delta X_{stage} = D\tan\Delta\theta_{stage} \ , \tag{8.2}$$

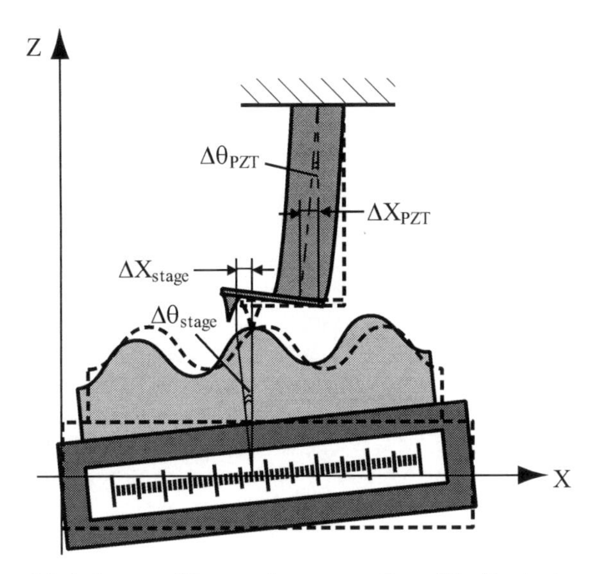

Figure 8.2. Influence of the angular error motion of the *Z*-actuator

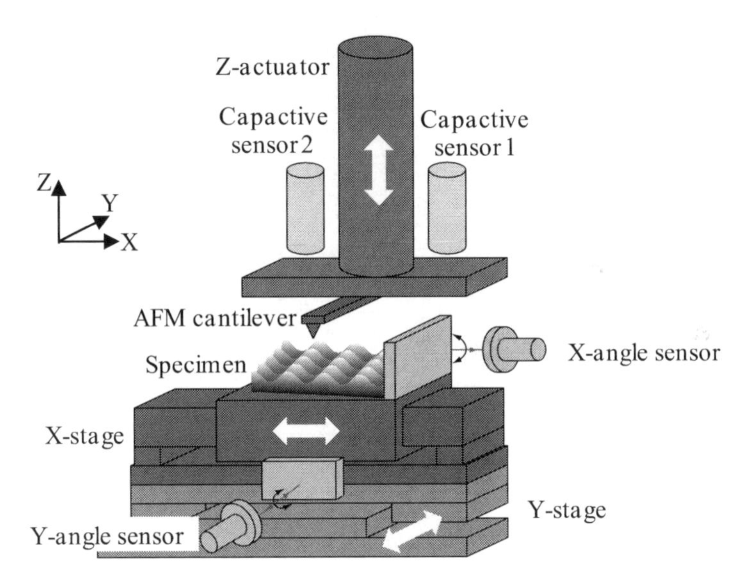

Figure 8.3. Schematic of the large area AFM with capacitive sensor compensation

where D is the Abbe offset, which is the distance between the scale of the stage and the probe tip of the AFM cantilever. It should be noted that although only errors in the X-coordinates are shown for simplicity, the same errors will occur in the Y-coordinates when scanning along the Y-direction. Since L and D are large in the case of large area measurement, it is necessary to compensate for the tilt error motions.

Figure 8.3 shows a schematic of the AFM designed for the measurement of large area surfaces through compensation not only for the displacement error of the

Z-actuator but also for the tilt error motions of scanning. A PZT actuator with a stroke of 100 μm is used as the Z-actuator for long stroke measurement in the Z-direction. In order to measure the displacement as well as the tilt motion of the PZT actuator, two capacitive sensors are aligned at two sides of the PZT actuator along the Z-direction. The PZT actuator is mounted on a manual stage (Z-stage), which is used as a coarse adjustment mechanism for the AFM tip position relative to the sample surface in the Z-direction. The specimen is moved by a linear motor-driven stage with an aerostatic bearing and a stepping motor-driven stage with a sliding contact bearing for scanning in the X- and Y-directions over an area of 50 mm × 40 mm. The tilt motions of the stages can be measured by using angle sensors described in Chapters 1 and 2 for compensation. A piezo-resistive cantilever [9, 10] is employed instead of the conventional optical force sensing device. This makes the AFM structure simple and compact. The cantilever is attached at the end of the PZT actuator. A photograph of the AFM probe unit is shown in Figure 8.4 and the overall system is shown in Figure 8.5.

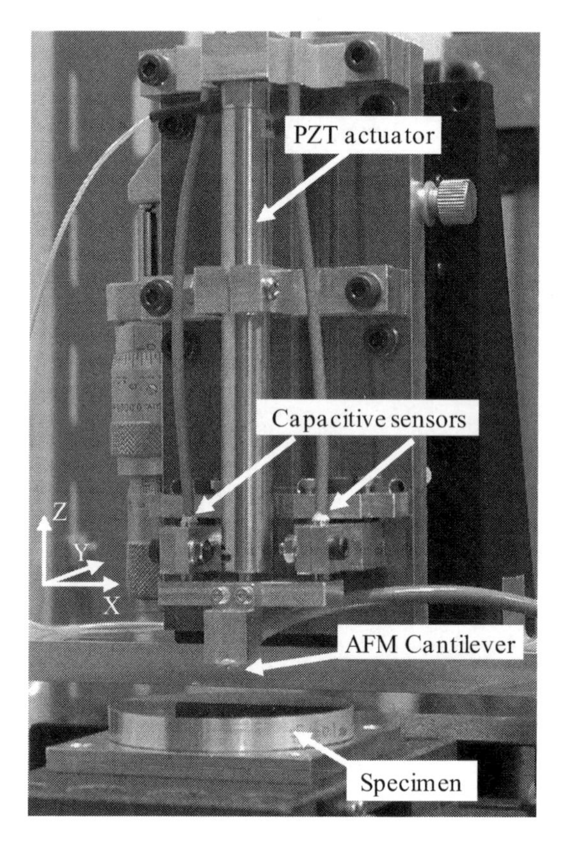

Figure 8.4. Photograph of the large area AFM with compensation of capacitive sensors

Figure 8.5. Photograph of the overall system

Figure 8.6 shows the block diagram of the electronics of the AFM system. The linear motor-driven air-bearing stage (*X*-stage) is controlled by a PC via RS-232C. The scanning position of *X*-stage is obtained by a linear encoder and recorded by the PC via a digital input (DI). The resolution of the linear encoder is approximately 0.28 nm. The scanning range of the *X*-stage is 50 mm. The stepping motor-driven stage (*Y*-stage) is driven by the PC via a digital-to-analog (DA) converter. The stage can be moved with a step of 10 nm by using a micro-step driver. The variation in the resistance of the cantilever corresponding to the bending of the cantilever, which is caused by the variation of the surface profile of the sample, is detected by a signal conditioner through a Wheatstone bridge. The difference between the signal conditioner output and the command value from the voltage reference, which is taken by a comparator, is integrated by a PI controller. The output of the PI controller is sent to the amplifier of the PZT actuator for feedback control of the actuator till the difference becomes zero, so that the gap between the AFM cantilever tip and the sample surface is kept constant. The surface profile height in the *Z*-direction is obtained from the mean of the two capacitive sensors outputs recorded by the PC via an analog-to-digital (AD) converter.

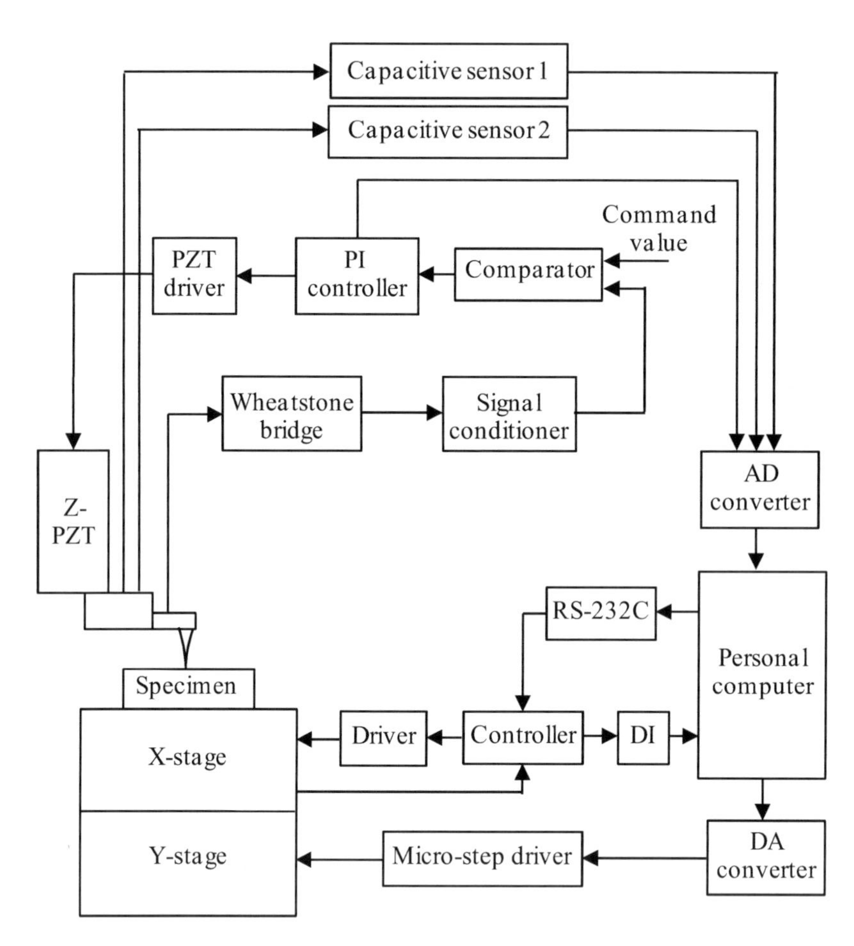

Figure 8.6. Block diagram of the electronics of the AFM system

Figure 8.7 shows photographs of the AFM cantilever with two piezo-resistive sensors. One sensor is for measurement of the deflection of the cantilever and the other is for the temperature compensation. The resistances of the sensors are assumed to be R_M and R_E, respectively. The initial resistances of R_M and R_E are the same (R_0). R_M changes in response to both the deflection of the AFM cantilever and the change of temperature. R_E only changes in response to the change of temperature. Let the change of R_M caused by the deflection of the AFM cantilever be ΔR_M. ΔR_M can be converted into the change of voltage by the Wheatstone bridge circuit shown in Figure 8.8 without the influence of temperature change. Assuming the voltage output of the bridge is Δe and the voltage of the electrical power supply is e_0, the relationship between Δe and ΔR_M can be expressed by:

$$\Delta e = \frac{\Delta R_M}{R_0}\frac{e_0}{4}. \tag{8.3}$$

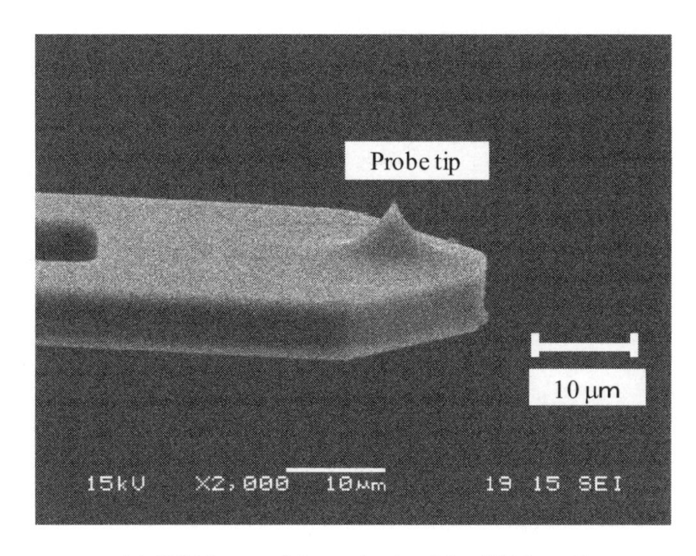

(a) SEM image of the probe tip of the AFM cantilever

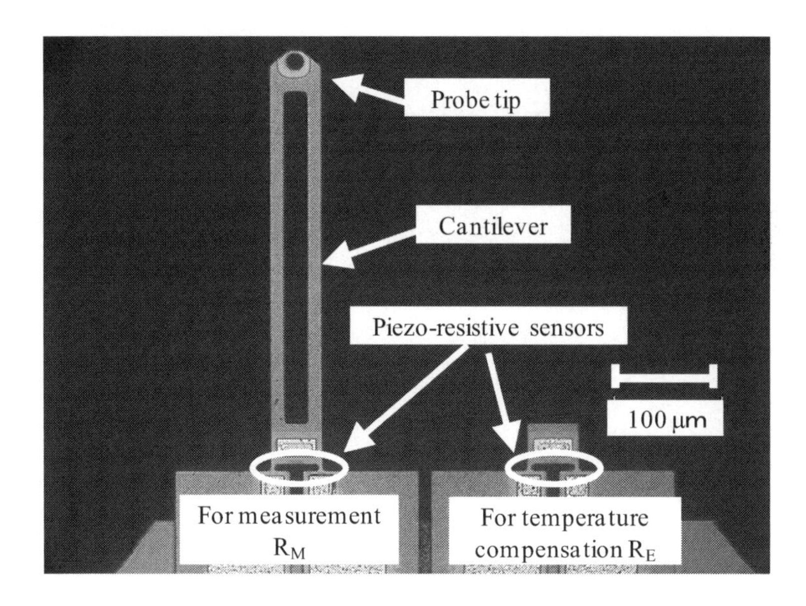

(b) Photograph of the AFM cantilever with piezo-resistive sensors

Figure 8.7. The AFM cantilever and probe tip

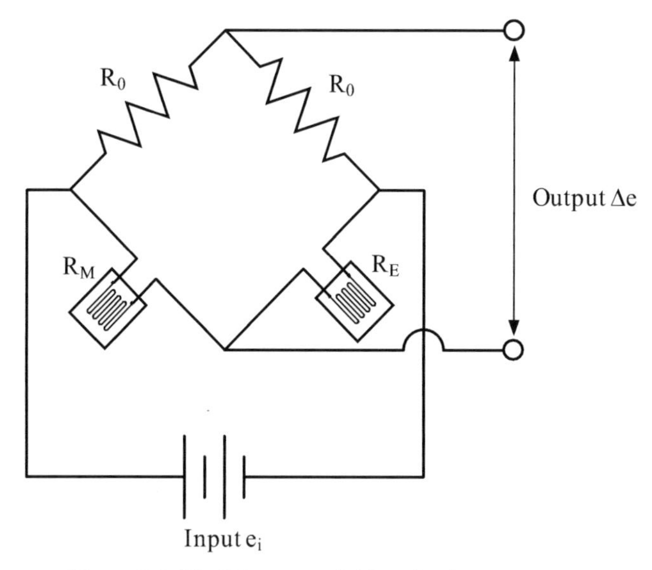

Figure 8.8. The Wheatstone bridge circuit

On the other hand, the relationship between ΔR_M and the deflection Δz of the cantilever can be written as:

$$\Delta z = \frac{\Delta R_M}{R_0} \cdot \frac{1}{\frac{3\Pi E t}{2h^2}}, \tag{8.4}$$

where Π (= 6.6×10^{-11} m^2/N) is the piezoelectric coefficient of the piezo-resistive sensor. E (= 1.9×10^{11} N/m^2), t (= 4 μm) and h (= 400 μm) are the Young's modulus, thickness and length of the AFM cantilever.

Let the stiffness of the AFM cantilever be k (= 3 N/m). The contact force F of the AFM probe tip and the measured surface can thus be expressed by:

$$F = k\Delta z = k\frac{\Delta R_M}{R_0} \cdot \frac{1}{\frac{3\Pi E t}{2h^2}}. \tag{8.5}$$

Figure 8.9 shows the measured relationship between the Z-directional displacement of the PZT actuator and the output of the piezo-resistive force sensor, which is called the force curve. As can be seen in the figure, the AFM cantilever was deflected when the probe tip approached the measured surface, resulting in a change of the repulsive force sensor. The working point of the cantilever was selected in the repulsive area, which was more stable than the attractive area. The contact force was chosen to be less than 0.5 μN in the experiment of surface profile measurement.

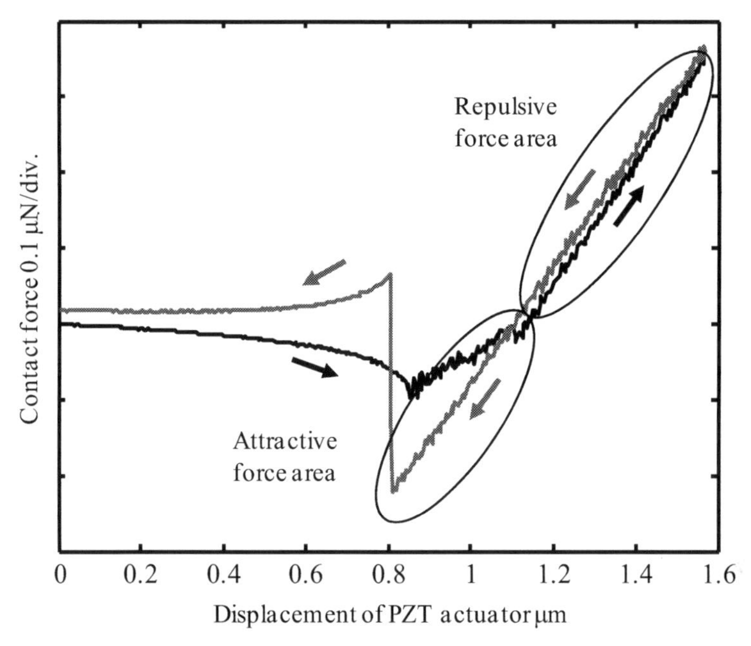

Figure 8.9. The force curve

Figure 8.10 shows the *Z*-directional displacement of the PZT actuator measured by the two capacitive sensors. The voltage applied to the PZT actuator was changed from −10 V to 125 V with a step of 0.01 V. The difference between the outputs of the two capacitive sensors is also shown in the figure. The maximum difference was approximately 5 μm with a hysteresis of approximately 1 μm, which corresponded to a maximum tilt error of 16 arcsec with a hysteresis of 5 arcsec. The error ΔX_{PZT} defined in Figure 8.2 is shown in Figure 8.11. The tilt error will cause a maximum error of approximately 8 μm in the *X*-direction, which should be compensated for accurate measurement of the surface profile.

The measurement result of a sinusoidal surface described in Chapter 3 is shown in Figure 8.12. The amplitudes and pitches in both the *X*- and *Y*-directions were 0.1 and 150 μm, respectively. The scanning area was 1000 μm (*X*) × 1000 μm (*Y*). The sampling numbers were 200 and the sampling intervals were 5 μm in both the *X*- and *Y*-directions. The scanning time was approximately 2 s for each of the *X*-lines and approximately 20 min for the entire imaging. Figure 8.12 (a) shows the result obtained by simply converting the applied voltage of the PZT actuator to the displacement based on the mean voltage-displacement sensitivity of the PZT actuator obtained from Figure 8.10. Figure 8.12 (b) shows the result evaluated from the mean of the outputs of the two capacitive sensors. Pitches of the sinusoidal structure were 150 μm in both Figures 8.12 (a) and 8.12 (b), which were the same as the designed value. On the other hand, however, the amplitude in Figure 8.12 (a) was approximately 0.3 μm, which was approximately three times larger than the designed value. In contrast, the result in Figure 8.12 (b) showed a good correspondence with the designed amplitude.

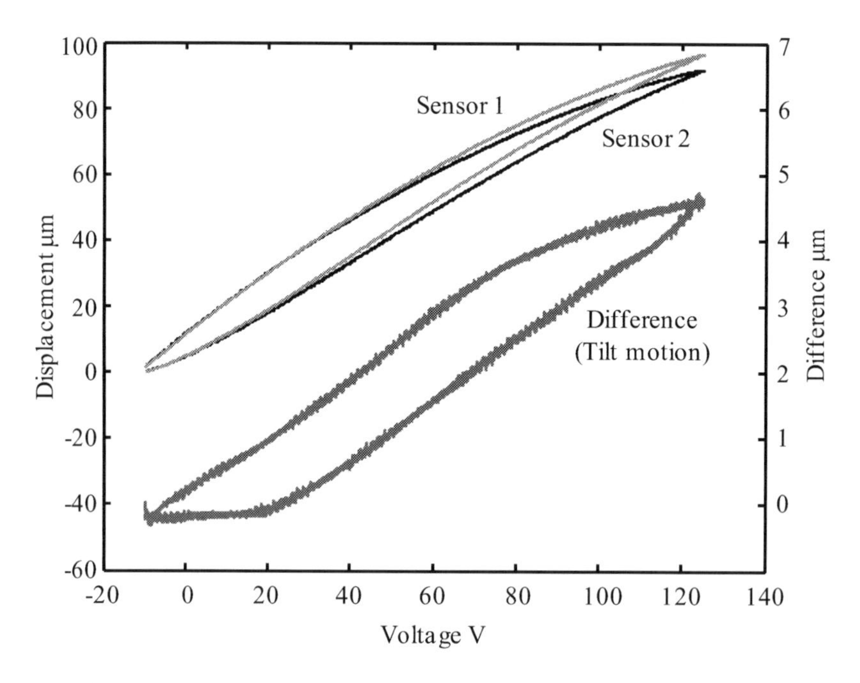

Figure 8.10. Displacement of the PZT actuator and difference between the outputs of the two capacitive sensors

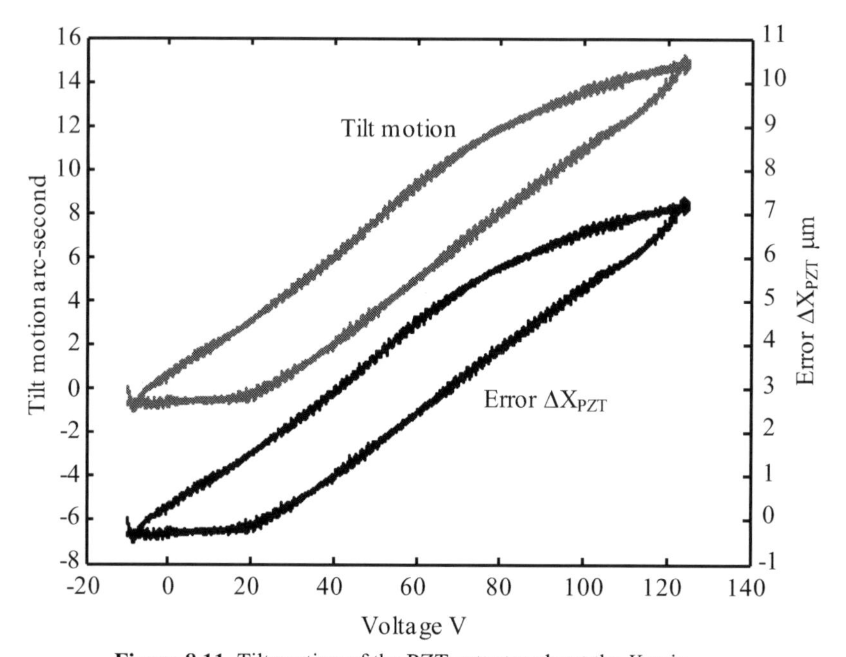

Figure 8.11. Tilt motion of the PZT actuator about the *Y*-axis

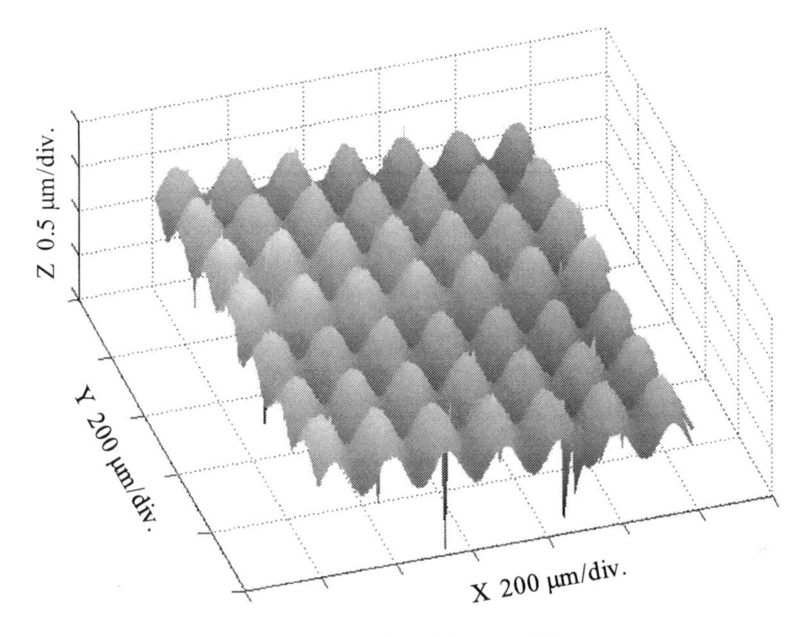

(a) Without compensation of the capacitive sensors

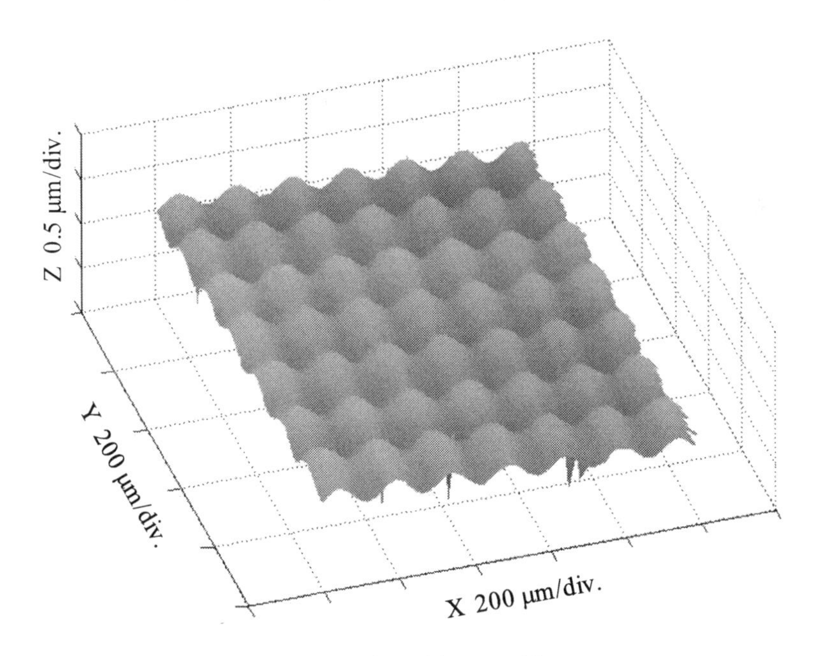

(b) With compensation of the capacitive sensors

Figure 8.12. Measurement results of a micro-structured surface by using the AFM

Figure 8.13 shows the result by an interference microscope [11], which is consistent to the result in Figure 8.12 (b). Figure 8.14 shows the result over a large area of 10 mm (*X*) × 10 mm (*Y*) [12].

The most serious problems inherent in the capacitive sensors are the influence from the electromagnetic noises and the limitation in the dynamic range. Figure 8.15 (a) shows the noise levels of the capacitive sensor using the internal power supply embedded in the amplifier of the sensor with a cut-off frequency of 1 kHz. The noise level was approximately 30 mV (corresponding to 150 nm). The noise was reduced to 10 mV (50 nm) when using a high-performance external power supply (Figure 8.15 (b)). Although a noise level of 2 mV (10 nm) was achievable when the cut-off frequency was set to be 100 Hz, this will reduce the measurement speed of the AFM. In the following section, a linear encoder with larger dynamic range and stronger robustness is employed instead of the capacitive sensor [13].

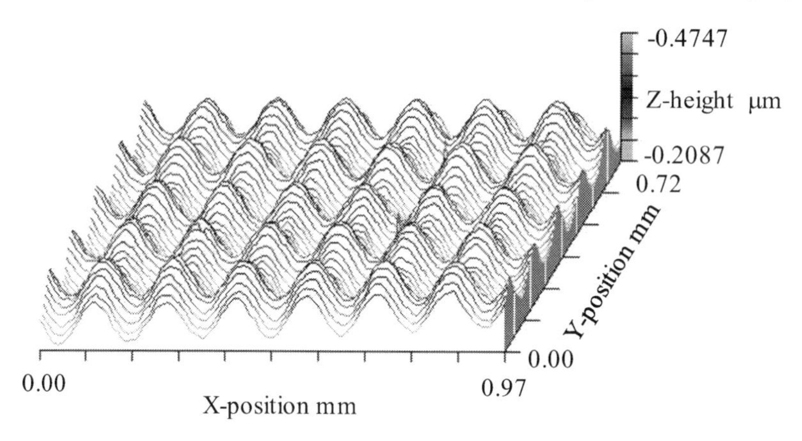

Figure 8.13. Measurement result by using a white light interference microscope

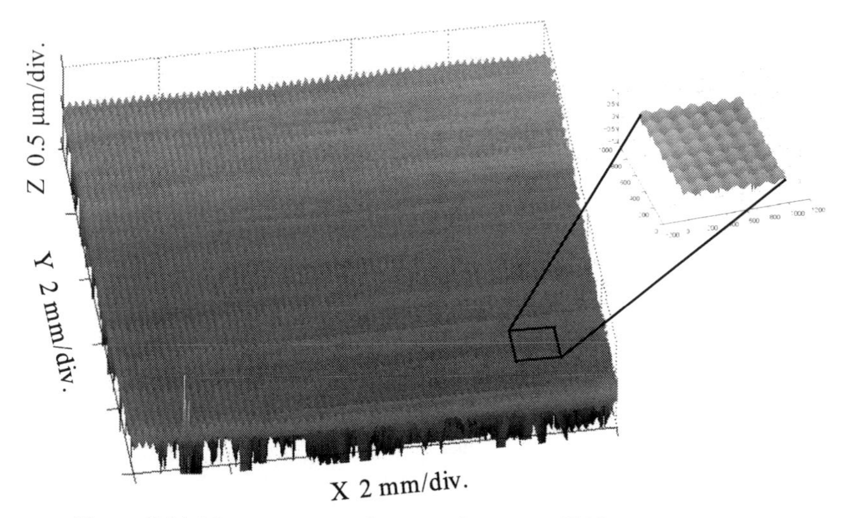

Figure 8.14. Measurement results over a large area (*X* 10 mm × *Y* 10 mm)

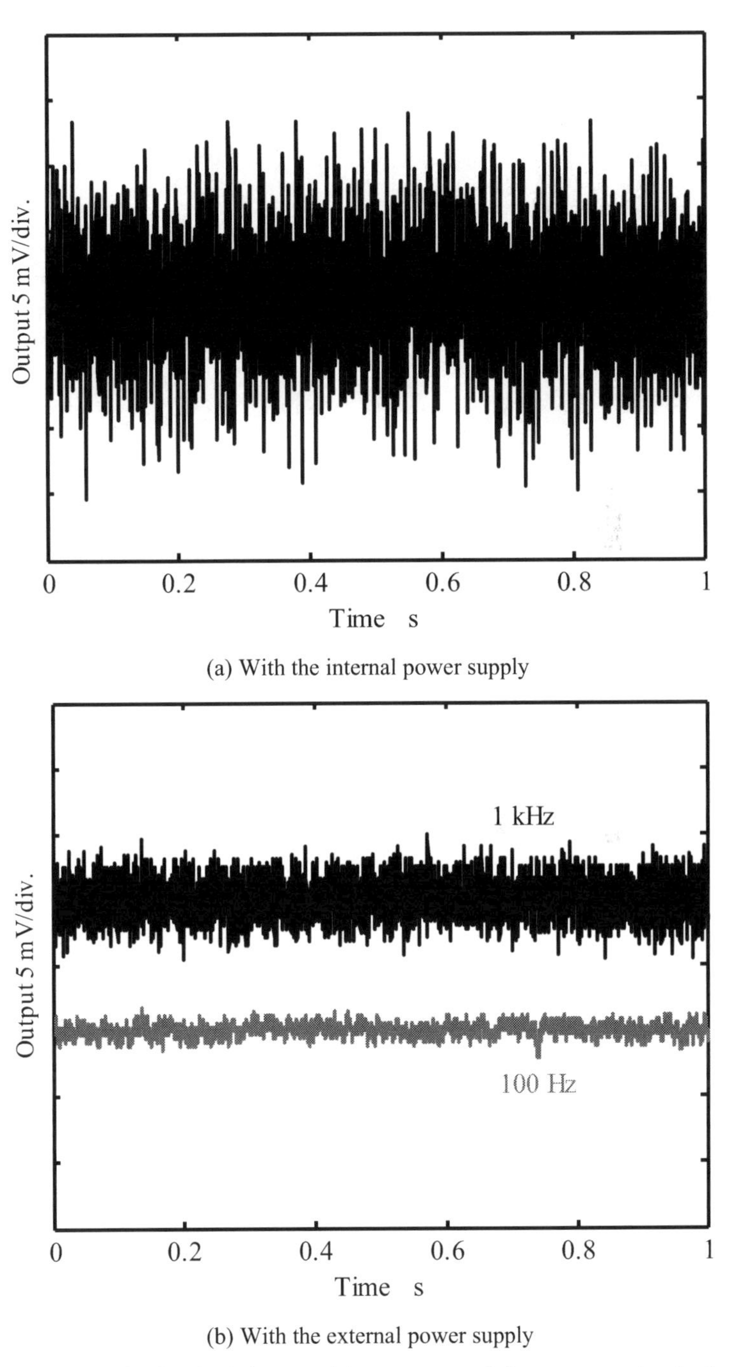

(a) With the internal power supply

(b) With the external power supply

Figure 8.15. Noise level of the capacitive sensor used for compensation in the AFM probe unit

8.3 Linear Encoder-compensated Scanning Probe Microscope

Figures 8.16 and 8.17 show the schematic and a photograph of the AFM probe unit in which a linear encoder is employed for compensation of the error from the PZT actuator. The probe unit consists of a silicon cantilever with a piezo-resistive sensor for probing the surface, a PZT actuator for servo controlling the cantilever and a linear encoder for measuring the displacement of the PZT actuator. The silicon cantilever and the encoder scale are mounted on the moving end of the PZT actuator. The PZT actuator and the read head of the linear encoder are fixed on the base of the AFM probe unit. When the probe unit is scanned over the surface for the profile measurement, the *Z*-directional position of the cantilever is servo controlled by the PZT actuator in such a way that the output of the piezo-resistive sensor is kept constant. The surface profile height can thus be obtained from the output of the linear encoder used to measure the displacement of the PZT actuator.

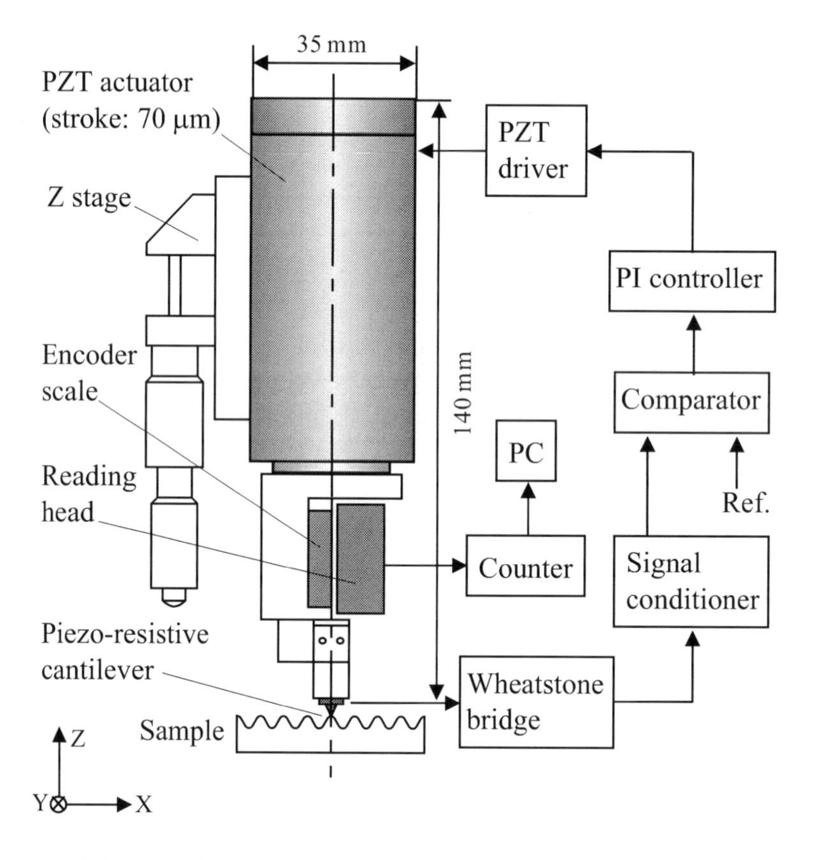

Figure 8.16. Schematic of the AFM probe unit with a linear encoder for error compensation

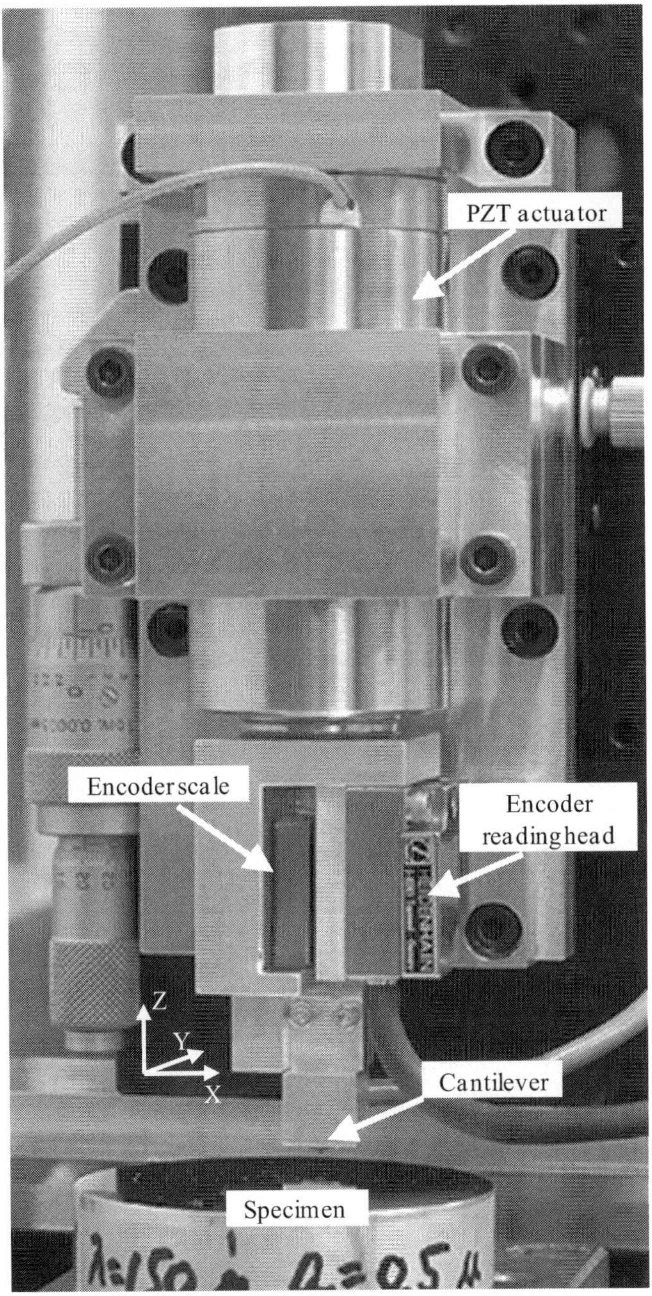

Figure 8.17. Photograph of the AFM probe unit with a linear encoder for error compensation

The maximum stroke of the PZT actuator is 70 μm. The size, signal period and measurement range of the linear encoder are 20 mm (*X*) × 15 mm (*Y*) × 20 mm (*Z*), 2 μm and 10 mm, respectively. The encoder scale is aligned along the axis of motion of the actuator so that the Abbe error can be avoided. The AFM probe unit is made compactly, which is approximately 35 mm (*X*) × 35 mm (*Y*) × 150 mm (*Z*).

Figure 8.18 shows the PZT actuator displacement measured by the linear encoder built in the AFM probe unit. The voltage applied to the PZT actuator was changed from −10 V to 150 V. It can be seen from the figure that the PZT actuator had a displacement of approximately 53 μm with an applied voltage of 150 V. The maximum non-linearity was approximately 6 μm, which can be compensated for with the linear encoder output. Figure 8.19 shows the PZT actuator displacement measured by the linear encoder in nanometer range. The voltage applied to the PZT actuator was changed with a step of 6 mV. It can be seen that the linear encoder was able to detect the step of approximately 1 nm. The dynamic range of the AFM probe unit was calculated to be approximately 94 dB.

Frequency response of the AFM probe unit is another important characteristic for measurement of large area micro-structured surfaces. The bandwidth of the PZT actuator was 15 kHz and that of the linear encoder was 3.3 kHz. Figure 8.20 shows the open-loop frequency response of the AFM probe unit. It can be seen that the AFM probe unit could work stably under a frequency of 1.8 kHz. Figure 8.21 shows the closed-loop stability of the AFM probe unit. The output of the linear encoder was sampled with an interval of 0.05 s during a term of 10 min. It can be seen that the stability was approximately 20 nm for a temperature change of approximately 0.2°C. A correlation between the AFM instability and the temperature change can be observed.

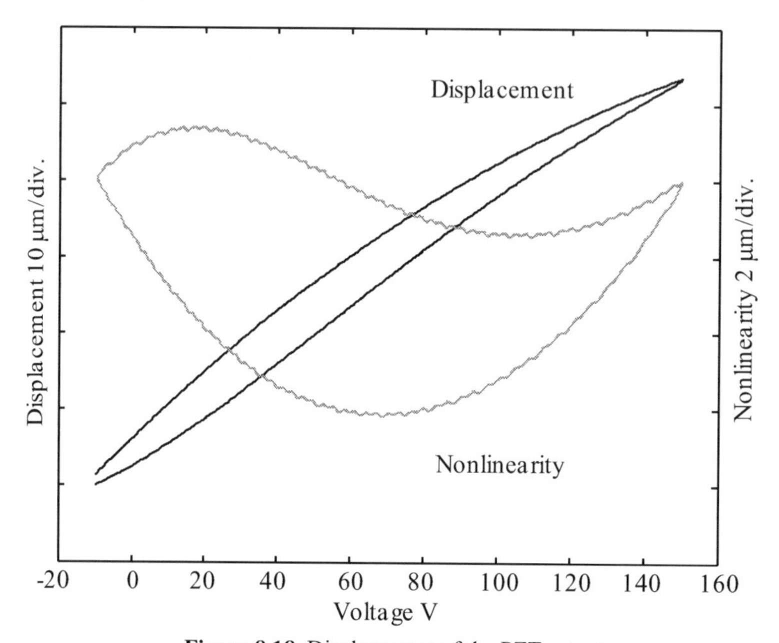

Figure 8.18. Displacement of the PZT actuator

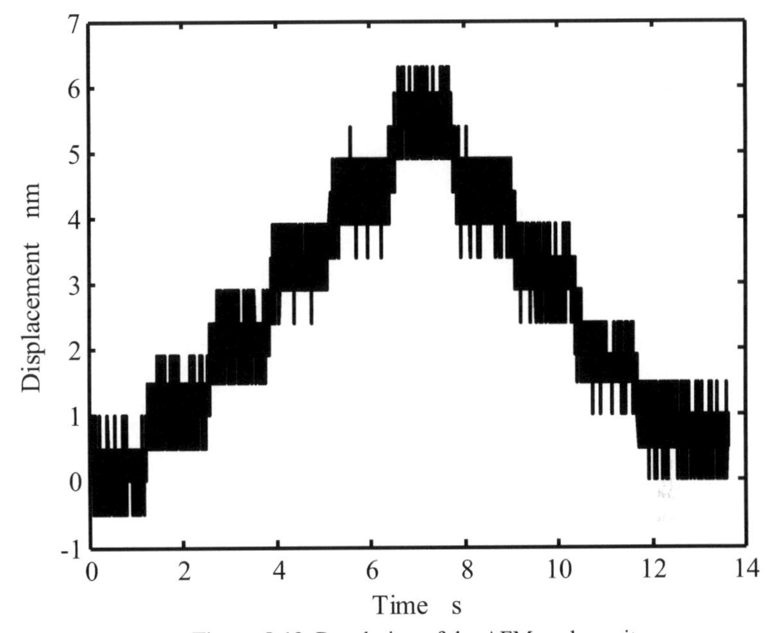

Figure 8.19. Resolution of the AFM probe-unit

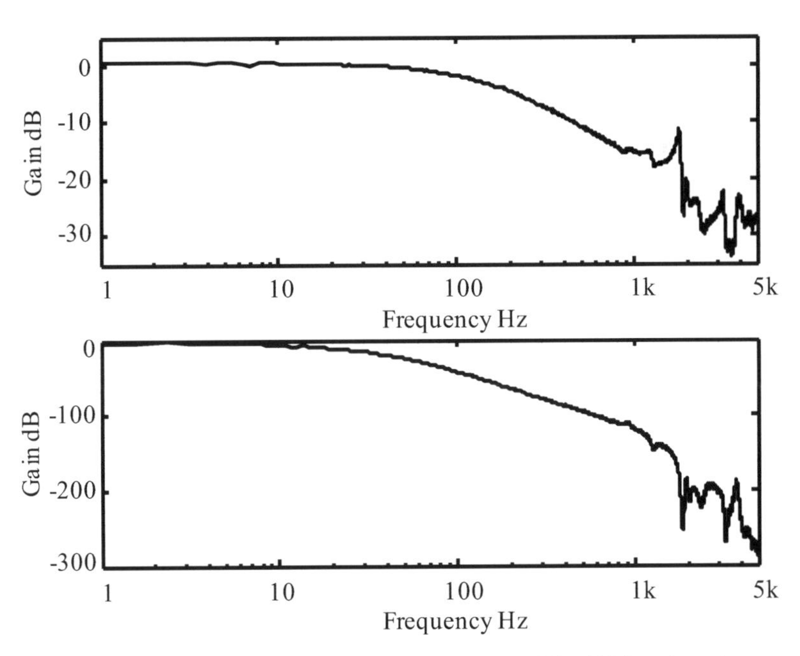

Figure 8.20. Open-loop frequency response of the AFM probe unit

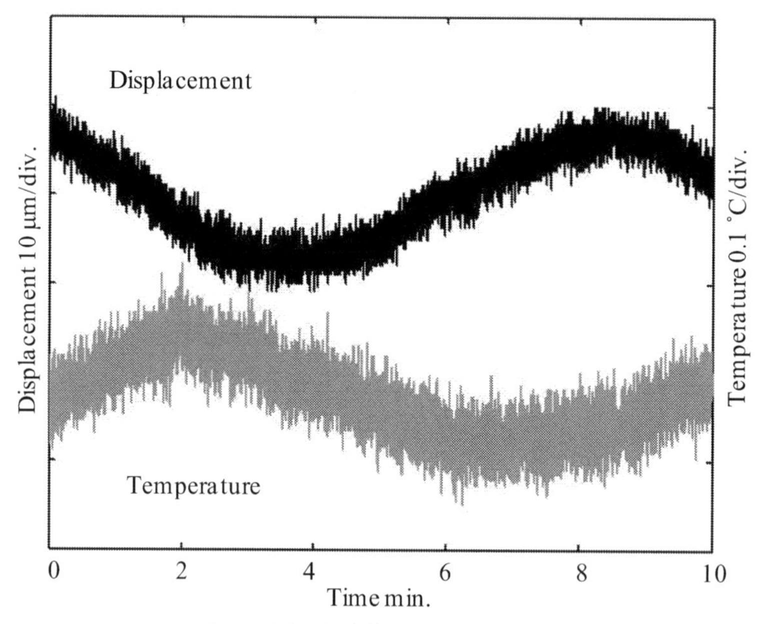

Figure 8.21. Stability of the AFM probe unit

Figure 8.22 shows a photograph of the measurement system composed of the AFM probe unit and *XY*-scanning stages. The AFM probe unit was kept stationary and the specimen was moved by the scanning stages in a raster scanning pattern. The *XY*-scanning stages were the same as that shown in Figure 8.5. The *X*-stage was a linear motor-driven stage with an air bearing for scanning the specimen at a constant speed. The *X*-stage had a positioning resolution of approximately 0.28 nm based on the feedback control with a linear encoder (*X*-encoder). The stepping motor-driven stage with a leadscrew was used to move the specimen in the *Y*-direction step-by-step. The movement of the *Y*-stage was measured by a liner encoder (*Y*-encoder) with a resolution of 1 nm. Figure 8.23 shows the *Y*-stage displacement measured by the *Y*-encoder when the *Y*-stage was moved from 0 to 10 mm. The *Y*-stage had a periodic positioning error corresponding to the lead of the leadscrew. The maximum positioning error was approximately 3.5 μm, which could be compensated by using the *Y*-encoder.

The *Z*-directional error motion of the *XY*-scanning stages was measured by scanning a capacitive sensor with respect to a reference mirror. As can be seen in Figures 8.24 and 8.25, the out-of-flatness of the reference mirror measured by an interferometer and the *Z*-directional error motion were 14 and 150 nm, respectively. The error motion, which directly influences the measurement result of the AFM, was compensated based on the result shown in Figure 8.25. Figures 8.26 and 8.27 show the AFM images of the same sinusoidal surface measured in the last section. The result shown in Figure 8.26 was obtained by simply converting the applied voltage of the PZT actuator to displacement based on the mean voltage-displacement sensitivity of the PZT actuator, just as with conventional AFMs.

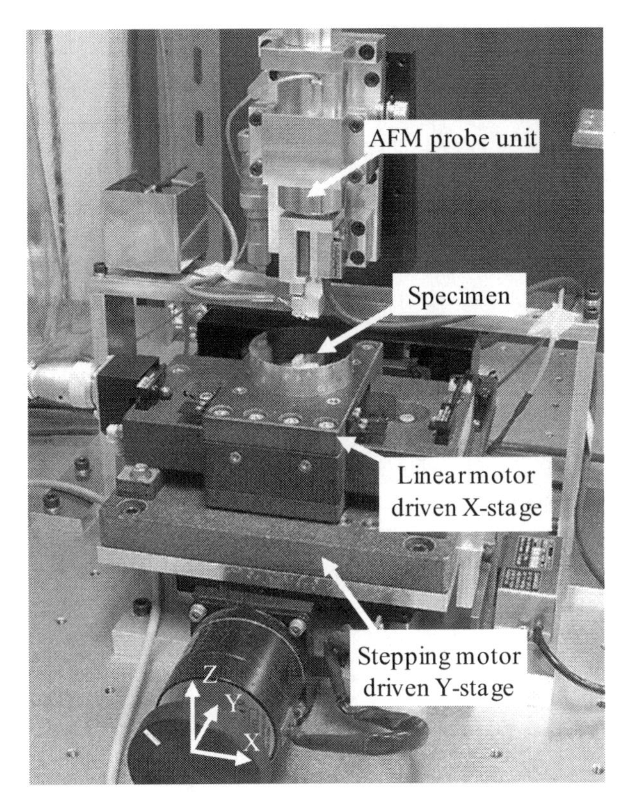

Figure 8.22. Photograph of the AFM measurement system

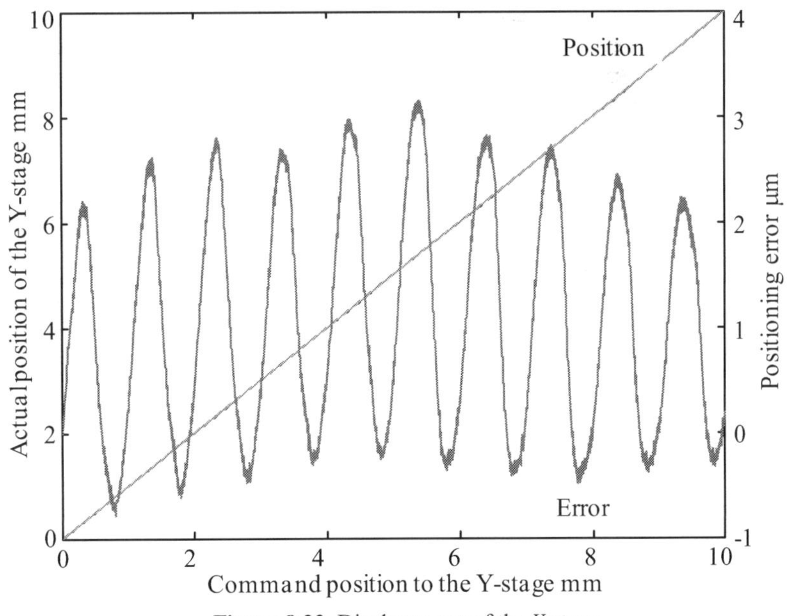

Figure 8.23. Displacement of the *Y*-stage

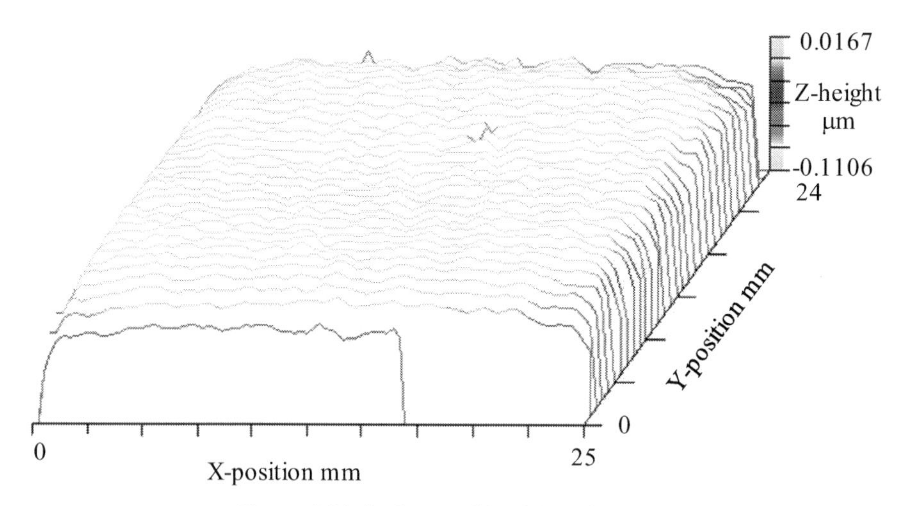

Figure 8.24. Surface profile of the reference mirror

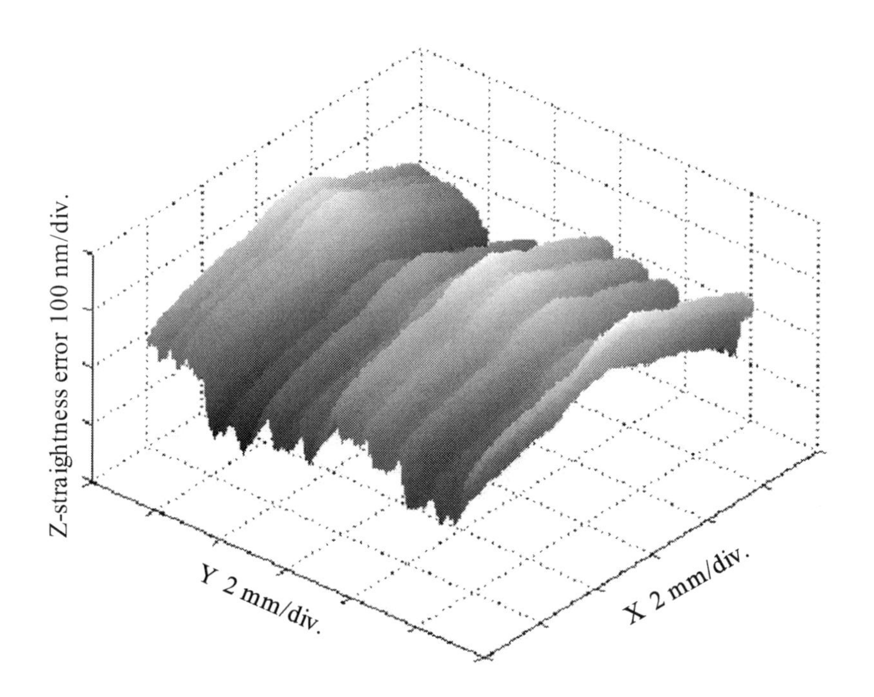

Figure 8.25. Measurement result of the *Z*-directional error motion of the *XY* scanning stages

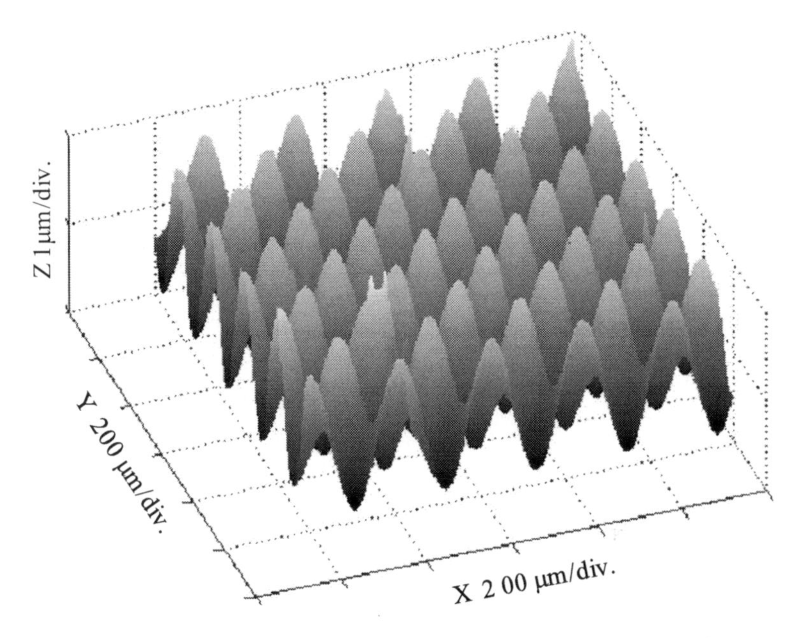

Figure 8.26. AFM image of the sinusoidal surface without compensation

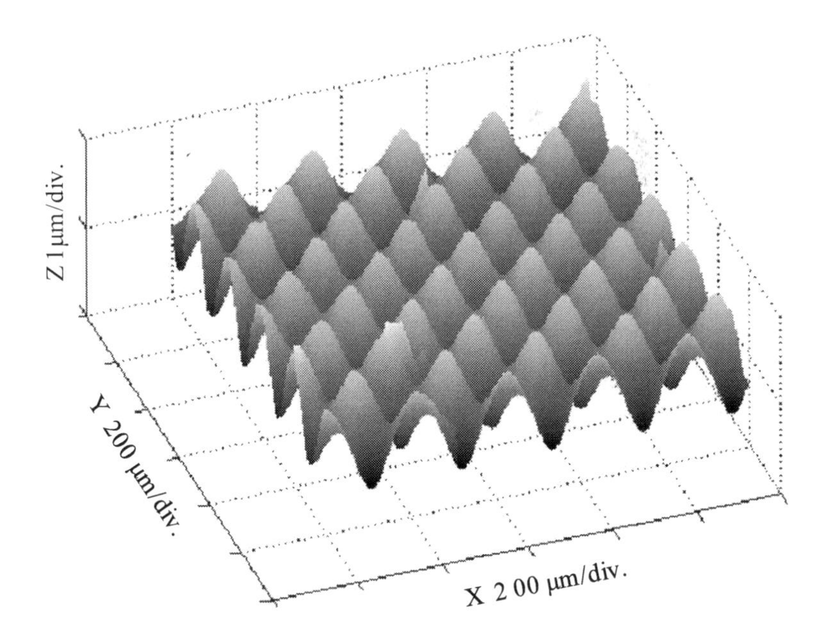

Figure 8.27. AFM image of the sinusoidal surface with compensation based on the output of the linear encoder

The result shown in Figure 8.27 was compensated based on the output of the *Z*-encoder. The scanning areas were 1 mm (*X*) × 1 mm (*Y*). The sampling numbers were 200 and the sampling intervals were 5 μm in both the *X*- and *Y*-directions. The average and the deviation of the sinusoidal amplitude in Figure 8.26 were 1.625 μm ± 0.028 μm (3σ), and those in Figure 8.27 were 0.987 μm ± 0.021 μm (3σ). The average of the amplitude in Figure 8.27 was much closer to the design value and the deviation was also smaller when compared with the non-compensated result shown in Figure 8.26. Figure 8.28 shows the compensated AFM image over a larger area.

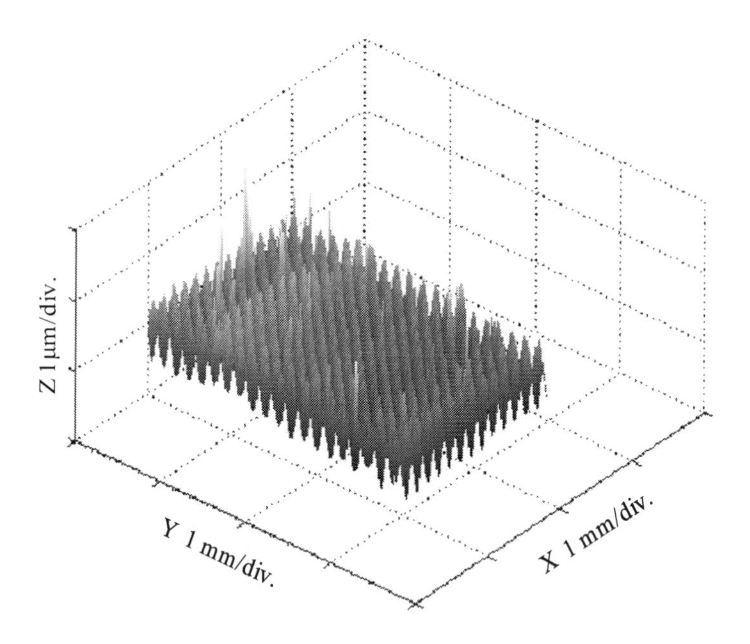

Figure 8.28. A large area AFM image with error compensation

8.4 Spiral Scanning Probe Microscope

The AFM systems described above have the same raster scanning mechanism as conventional AFMs. As can be seen in Figure 8.29 (a), since multiple scans are necessary in the raster scanning mechanism, it makes a large area measurement time-consuming. In this section, a spiral scanning mechanism is described. As shown in Figure 8.29 (b), since only one scan along the radial direction is needed in the spiral scanning mechanism, the measurement time can be shortened. On the other hand, there are two challenges for the spiral scanning mechanism when compared with the raster scanning mechanism. One is that the interval between measurement points becomes large in the outer area of the specimen as shown in Figure 8.30. The other issue is that the center of the measurement coordinate should be aligned with the center of the spindle as shown in Figure 8.31.

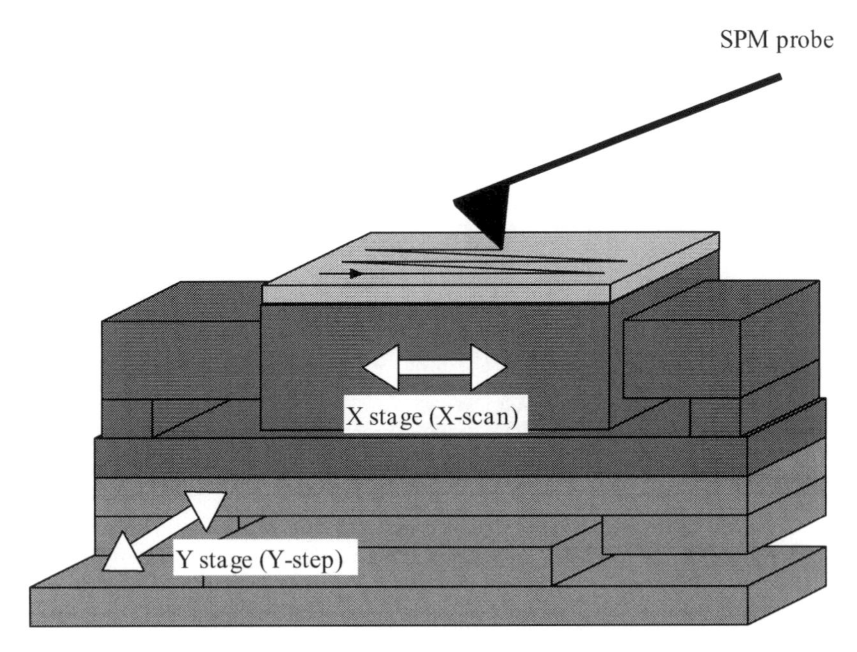

(a) Raster scanning (multiple scans in *X*-directions)

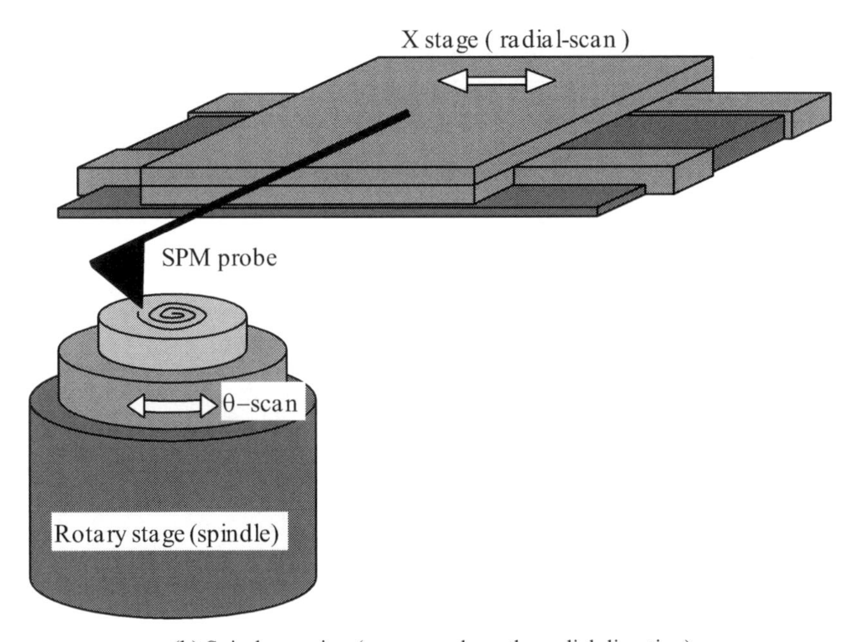

(b) Spiral scanning (one scan along the radial direction)

Figure 8.29. Raster and spiral scanning mechanisms

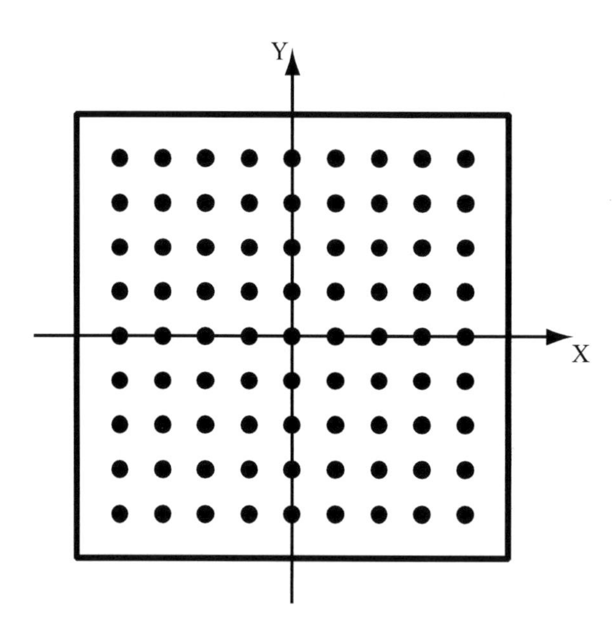

(a) Raster scanning
(uniform intervals between measurement points /uniform lateral resolution)

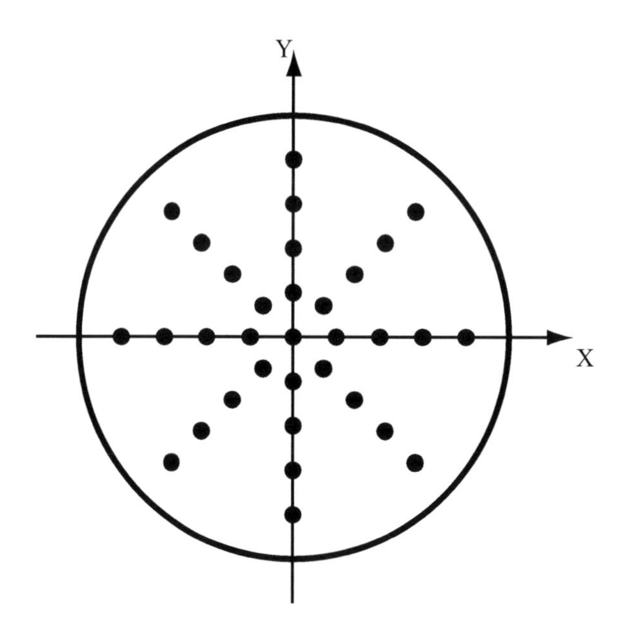

(b) Spiral scanning
(larger intervals between measurement points /lower later resolution in outer area)

Figure 8.30. Intervals between measurement points

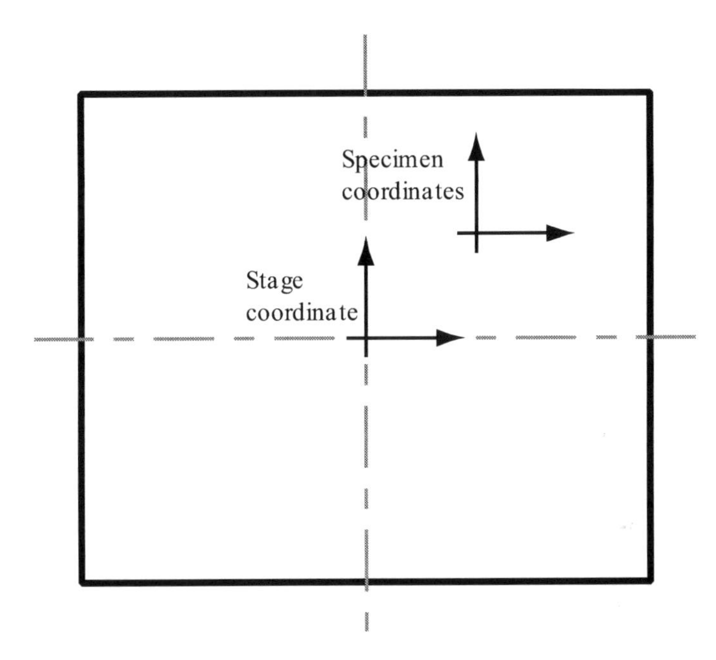

(a) Raster scanning (no need for alignment of coordinate system)

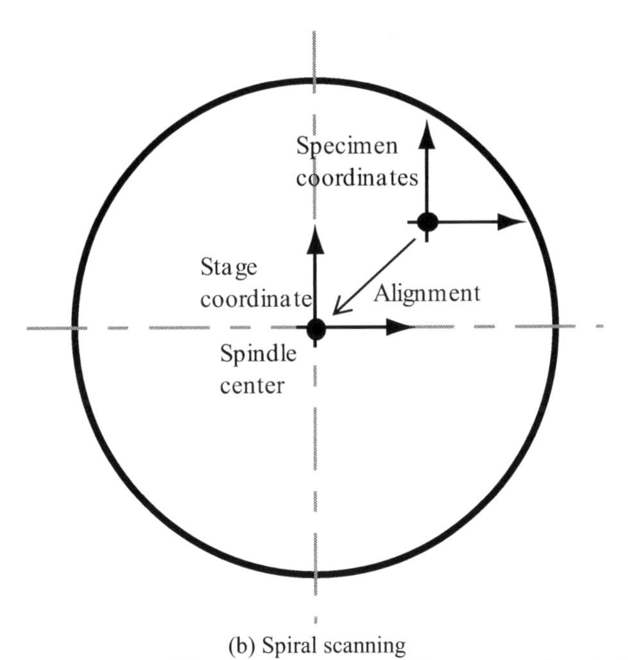

(b) Spiral scanning
(necessary for alignment of the coordinate system with respect to the spindle center)

Figure 8.31. Alignment of coordinate system

Figure 8.32 shows a schematic of the AFM system developed with the raster scanning mechanism. The specimen is vacuum-chucked on the spindle with an aerostatic bearing (air-spindle). The *X*-slide, which is also with an aerostatic bearing (air-slide), is used to mount the AFM probe unit. The rotation of the spindle and the movement of the *X*-slide are synchronized so that the surface of the specimen can be scanned by the AFM probe unit in a spiral scanning pattern. Assume that the *X*-slide starts to move from the center position of the spindle. The polar coordinates and the Cartesian coordinates of the cantilever tip of the AFM

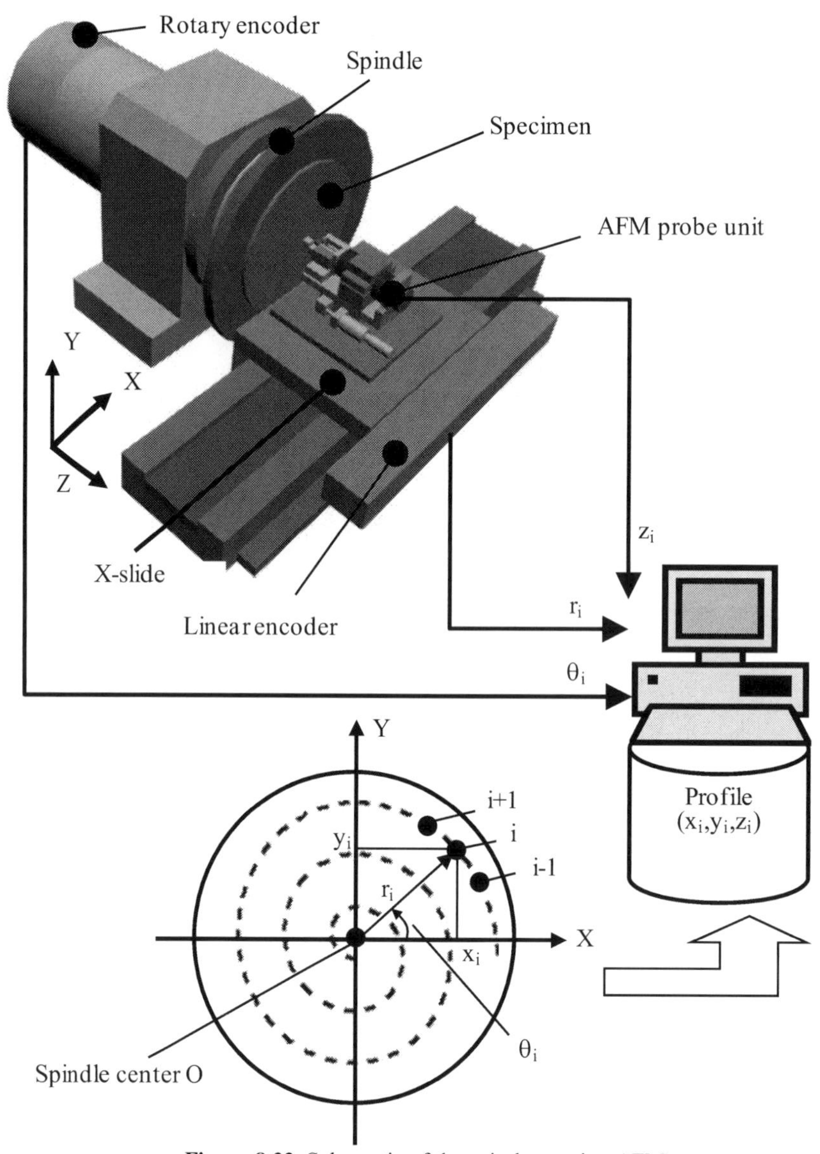

Figure 8.32. Schematic of the spiral scanning AFM

probe unit in the XY-plane can be expressed by the following equations, respectively:

$$(r_i, \theta_i) = \left(\frac{Fi}{UT}, 2\pi \frac{i}{U} \right), \; (i = 0, 1, ..., N-1), \tag{8.6}$$

$$(x_i, y_i) = (r_i \cos\theta_i, r_i \sin\theta_i), \; (i = 0, 1, ..., N-1), \tag{8.7}$$

where F is the feed rate of the X-slide in μm per minute, U is the pulse number of the rotary encoder of the spindle in pulses per revolution, T is the rotational speed of the spindle in rpm (revolution per minute), i is the ith rotary encoder pulse, and N is the total pulse number of the rotary encoder over the scanning range, respectively. The rotary encoder pulse is used as the external trigger signal for the data acquisition board of the personal computer. The rotary encoder output (θ_i) of the spindle, the linear encoder output (r_i) of the X-slide, the linear encoder output (z_i) of the AFM probe unit are simultaneously taken into the personal computer in response to the external trigger. The three-dimensional (3D) surface profile of the specimen can thus be obtained from the (x_i, y_i, z_i) coordinates in the Cartesian coordinate system.

Figure 8.33 shows a photograph of the spiral scanning AFM. The spindle can

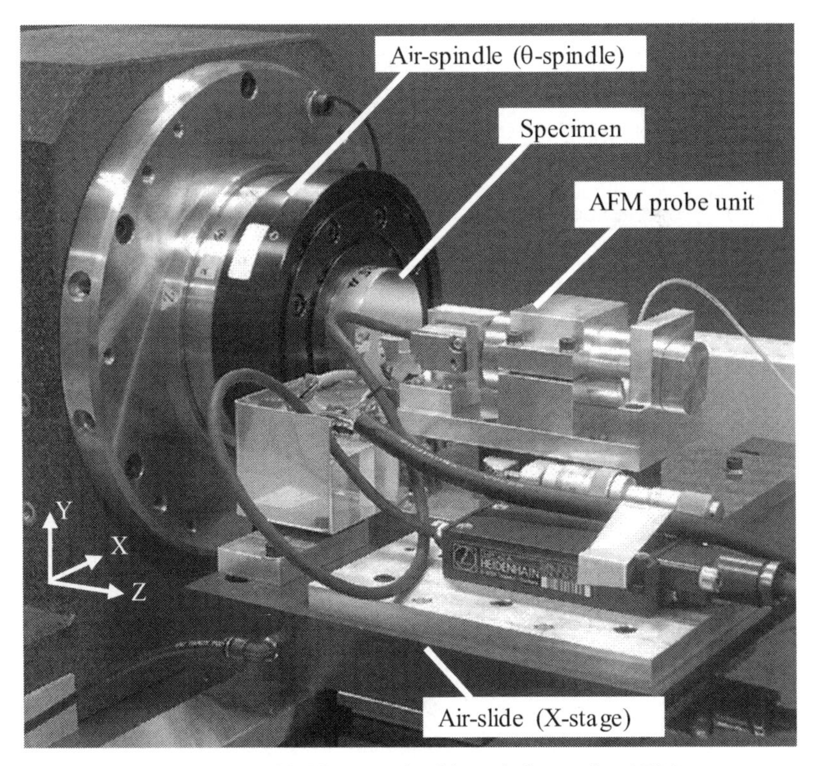

Figure 8.33. Photograph of the spiral scanning AFM

mount a specimen with a diameter of 130 mm (optionally 300 mm). The moving range of the X-slide is 300 mm. To respond to the first challenge shown in Figure 8.30, a rotary encoder with a pulse number of 1.5×10^8 corresponding to an angular resolution of 0.009 arcsec is employed. This realizes a lowest lateral resolution of 2 nm for the AFM over a measurement area of 100 mm in diameter. The linear encoder of the X-slide has a resolution of 0.28 nm.

Figure 8.34 shows a photograph of the AFM probe unit with an alignment stage in the Y-direction. The AFM probe unit is basically the same as that shown in Figure 8.17. To respond to the second challenge of the spiral scanning AFM shown in Figure 8.31, a small stage is mounted between the AFM cantilever and the PZT actuator so that the position of the cantilever tip can be adjusted in the Y-direction toward the center of the spindle.

Figure 8.35 shows an analysis of the influence from the alignment error of the cantilever tip with respect to the spindle center, which is the second challenge to the spiral AFM shown in Figure 8.31. As can be seen in the figure, if the position of the AFM cantilever tip (O') is not aligned with the spindle center (O), which is the origin of the coordinate system for the spiral scanning measurement, the actual measuring point (P_1) will be different from the ideal measuring point (P_0), causing the distortion of the measured surface profile. δ_X and δ_Y are the X-directional and Y-directional components of the alignment error, respectively. Assume that the coordinate of P_0 is (x, y) and that of P_1 is (x_1, y_1). If the alignment error is small enough, x_1 and y_1 can be expressed as follows:

$$x_1 = x + \frac{1}{\sqrt{x^2 + y^2}}\left(\delta_X \cdot x - \delta_Y \cdot y\right), \tag{8.8}$$

$$y_1 = y + \frac{1}{\sqrt{x^2 + y^2}}\left(\delta_Y \cdot x + \delta_X \cdot y\right). \tag{8.9}$$

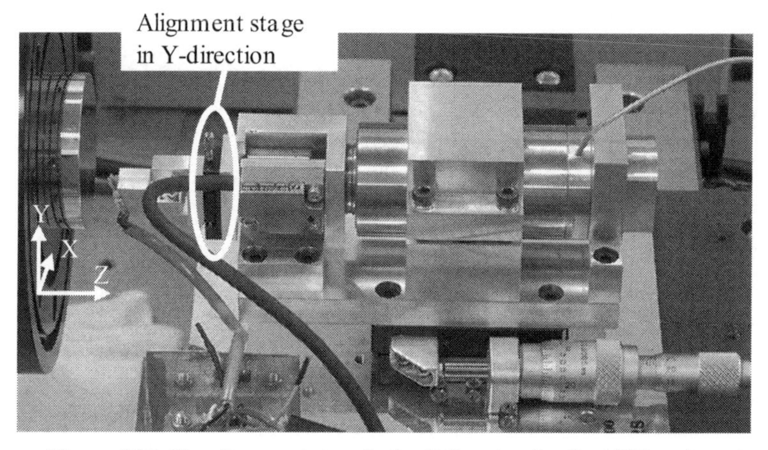

Figure 8.34. The alignment stage in the Y-direction for the AFM probe unit

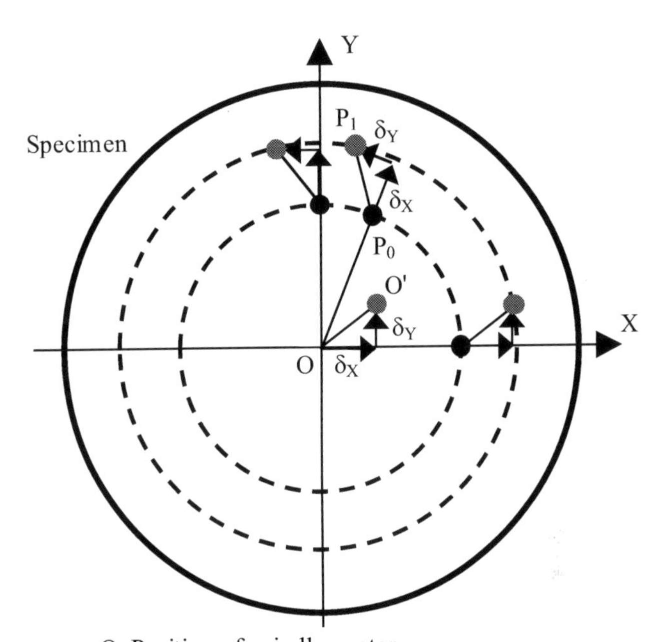

O: Position of spindle center
O' : Position of AFM cantilever tip
P_0 (x, y) : Ideal measuring point
P_1 (x_1, y_1) : Actual measuring point
δ_X: Center-alignment error in X-direction
δ_Y: Center-alignment error in Y-direction

Figure 8.35. Influence of the alignment error of the AFM cantilever tip with respect to the spindle center

Figures 8.36 and 8.37 show the differences between $P_0(x, y)$ and $P_1(x_1, y_1)$ caused by δ_X and δ_Y, respectively. As can be seen in the figures, δ_X causes position errors along the radial direction and δ_Y causes position errors along the circumferential direction. The error vector has the same amplitude as the alignment error.

Figure 8.38 shows a simulation result of profile distortions caused by the center-alignment error. The surface profile of the specimen is a sinusoidal *XY*-grid superimposed with *X*- and *Y*-directional sine waves. The amplitudes of the *X*- and *Y*-directional sine waves are set to be 0.25 μm, which results in a peak-to-valley grid amplitude of 1 μm. The corresponding spatial pitches in the two directions are 150 μm. In the simulation, both the error components δ_X and δ_Y are set at 50 μm. Figure 8.38 (a) is a three-dimensional expression of the simulation result of the surface profile and Figure 8.38 (b) is a top view of the result. Large profile distortions can be observed in the figures. To realize accurate measurement of the surface profile by the spiral AFM, it is necessary to accurately conduct the alignment of the AFM cantilever tip with respect to the spindle center.

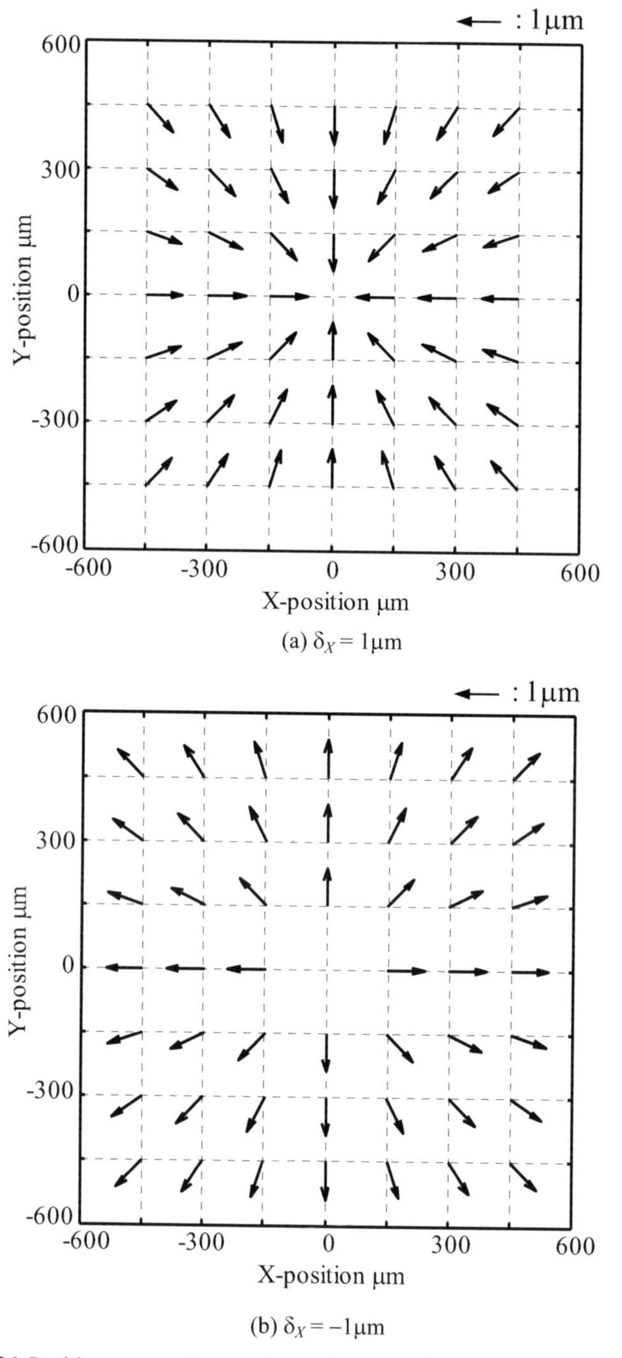

(a) $\delta_X = 1\mu m$

(b) $\delta_X = -1\mu m$

Figure 8.36. Position errors of measuring points caused by the *X*-directional component of center-alignment error (δ_X)

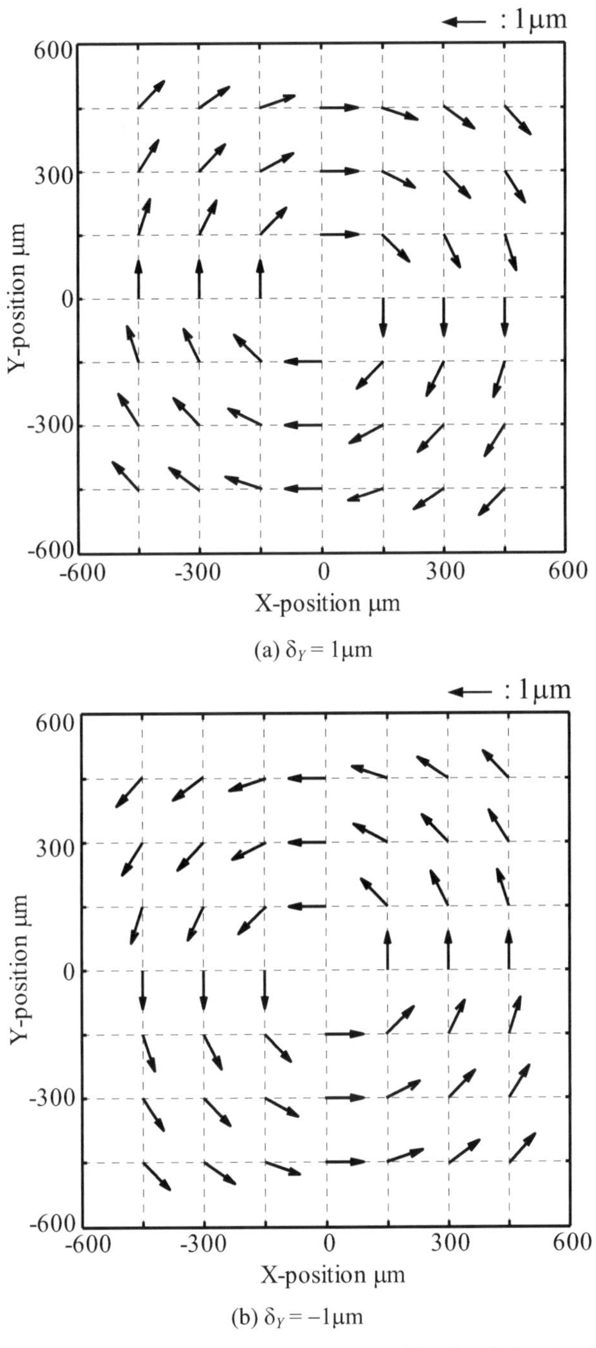

Figure 8.37. Position errors of measuring points caused by the *Y*-directional component of center-alignment error (δ_Y)

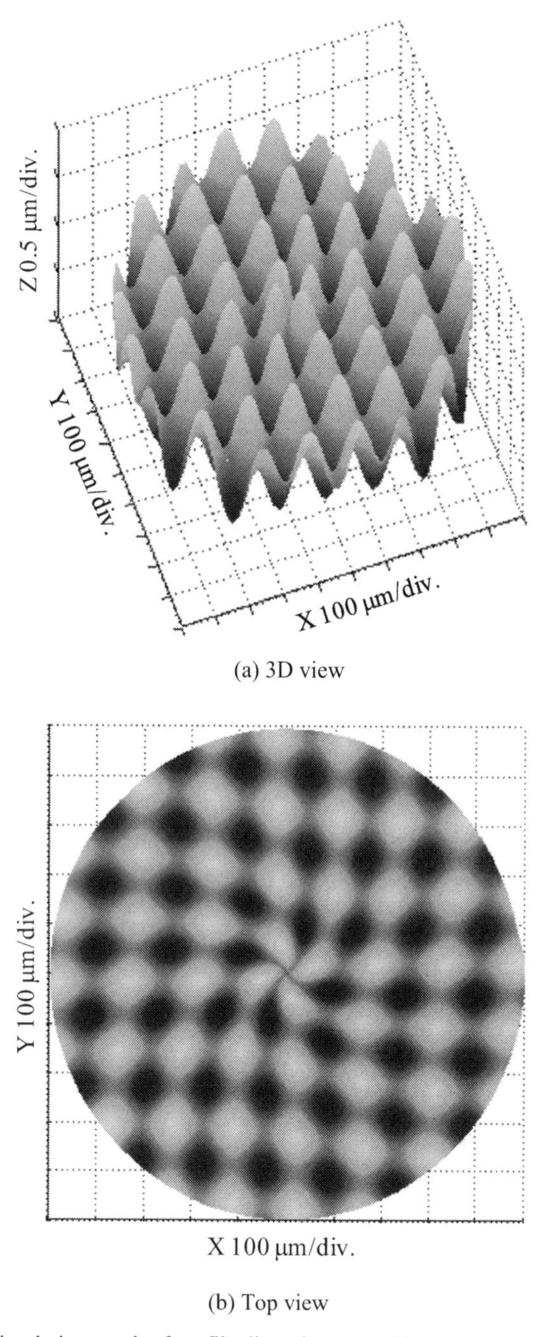

Figure 8.38. Simulation result of profile distortion caused by the center-alignment error

Figure 8.39 shows the experimental setup for the center-alignment. An aluminum artifact with a micro-bump at the center position was employed. The artifact was first aligned to the spindle center by using a high-magnification microscope to monitor the micro-bump of the artifact rotated by the spindle. Then the AFM probe unit was moved by the *X*-slide to scan the micro-bump which was kept stationary. The *Y*-position of the AFM tip was changed by using the alignment stage of the AFM probe unit for multiple *X*-scans at different *Y*-positions of the micro-bump. The results were employed for evaluating the position of the AFM cantilever tip relative to the spindle center, based on which accurate center-alignment could be carried out. A capacitive sensor was employed to monitor the *Y*-position of the cantilever holder. The alignment in the *X*-direction was carried out by moving the AFM-head by the *X*-slide. It should be noted that the center-alignment is necessary only at the time exchanging the AFM cantilever.

After the center-alignment, the artifact was exchanged with a sinusoidal *XY*-grid specimen. The specimen had same specifications as those of the simulation shown in Figure 8.38. Figure 8.40 shows the measured surface profile at the center part of the grid specimen over an area of 1 mm in diameter. The feed speed of the *X*-slide was 0.1 μm/s, and the rotation speed of the spindle was 3.6°/s. The movement length of the *X*-slide was 0.5 mm. Figures 8.40 (a) and 8.40 (b) show the 3D view and top view of the profile, respectively. No profile distortions due to the center-alignment error were observed in the result. The spike components of the surface were caused by fabrication errors or contaminations.

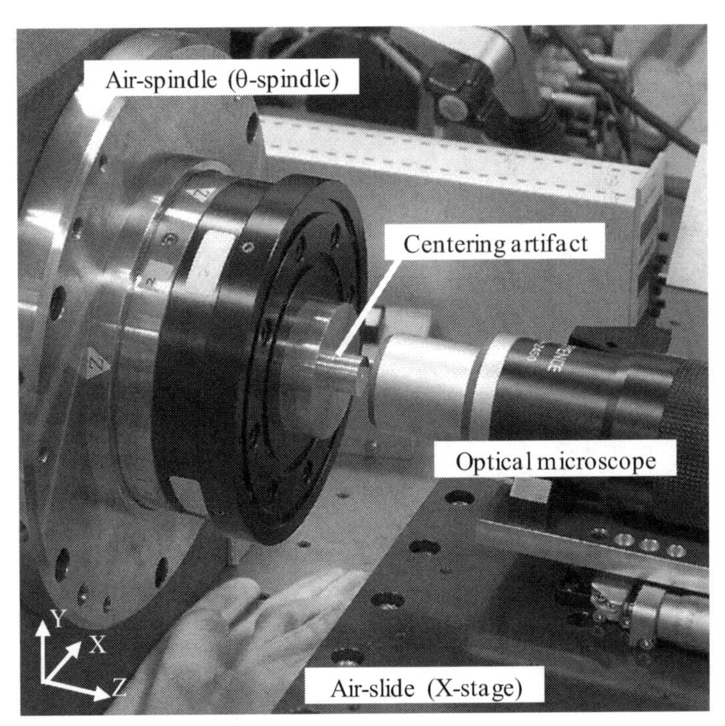

Figure 8.39. The setup for center-alignment

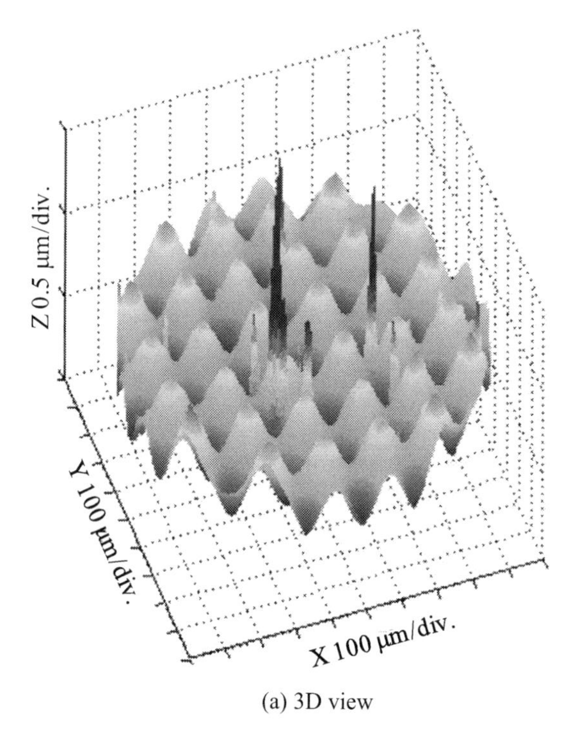

(a) 3D view

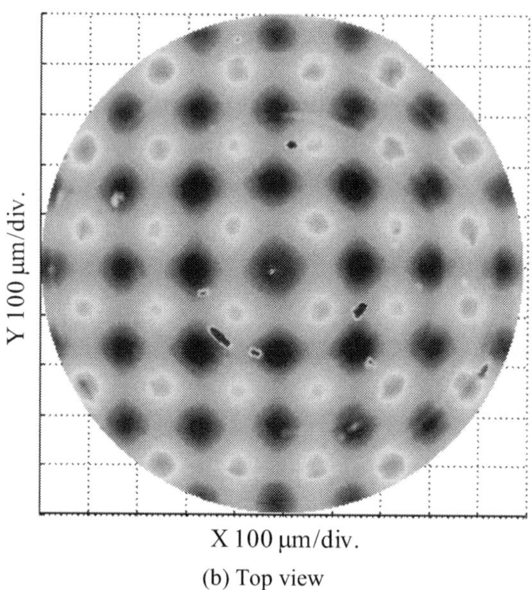

(b) Top view

Figure 8.40. Measured profile of the sinusoidal surface by the spiral AFM after center-alignment

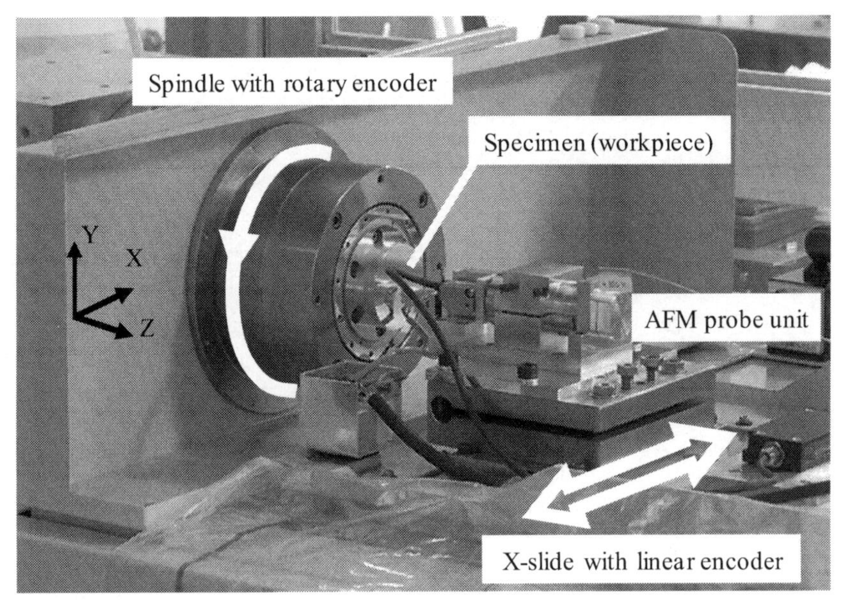

Figure 8.41. A spiral scanning AFM for surface profile measurement on a diamond turning machine

The AFM probe unit was mounted on the diamond turning machine used to fabricate the *XY* sinusoidal grid (Figure 8.41) so that the grid could be measured on the machine. The cross-slide (*X*-slide) and the workpiece spindle of the machine were employed for construction of the spiral scanning AFM [14]. Because the surface profile can be evaluated by the AFM on the machine, the measurement result can be easily fed back to the machining process. The linear encoder integrated in the AFM probe unit can realize accurate measurement of the *Z*-directional surface profile in the presence of electromagnetic noise, which is associated with the machine. Figure 8.42 shows an image of the *XY*-grid taken by the spiral AFM on the machine.

8.5 Summary

Two AFM probe units with compensation of the *Z*-directional error of the AFM probe unit have been presented. The first AFM probe unit employs a PZT actuator with a stroke of 100 μm as the *Z*-actuator to realize a large measurement range in the *Z*-direction. Two capacitive sensors are aligned on the two sides of the PZT actuator along the *Z*-direction. The displacement as well as the tilt motion of the actuator can be accurately measured and compensated for based on the sensor outputs. The second AFM probe unit employs a linear encoder for measurement and compensation of the error motion from the PZT actuator. The actuator used in the AFM probe unit has a stiffness of 100 N/μm and a stroke of 53 μm. The high

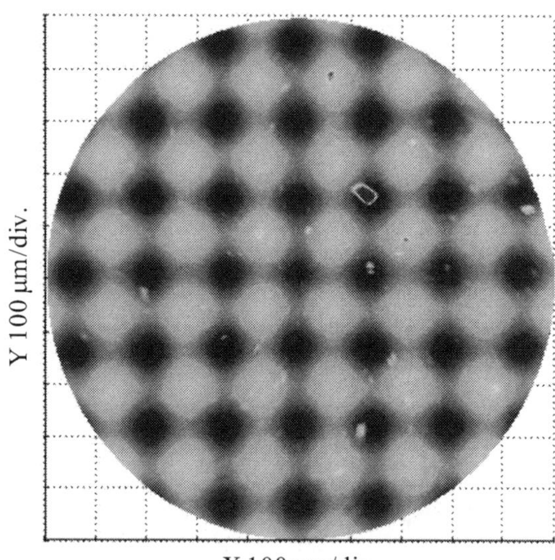

Figure 8.42. An AFM image of the *XY*-grid taken on the diamond turning machine

stiffness of the PZT actuator makes it possible to mount a scale of a linear encoder together with the AFM cantilever. The displacement of the PZT actuator, which corresponds to the surface profile height, can be accurately measured by the linear encoder with a resolution of 1 nm.

A spiral scanning mechanism has been presented instead of the raster scanning mechanism which is used in conventional AFMs. A large area spiral scanning AFM has been constructed by using the linear encoder-compensated AFM probe unit and a spiral scanning mechanism. The scanning mechanism is composed of a linear stage with a stroke/resolution of 300 mm/0.24 nm and a spindle with a spindle size/rotary resolution of 130 mm in diameter/0.009 arcsec. The sample is mounted on the spindle and the AFM probe unit is mounted on the linear stage. The importance of alignment of the AFM cantilever tip with the spindle center has been verified. A center-alignment method by using an optical microscope and a centering artifact has been presented.

The linear encoder compensated AFM probe unit has also been mounted on a diamond turning machine for on-machine measurement of large area micro-structured surface profiles. It has been verified that the strong robustness of the linear encoder to electromagnetic noise is important and effective for use of the AFM probe unit on a machine tool.

References

[1] Evans CJ, Bryan JB (1999) "Structured", "textured" or "engineered" surfaces. Ann CIRP 48(2):541–556

[2] Kiyono S, Gao W, Zhang S, Aramaki T (2000) Self-calibration of a scanning white light interference microscope. Opt Eng 39(10):2720–2725

[3] Binning G, Quate CF, Gerber CH (1986) Atomic force microscope. Phys Rev Lett 56:930–933

[4] Holmes M, Hocken RJ, Trumper DL (2000) The long-range scanning stage: a novel platform for scanned-probe microscopy. Precis Eng 24(3):191–209

[5] Gao W, Nomura M, Kewon H, Oka T, Liu QG, Kiyono S (1999) A new method for improving the accuracy of SPM and its application to AFM in liquid. JSME Int J Ser C 42(4):877–883

[6] Kweon H, Gao W, Kiyono S (1998) In situ self-calibration of atomic force microscopy. Nanotechnology 9:72–76

[7] Gonda S, Doi T, Kurosawa T, Tanimura Y, Hisata N, Fujimoto TH, Yukawa H (1999) Real-time interferometrically measuring atomic force microscope for direct calibration of standards. Review of Scientific Instruments 70(8):3362–3368

[8] Dai G, Pohlenz F, Danzebrink HU, Xu M, Hasche K, Wilkening G (2004) Metrological large range scanning probe microscope. Rev Sci Instrum 75(4):962–969

[9] Tortonese M, Barrett RC, Quate CF (1993) Atomic resolution with an atomic force microscope using piezoresistive detection. Appl Phys Lett 62(8):834–836

[10] Seiko Instruments Inc. (2010) Reference data of piezo-resistive cantilever. Seiko Instruments Inc., Japan

[11] Zygo Corporation (2010) http://www.zygo.com. Accessed 1 Jan 2010

[12] Aoki J, Gao W, Kiyono S, Ono T (2005) A high precision AFM for nanometrology of large area micro-structured surfaces. Key Eng Mater 295–296:65–70

[13] Aoki J, Gao W, Kiyono S, Ochi T, Sugimoto T (2005) Design and construct of a high-precision and long-stoke AFM probe-unit integrated with a linear encoder. Nanotechnol Precis Eng 3(1):1–7

[14] Gao W, Aoki J, Ju BF, Kiyono S (2007) Surface profile measurement of a sinusoidal grid using an atomic force microscope on a diamond turning machine. Precis Eng 31:304–309

9

Automatic Alignment Scanning Probe Microscope System for Measurement of 3D Nanostructures

9.1 Introduction

Nanostructures with dimensions on the order of 1 to 100 nm represented by the nanoedge of a single point diamond cutting tool can be made by nanomanufacturing. The diamond cutting tool is used in ultra-precision cutting to fabricate precision workpieces [1–5]. As shown in Figure 9.1, the tool has a very sharp edge with a radius on the order of 10 to 100 nm. The micro/nanowear of the tool edge poses a large problem because it influences the quality of the machined surface [6]. Diamond cutting has also been used to generate three-dimensional (3D) micro-structured surfaces. In such cases, the fabrication accuracy is influenced not only by the tool edge sharpness, but also by the local 3D profile of the tool edge [7]. Therefore, it is important to conduct precision nanometrology of the edge wear as well as the 3D edge profile.

The tool edge is conventionally monitored by an optical microscope or a scanning electron microscope (SEM) [8–16]. The optical microscope is easy to use. Its resolution, however, is limited to the micrometer range by the optical diffraction

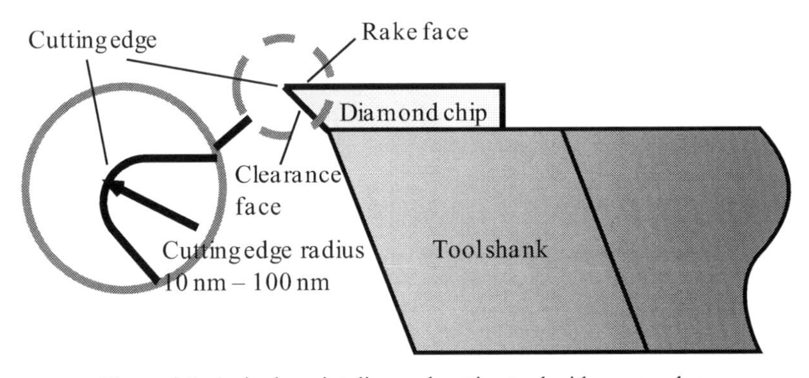

Figure 9.1. A single-point diamond cutting tool with a nanoedge

phenomena. The SEM is excellent for the nanometric resolution, but the measurement has to be carried out in a vacuum chamber. In addition, an SEM image is basically a 2D projection of a 3D object, and is not proper for measurement of the 3D tool edge profile.

Differing from optical microscopes and SEMs, scanning probe microscopes (SPM), represented by atomic force microscopes (AFMs), are effective for 3D profile measurement of nanostructures because SPMs have nanometer resolution in the *X*-, *Y*-, *Z*-directions [17, 18]. Although it is possible to use the AFM for characterization of edges of diamond tools [19], it is difficult to align the AFM probe tip with the tool edge. The alignment in a conventional AFM is carried out with the visual feedback from an optical microscope. As shown in Figure 9.2, however, the optical microscope focuses on the back of the AFM cantilever, not directly on the interface between the AFM probe tip and the tool edge. The extremely small working distance and the low resolution of the optical microscope are the other limitations for AFM tip and tool edge alignment.

In this section, an optical probe is presented for automatic alignment of the AFM probe tip with the tool edge top.

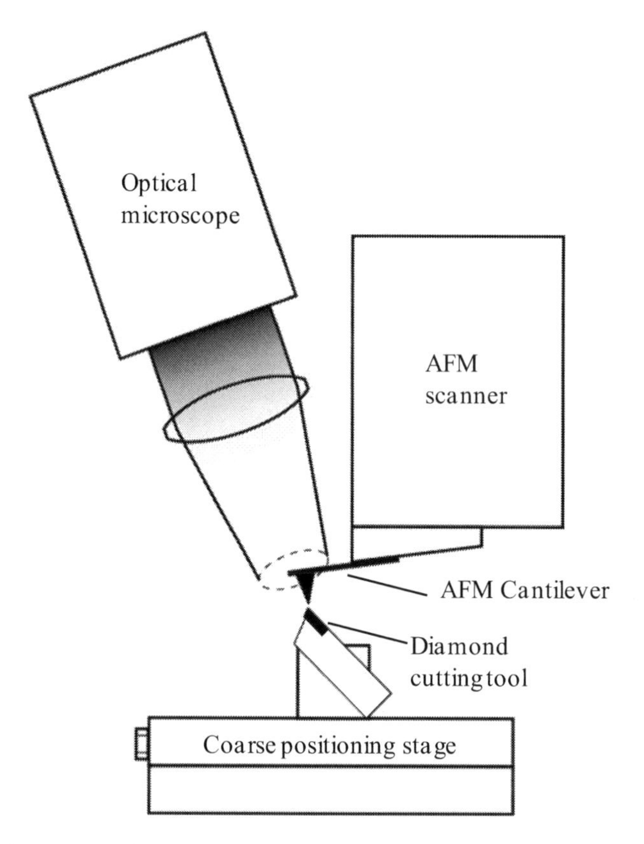

Figure 9.2. A conventional AFM with the visual feedback from an optical microscope for alignment of the AFM tip with the specimen

9.2 Optical Probe for Automatic Alignment

9.2.1 Alignment Principle

Figure 9.3 shows a schematic of the optical probe for aligning the AFM tip with the tool edge top. A collimated laser beam from a laser source is focused by an objective lens to generate a small beam spot at the beam waist. The laser beam is then received by a photodiode after passing through a collecting lens. The *XYZ*-coordinate system is set as shown in the figure. The origin of the coordinate system is defined as the center of the spot at the beam waist, which is called the reference point for the alignment of the AFM tip and the top of the cutting tool edge. The light diffraction at the AFM tip and the tool top is not considered and all the light beams collected by the collecting lens are assumed to be received by the PD.

Assume that the sectional profile of the beam in the *XZ*-plane is circular. The radius of the beam can be expressed by [20]:

$$a(y) = a_0\sqrt{1+\left(\frac{\lambda y}{\pi a_0^2}\right)^2}\,, \tag{9.1}$$

where a_0 is the beam radius at the beam waist. The corresponding area of the sectional beam is given by:

$$A(y) = \pi a(y)^2\,. \tag{9.2}$$

Figures 9.4 and 9.5 show the tool edge and the AFM probe in the laser beam. When the AFM probe or the tool edge is brought into the laser beam, because a part of the laser beam will be blocked, the optical power of the beam received by the photodiode will change with the position of the AFM probe or the tool edge in the laser beam. Assume that (x_1, y_1, z_1) and (x_2, y_2, z_2) are the coordinates of the tool edge top and the AFM probe tip, respectively. Let the areas not blocked by the tool edge and the AFM tip be denoted by $A_1(x_1,y_1,z_1)$ and $A_2(x_2,y_2,z_2)$, and the corresponding optical powers received by the photodiode be Q_{tool} and Q_{AFM}, respectively. Assume that the optical power of the laser source is Q. The relative outputs of the photodiode, defined as the ratios of Q_{tool} to Q and Q_{AFM} to Q, respectively, can be expressed as follows:

$$s_{tool}(x_1,y_1,z_1) = \frac{Q_{tool}(x_1,y_1,z_1)}{Q}\times 100\% = \frac{A_1(x_1,y_1,z_1)}{A(y_1)}\times 100\%\,, \tag{9.3}$$

$$s_{AFM}(x_2,y_2,z_2) = \frac{Q_{AFM}(x_2,y_2,z_2)}{Q}\times 100\% = \frac{A_2(x_2,y_2,z_2)}{A(y_2)}\times 100\% \tag{9.4}$$

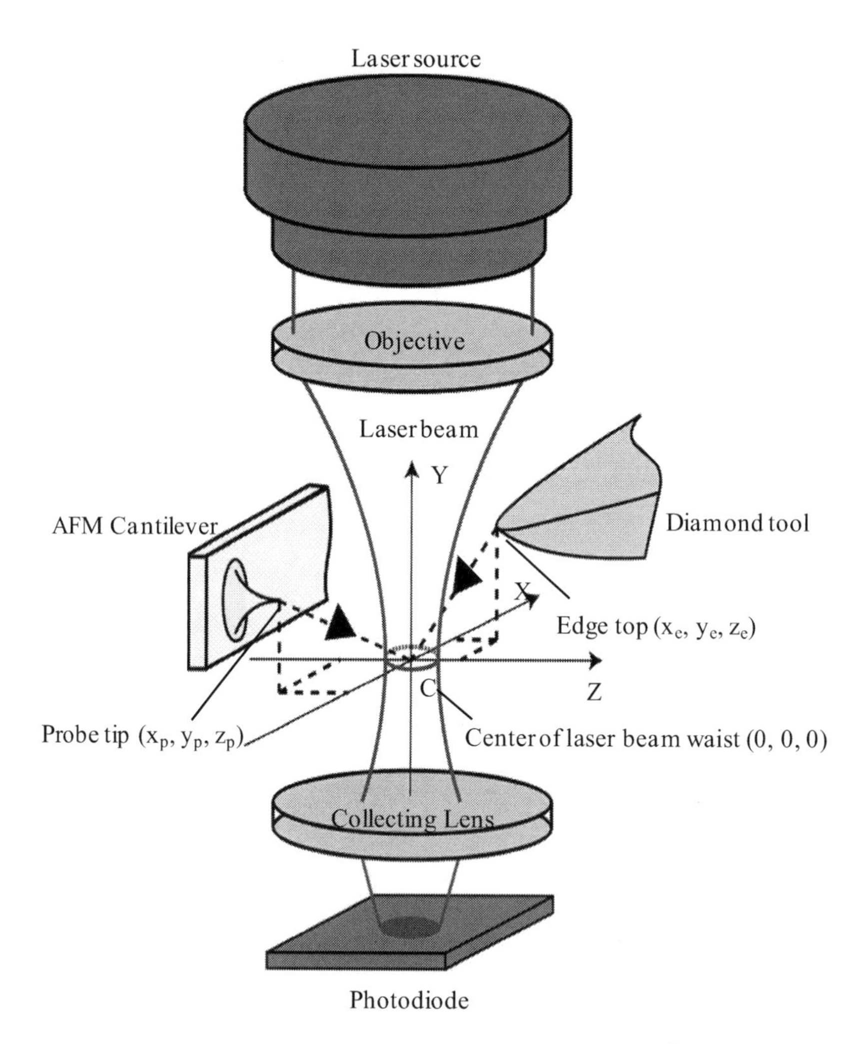

Figure 9.3. Schematic of the optical probe for automatic alignment of the AFM probe tip with the cutting tool edge top

where

$$Q_{tool}(x_1, y_1, z_1) = Q\frac{A_1(x_1, y_1, z_1)}{A(y_1)}, \tag{9.5}$$

$$Q_{AFM}(x_2, y_2, z_2) = Q\frac{A_2(x_2, y_2, z_2)}{A(y_2)}. \tag{9.6}$$

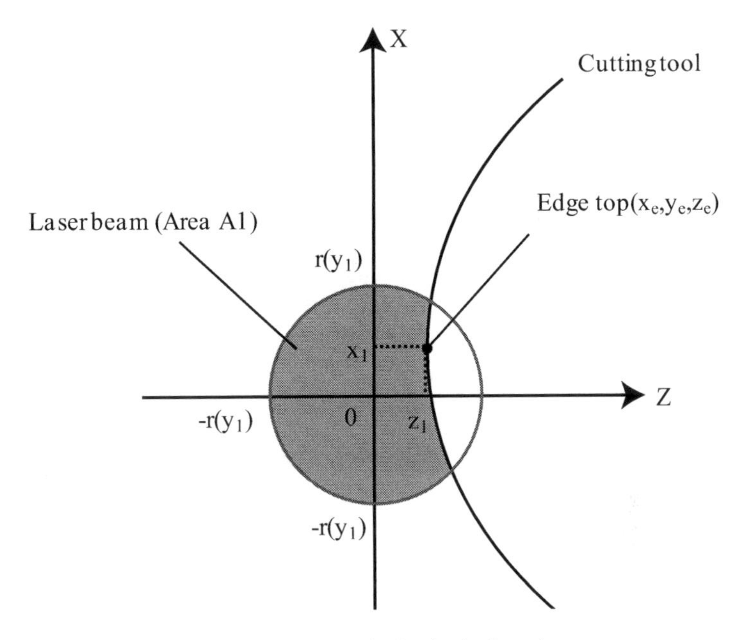

Figure 9.4. The tool edge in the laser beam

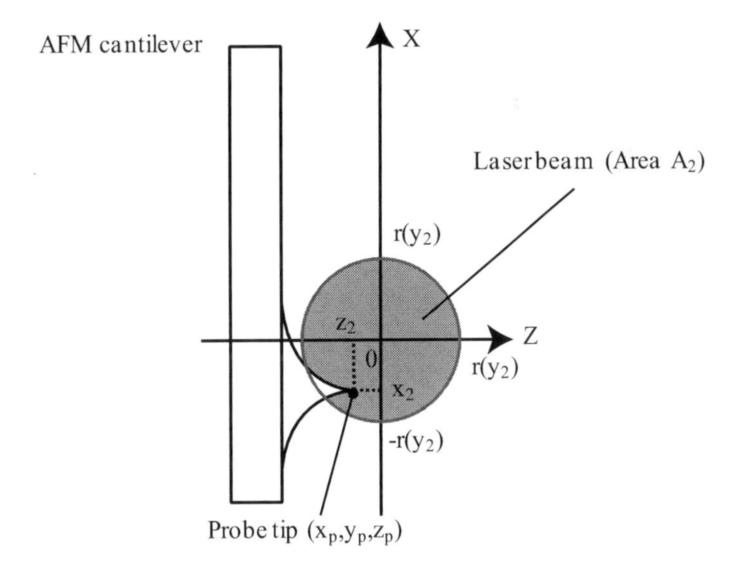

Figure 9.5. The AFM probe in the laser beam

It can be seen that the relative photodiode output and the optical power received by the photodiode are functions of the position of the tool edge top or AFM probe tip, which can be utilized for the alignment of the AFM probe tip with the tool edge top.

9.2.2 Alignment Procedure

Step 1: Alignment of the tool edge top along the *X*-direction

The alignment starts by positioning the tool edge top at the center of the laser beam along the *X*-direction (Figure 9.6). Figure 9.7 shows a result of the computer simulation for investigating the relationship between the photodiode output and the position of the tool edge top when the alignment is carried out along the *X*-direction. The relationship between the photodiode output $s_{tool}(x_e, y_e, z_e)$ and x_e is shown in Figure 9.7. The diameter of the beam waist is set to 10 μm. The radius of the round nose tool is 1 mm. The geometries of the tool and the laser beam are the same for all the simulations shown below.

In Figure 9.7 (a), the *Z*-position z_e of the tool edge top is set to −2.5 μm, 0 μm, and +2.5 μm, while the *Y*-position y_e of the tool edge top is kept at 0. Although the curve of s_{tool} versus x_e shifted vertically with varying z_e, the minimum value of $s_{tool}(x_e, y_e, z_e)$ with respect to x_e is always at the center of the beam (x_e = 0). In Figure 9.7 (b), y_e is successively set to −50 μm, 0 μm, and +50 μm, while z_e is kept at 0. Although the minimum value of $s_{tool}(x_e, y_e, z_e)$ with respect to x_e also occurs at $x_e = 0$, the curve does not shift vertically with varying y_e.

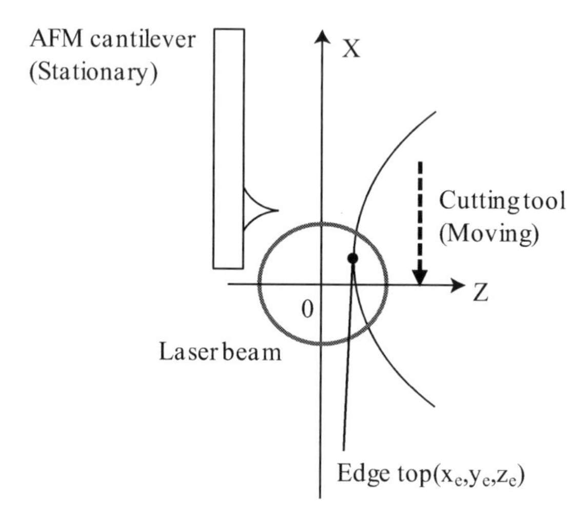

Figure 9.6. Step 1: Alignment of the tool edge top along the *X*-direction

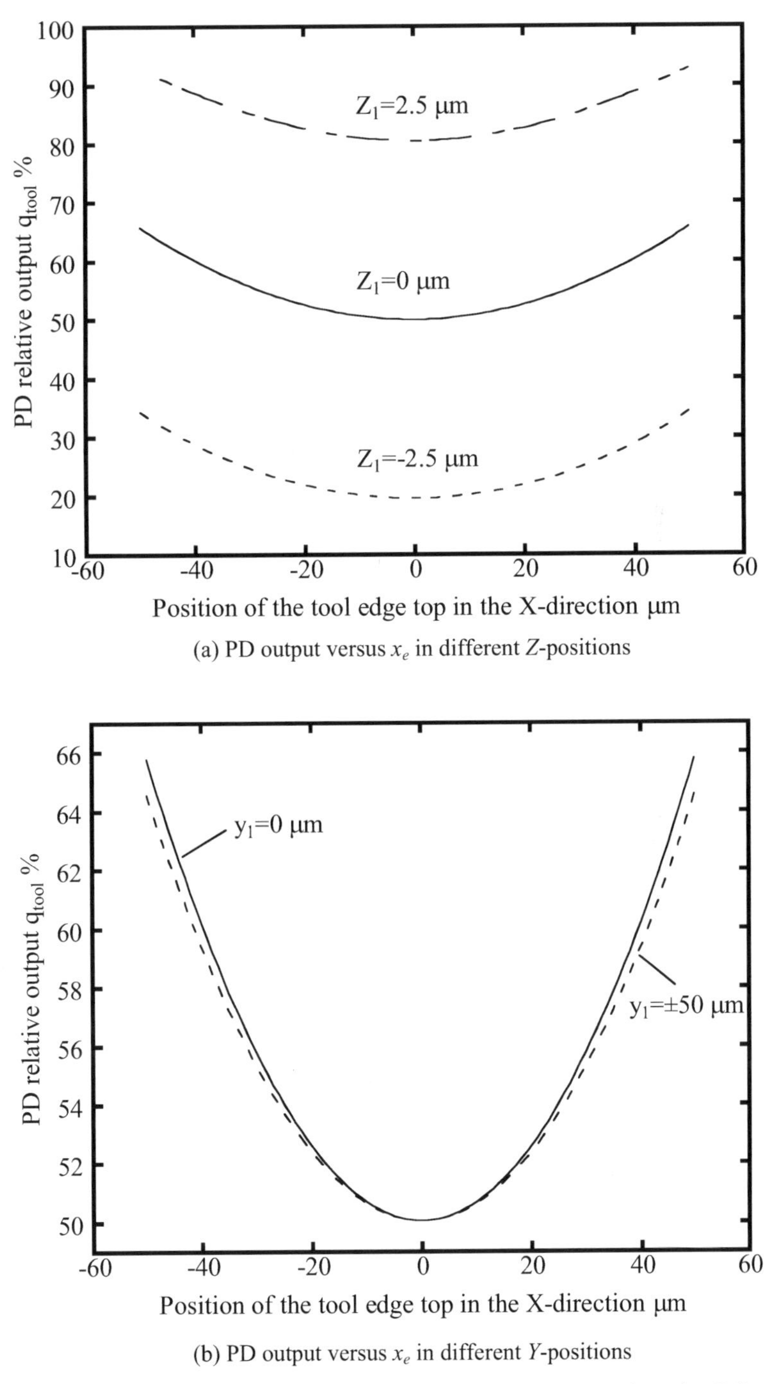

(a) PD output versus x_e in different Z-positions

(b) PD output versus x_e in different Y-positions

Figure 9.7. Simulation results for alignment of the tool edge top along the X-direction

It can be seen from the results in Figure 9.7 that the position of the tool edge top can be aligned to the *X*-directional center of the laser beam by moving the tool edge along the *X*-direction to track the minimum of the photodiode output. The initial values of y_e and z_e do not influence the alignment in the *X*-direction.

Step 2: Alignment of the tool edge top along the *Y*-direction

The second step of the alignment is to align the tool edge top along the *Y*-direction as shown in Figure 9.8.

Simulation results of aligning the tool edge top in the *Y*-direction are shown in Figures 9.9. x_e is assumed to be zero in Figure 9.9 (a). It can be seen in Figure 9.9 (a) that the photodiode output $s_{tool}(x_e, y_e, z_e)$ does not change with y_e for $z_e = 0$. On the other hand, $s_{tool}(x_e, y_e, z_e)$ becomes a function of y_e for $z_e \neq 0$, which reaches a minimum at $z_e > 0$ and a maximum at $z_e < 0$. Based on this result, the tool edge top can be aligned to the *Y*-directional center of the laser beam waist by tracking the maximum and/or minimum value of the PD output through moving the tool edge along the *Y*-direction. The value of z_e can be estimated to be positive when $s_{tool}(x_e, y_e, z_e)$ is at its maximum and be negative when $s_{tool}(x_e, y_e, z_e)$ is at its minimum value.

The simulation results of $s_{tool}(x_e, y_e, z_e)$ versus y_e for different x_e are shown in Figure 9.9 (b). z_1 is set to be −2.5 μm. It can be seen that $s_{tool}(x_e, y_e, z_e)$ always shows a minimum at y_e = 0 μm regardless of the initial x_e. This result can be utilized for the alignment of the tool edge top in the *Y*-direction.

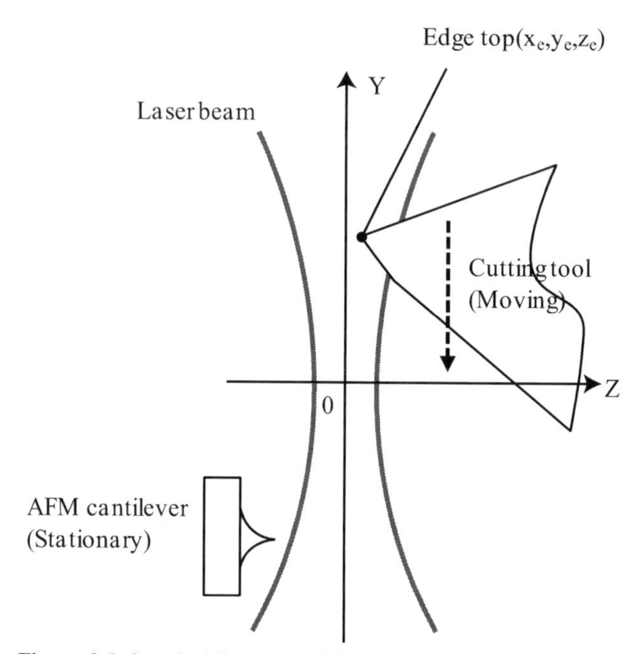

Figure 9.8. Step 2: Alignment of the tool edge top along the *Y*-direction

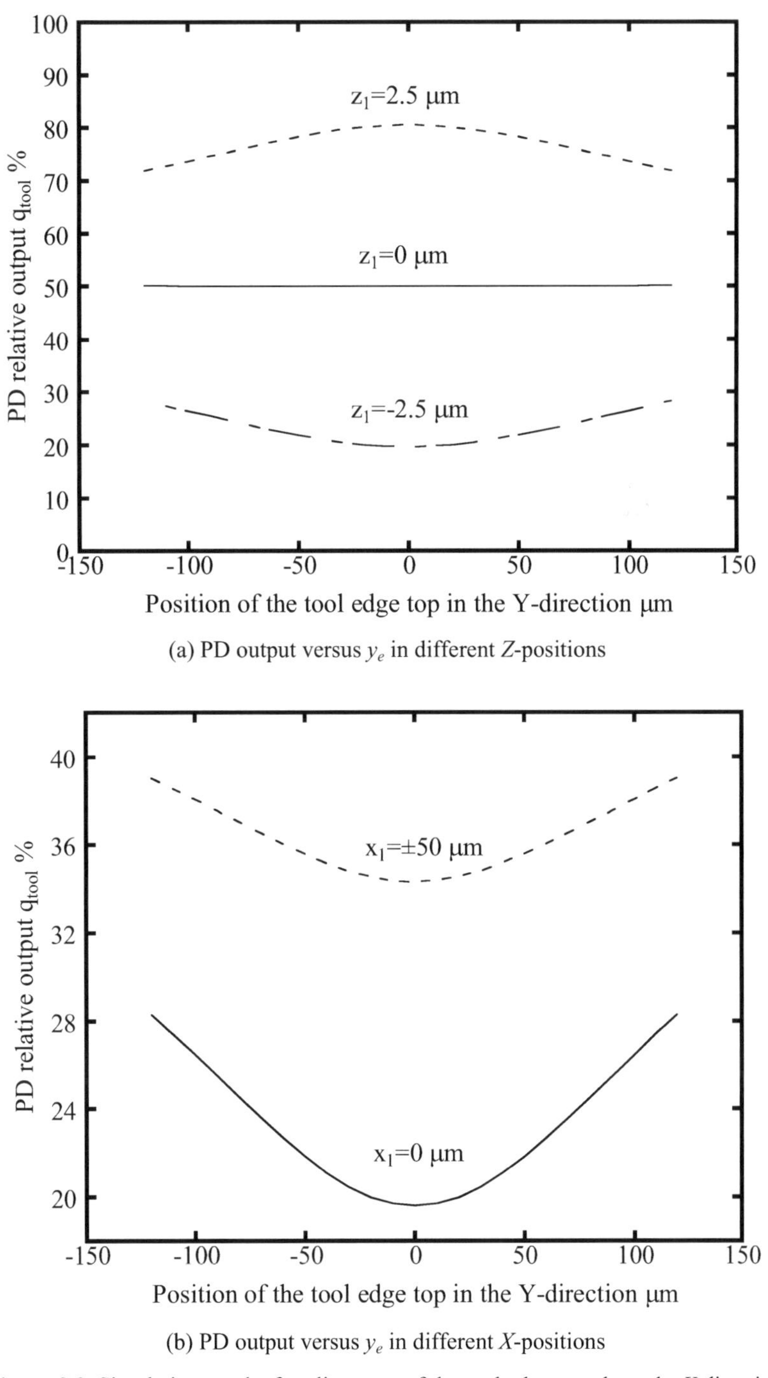

(a) PD output versus y_e in different Z-positions

(b) PD output versus y_e in different X-positions

Figure 9.9. Simulation results for alignment of the tool edge top along the *Y*-direction

Step 3: Alignment of the tool edge top along the Z-direction

The third step of the alignment is to align the tool edge top along the Z-direction as shown in Figure 9.10.

Figure 9.11 shows the simulation results of the photodiode output $s_{tool}(x_e, y_e, z_e)$ versus the Z-directional position z_e of the top of the tool edge. Figure 9.11 (a) shows the results obtained by changing x_e while y_e is kept zero. The curve of $s_{tool}(x_e, y_e, z_e)$ increases with z_e and shifts vertically with different x_e. The results when changing y_e while keeping x_e to be zero are shown in Figure 9.11 (b). Similar to that in Figure 9.11 (a), the curve of $s_{tool}(x_e, y_e, z_e)$ in Figure 9.11 (b) also increases with z_e, but there is no vertical shift when y_e changes. Obviously, $s_{tool}(x_e, y_e, z_e)$ cannot be directly used for positioning the tool edge to the beam center because the $s_{tool}(x_e, y_e, z_e)$ in Figures 9.11 (a) and 9.11 (b) does not have the maximum or minimum value with respect to z_e at the center of the laser beam.

To identify the tool edge top along the Z-direction, the derivative $s'_{tool}(x_e, y_e, z_e)$ of the curve $s_{tool}(x_e, y_e, z_e)$ in Figure 9.11 with respect to z_e is evaluated. Figure 9.12 shows the results. In Figure 9.12 (a), $s'_{tool}(x_e, y_e, z_e)$ has the maximum value at $z_e = 0$ when the tool edge top is at the X-directional center of the laser beam ($x_e = 0$). If x_e is not equal to zero, however, the Z-axis position of the maximum value of $s'_{tool}(x_e, y_e, z_e)$ will be off from the laser beam center, which makes the alignment along the Z-direction difficult. On the other hand, $s'_{tool}(x_e, y_e, z_e)$ always reaches its maximum at $z_e = 0$ in Figure 9.12 (b), where y_e can have different values, as long as $x_e = 0$. The results indicate that the derivative of the photodiode output can be utilized for positioning the tool edge top to the Z-directional center of the laser beam if the alignment in the X-direction has been accomplished.

Then, the tool is pulled out from the beam to a certain position along the Z-direction after the position of the tool edge relative to the beam center is recorded.

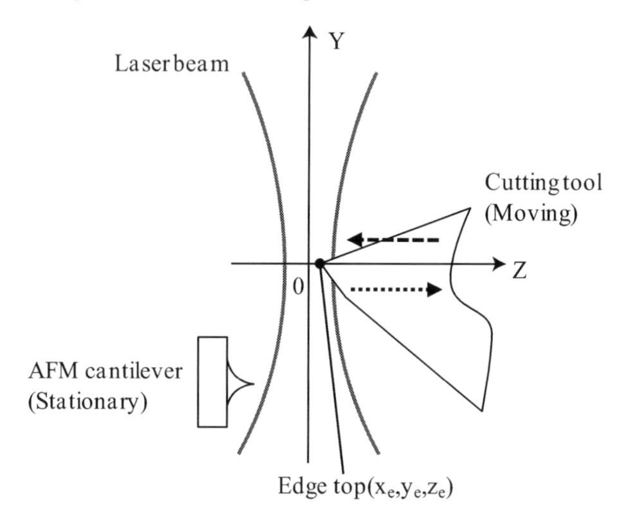

Figure 9.10. Step 3: Alignment of the tool edge top along the Z-direction

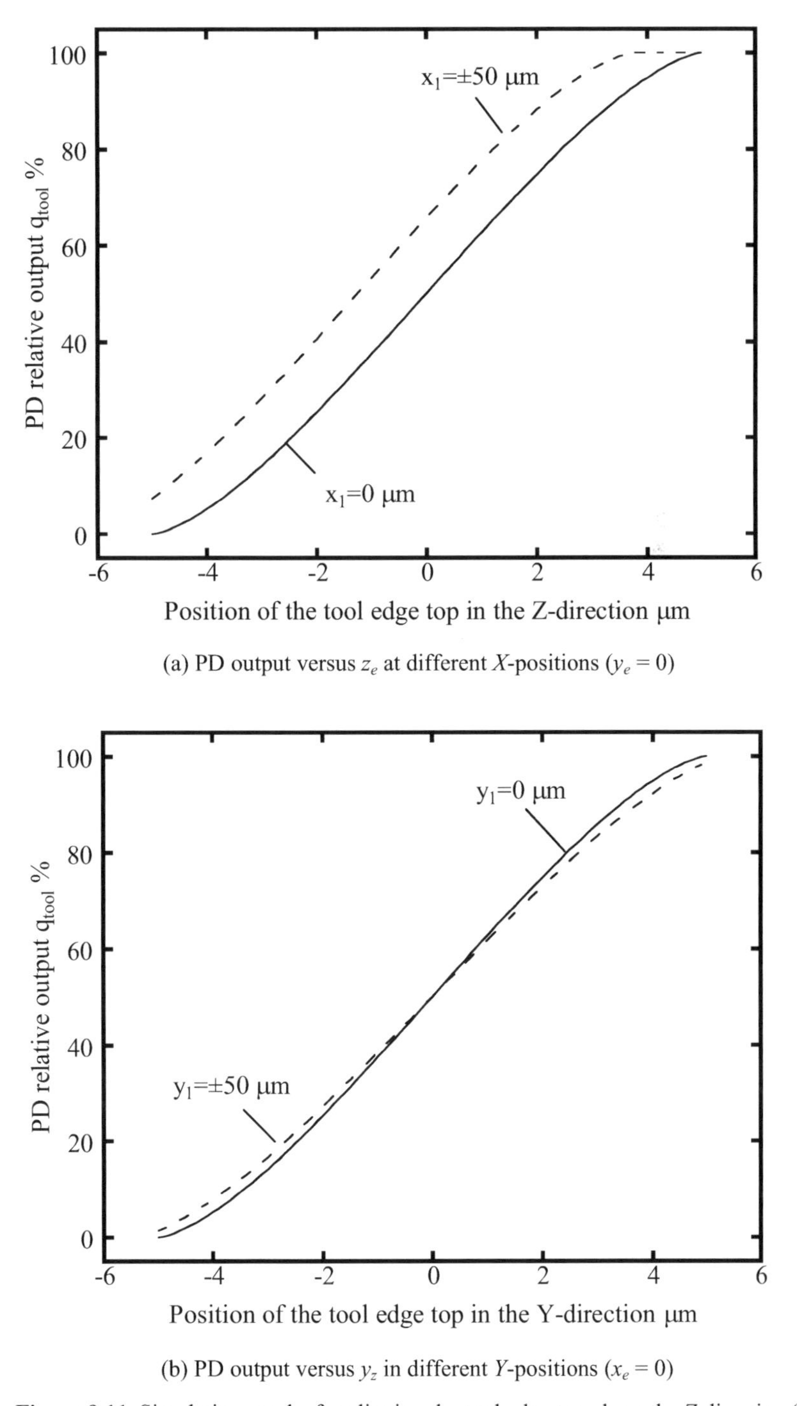

(a) PD output versus z_e at different X-positions ($y_e = 0$)

(b) PD output versus y_z in different Y-positions ($x_e = 0$)

Figure 9.11. Simulation results for aligning the tool edge top along the Z-direction (1)

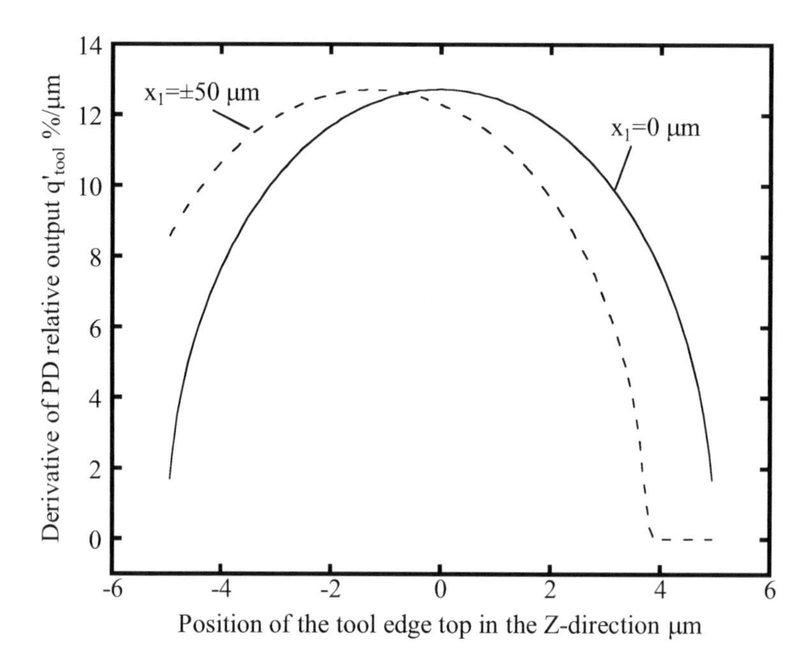

(a) Derivative of PD output versus z_e at different X-positions ($y_e = 0$)

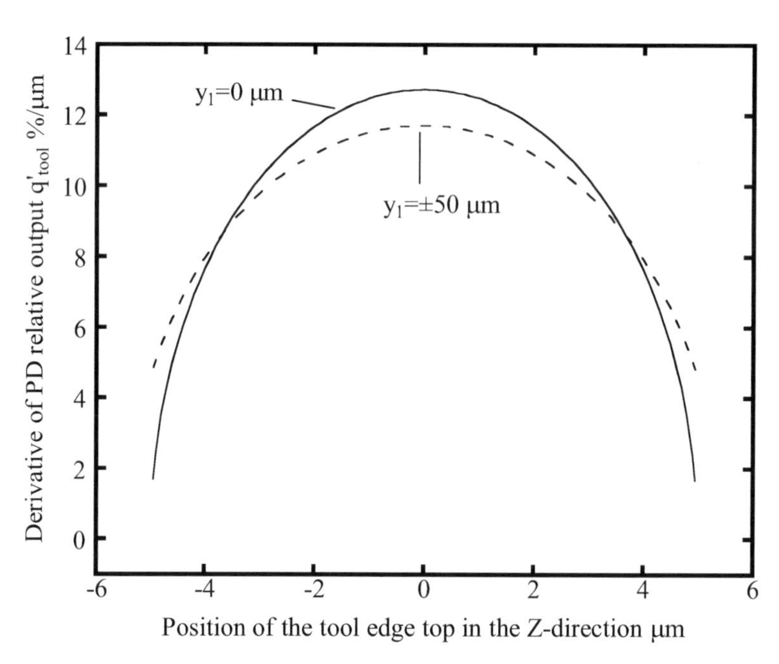

(b) Derivative of PD output versus y_z in different Y-positions ($x_e = 0$)

Figure 9.12. Simulation results for aligning the tool edge top along the Z-direction (2)

Steps 4 and 5: Alignment of the AFM probe tip along the *X*- and *Y*-directions

After the alignment of the tool edge top has been carried out, the tool is kept stationary and the AFM probe tip is aligned to the center of the laser beam waist along the *X*-direction shown in Figure 9.13 and then the *Y*-direction shown in Figure 9.14. The simulation results are shown in Figures 9.15 and 9.16, respectively. The AFM probe is assumed to have a triangular shape with a width of 10 μm and a height of 5 μm. It can be seen that the results are similar to those of the tool edge top although the size and the geometry of the AFM probe tip are quite different. The same alignment methods can thus be applied to both the tool edge top and the AFM probe tip.

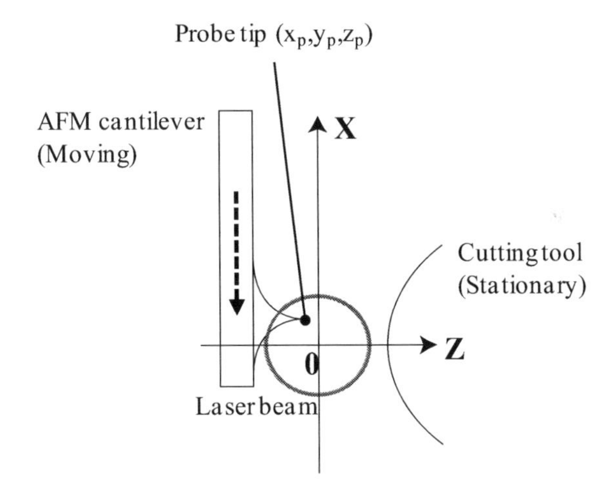

Figure 9.13. Step 4: Alignment of the AFM probe tip along the *X*-direction

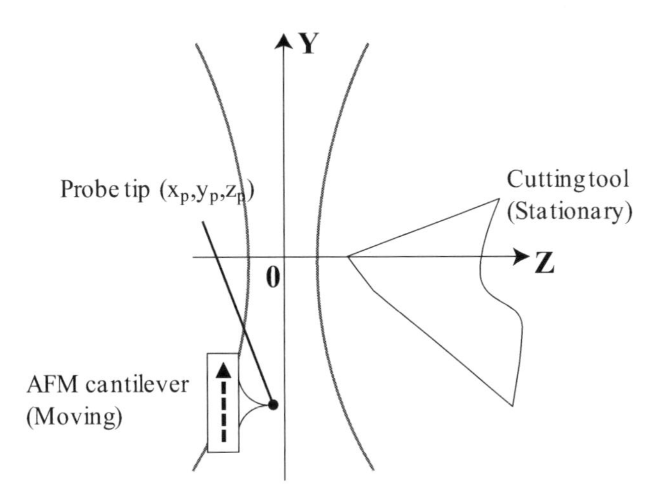

Figure 9.14. Step 5: Alignment of the AFM probe tip along the *Y*-direction

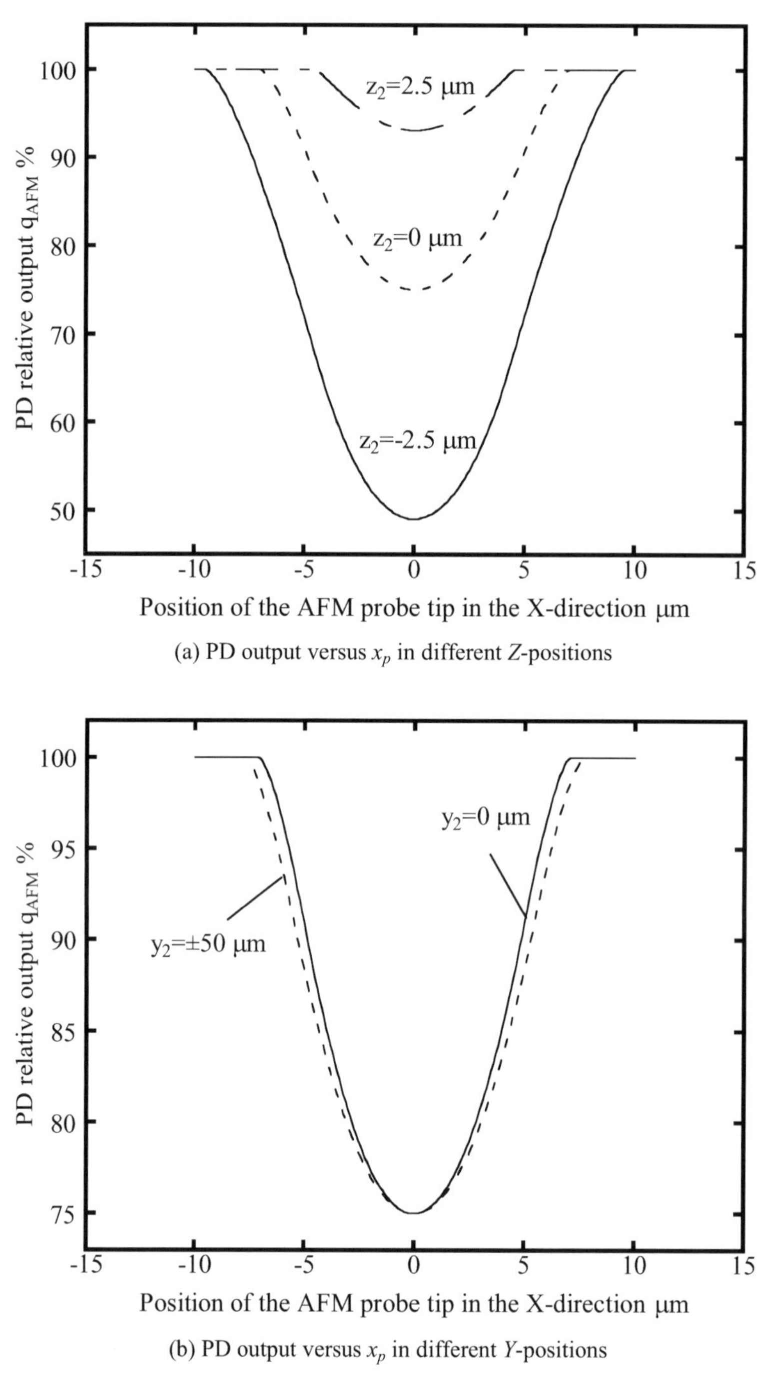

(a) PD output versus x_p in different Z-positions

(b) PD output versus x_p in different Y-positions

Figure 9.15. Simulation results for alignment of the AFM probe tip along the X-direction

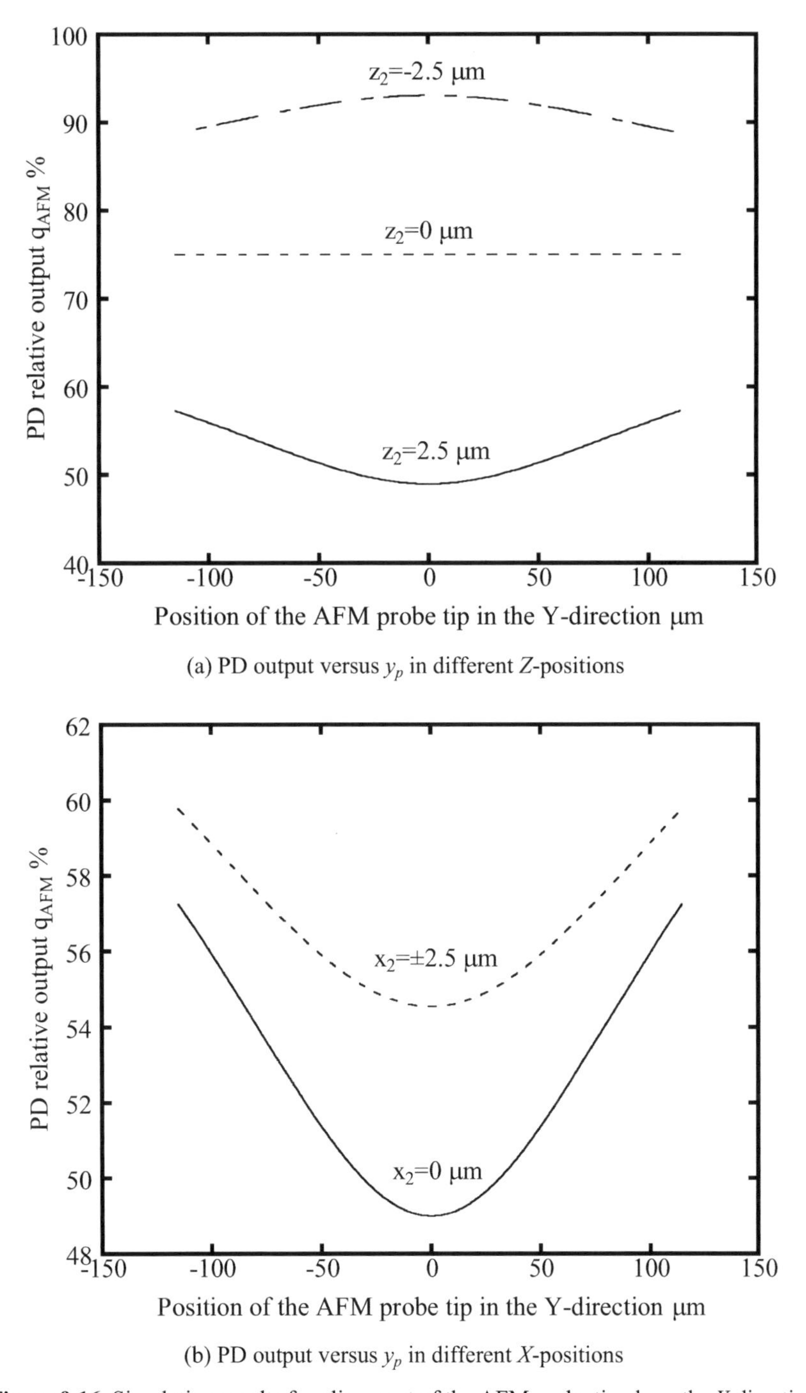

(a) PD output versus y_p in different Z-positions

(b) PD output versus y_p in different X-positions

Figure 9.16. Simulation results for alignment of the AFM probe tip along the Y-direction

Step 6: Making contact between the AFM probe tip and the tool edge top along the *Z*-direction

The tool edge top is returned to the center of the beam waist. Then the AFM probe tip is moved along the *Z*-direction to contact the top of the tool edge. The contacting process is the same as that carried out in a conventional AFM as shown in Figures 9.17 and 9.18.

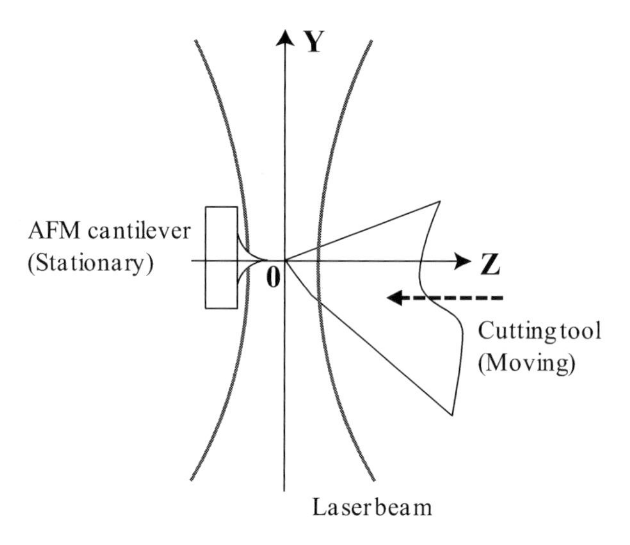

Figure 9.17. Step 6.1: Returning the tool

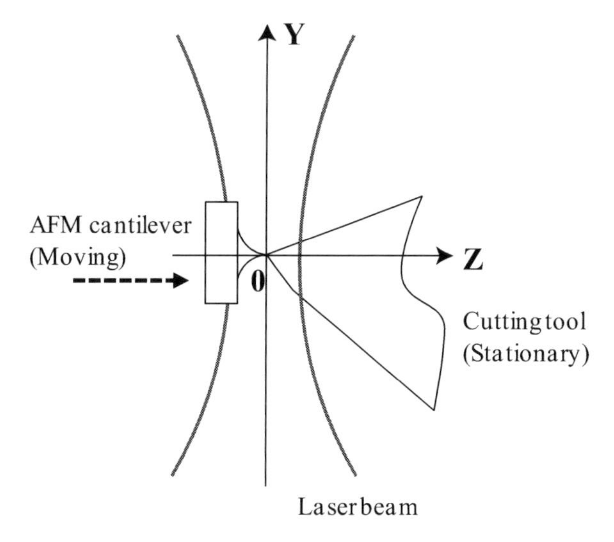

Figure 9.18. Step 6.2: Moving the AFM cantilever to make the contact

9.3 Instrumentation of the Automatic Alignment AFM

9.3.1 Instrument Design

Figure 9.19 shows a schematic of the automatic alignment AFM designed based on the alignment concept described in Section 9.2. The instrument is composed of four main parts: the AFM unit, the optical probe unit, the cutting tool unit and the electronics unit (note that the electronics unit is not shown in Figure 9.19). Each of the units is described as follows.

1. The AFM unit

The AFM unit is composed of an AFM scanner, an AFM cantilever and an alignment table. The AFM cantilever is set on the top of the AFM scanner, which is mounted on the alignment table. The AFM scanner is constructed by using three PZT actuators so that it can make *XY*-directional scanning and *Z*-directional servo control of the AFM cantilever. Each of the PZT actuators is servo controlled based

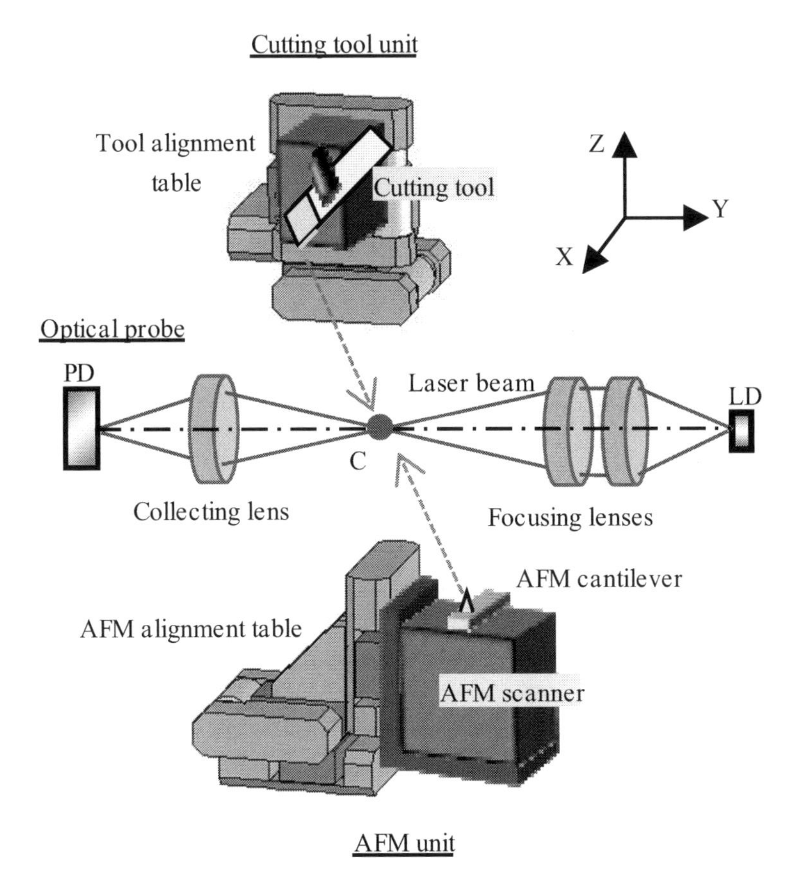

Figure 9.19. Schematic of the automatic alignment AFM

on the feedback signal from a capacitive-type displacement sensor embedded in the actuator. The stroke and resolution of the sensor is 100 μm and 1 nm, respectively. The closed-loop linearity of the motion of the PZT actuator is 0.02% of the travel range. The resonant frequency is approximately 790 Hz.

The piezo-resistive AFM cantilever is mounted on the top surface of the AFM scanner. The probe of the cantilever has a cone shape. The length of the cone probe has a length of 5 μm, a cone angle of 30º and a tip radius of less than 20 nm. The probe tip of the AFM cantilever can be moved by the AFM alignment table for the alignment in the *X*-, *Y*- and *Z*-directions, respectively. The table is driven by three linear stages actuated by servo motors. Each stage has a travel stroke of 15 mm, a maximum speed of 1.5 mm/s and a minimum incremental motion of 50 nm.

2. The optical probe unit

Figure 9.20 shows a schematic of the optical probe assembly to generate a focused light spot at the point *C*, which is employed as a reference point for the

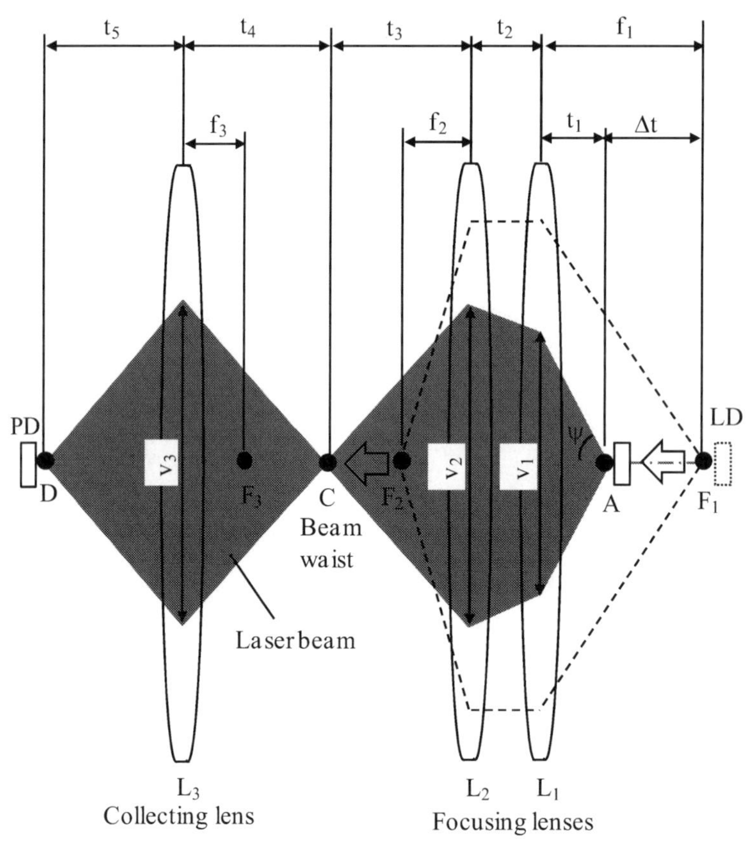

L1: collimation lens, L2: objective lens, L3: condenser lens

Figure 9.20. Design of the optical probe

alignment of the probe tip of the AFM cantilever and the edge top of the cutting tool. The laser beam from a laser diode (LD) is focused by using the lenses L_1 and L_2. The lens L_3 is employed for collecting the laser beam to a photodiode (PD). L_1, L_2, and L_3 are a collimation lens, an objective lens and a condenser lens, respectively.

As a design goal, the working distances (t_3, t_4) between the point C and the lenses L_2 and L_3 are set to be larger than 30 mm, which is required for mounting the micro-tool and the AFM unit. The focal lengths of the lenses are assumed to be f_1, f_2 and f_3, respectively. The LD is placed at the focal position F_1 of the lens L_1 in the typical arrangement of a focusing optics so that the laser beam from the LD can be collimated by L_1 and then focused at the focal position F_2 of the lens L_2. This is the simplest design for focusing a laser beam from a laser diode. However, the working distance t_3 is limited by the focal length of the objective lens L_3, which is generally shorter than 30 mm for getting a numerical aperture (NA) large enough. To increase the working distance t_3, the position of the LD is moved toward the lens L_1 from the point F_1 to the point A as shown in Figure 9.20. Letting the distance between F_1 and A be Δt, t_3 can be expressed as follows:

$$t_3 = \frac{f_1^2 f_2 - (f_1 - t_2) f_2 \Delta t}{f_1^2 - (f_1 + f_2 - t_2)\Delta t}, \tag{9.7}$$

where t_2 is the distance between lenses L_1 and L_2. The working distance t_3 is designed to be 32.9 mm by determining f_1, f_2, t_2 and Δt to be 20 mm, 25 mm, 10 mm and 3.5 mm, respectively.

The working distance t_4 is designed to be 30 mm by determining f_3 to be 15 mm based on the following relationship:

$$t_4 = t_5 = 2 f_3. \tag{9.8}$$

On the other hand, the diameter of each of the lenses should be designed to be larger than the beam diameter formed on the lens. The beam diameters v_1, v_2, v_3 on the lenses can be written as:

$$v_1 = 2t_1 \tan\psi, \tag{9.9}$$

$$v_2 = \frac{2(f_1^2 - f_1\Delta t + t_2\Delta t)\tan\psi}{f_1}, \tag{9.10}$$

$$v_3 = \frac{2(f_1^2 - f_1\Delta t + s_2\Delta t)t_4 \tan\psi}{f_1 t_3}, \tag{9.11}$$

where ψ is the beam divergence angle of the laser diode. v_1, v_2 and v_3 are evaluated to be 4.9 mm, 5.5 mm and 5.0 mm, respectively, where ω is 8.5°. The corresponding lens diameters are 12.5 mm, 7 mm and 15 mm, respectively, which are larger than v_1, v_2 and v_3. Figure 9.21 shows the light intensity of the focused laser beam at the beam waist (point *C*) measured by a beam profiler. The beam diameter was measured to be approximately 11 μm at point *C*.

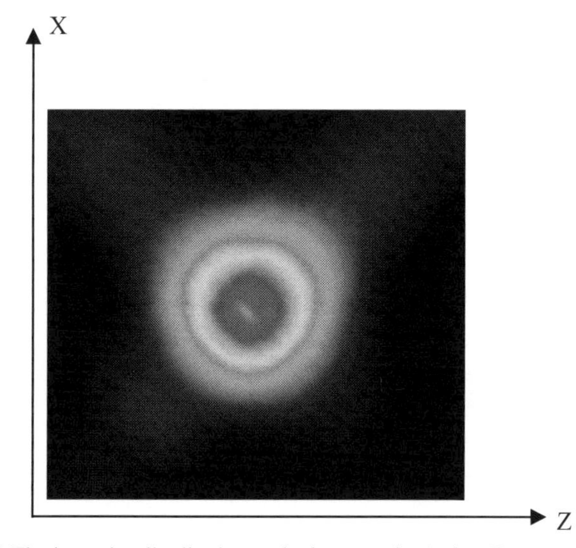

(a) The intensity distribution at the beam waist (point *C*)

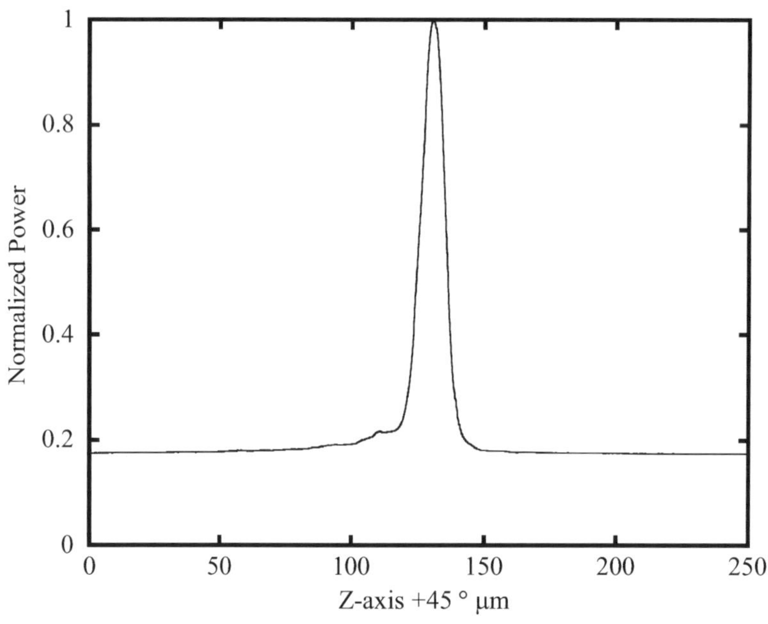

(b) Sectional profile of the beam intensity at point *C*

Figure 9.21. Intensity distribution of the focused laser beam

3. The cutting tool unit

The same type alignment table is employed in the cutting tool unit for the alignment of the edge top of the cutting tool clamped on the alignment table in Figure 9.19. The clamping mechanism for the cutting tool is designed in such a way that the tool can be easily mounted and removed from the cutting tool unit.

4. The electronics unit

The electronics unit is composed of two parts. One is the electronics for the optical probe and the other is for the profile measurement of the cutting edge by the AFM. As can be seen in Figure 9.22, the intensity of the laser is modulated by a sinusoidal signal with a certain frequency (10 kHz in the following experiment) from a function generator. The output of the photodiode, which has an AC component with the same frequency, is detected by a lock-in amplifier. The amplitude of the AC component in the photodiode output is a function of the positions of the AFM probe tip and/or the tool edge top. The output of the function generator is employed as the reference signal for the lock-in amplifier so that the amplitude of the AC component in the photodiode output can be detected in the presence of electronic noises. The output of the lock-in output is acquired by a personal computer via an analog-to-digital interface (ADC).

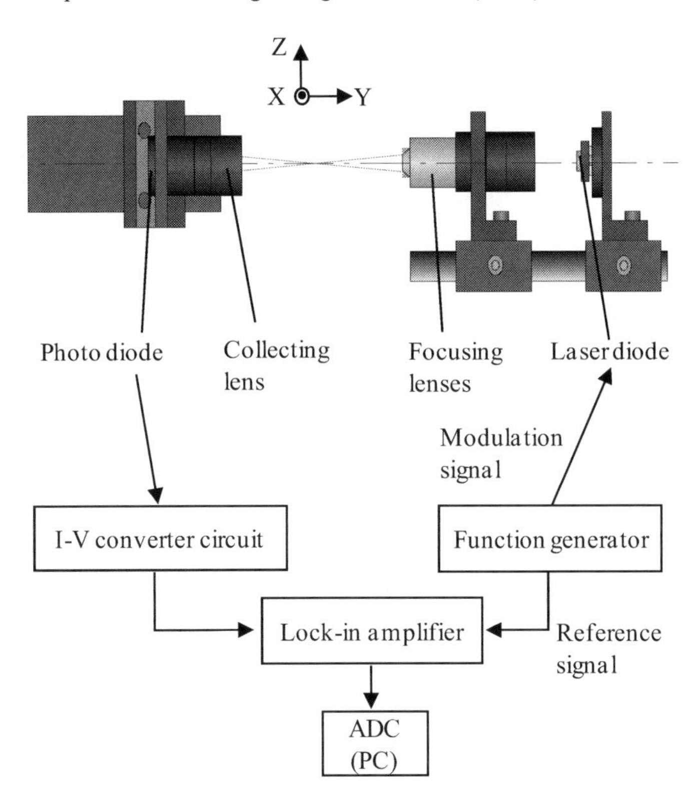

Figure 9.22. Electronics for the optical probe

Figure 9.23 shows the electronics for the AFM probe unit. The deflection of the AFM cantilever is converted to a resistance signal by the piezo-resistive sensor embedded in the cantilever, then to a voltage signal by the bridge circuit. The output of the bridge circuit is employed as the feedback signal for the PID controller of the *Z*-PZT in such a way that the deflection of the AFM cantilever is kept stationary. The displacement of the *Z*-PZT stage, which corresponds to the *Z*-directional height information of the tool edge profile, is detected by the capacitive sensor embedded in the stage. The voltage output of the capacitive sensor is input to a personal computer (PC) via an analog to digital converter. The reference signal for the PID controller is determined based on the contacting force between the AFM cantilever and the tool edge surface. The scanning of the AFM cantilever in the *X*- and *Y*-directions is carried out by moving the *X*- and *Y*-PZT stages, which

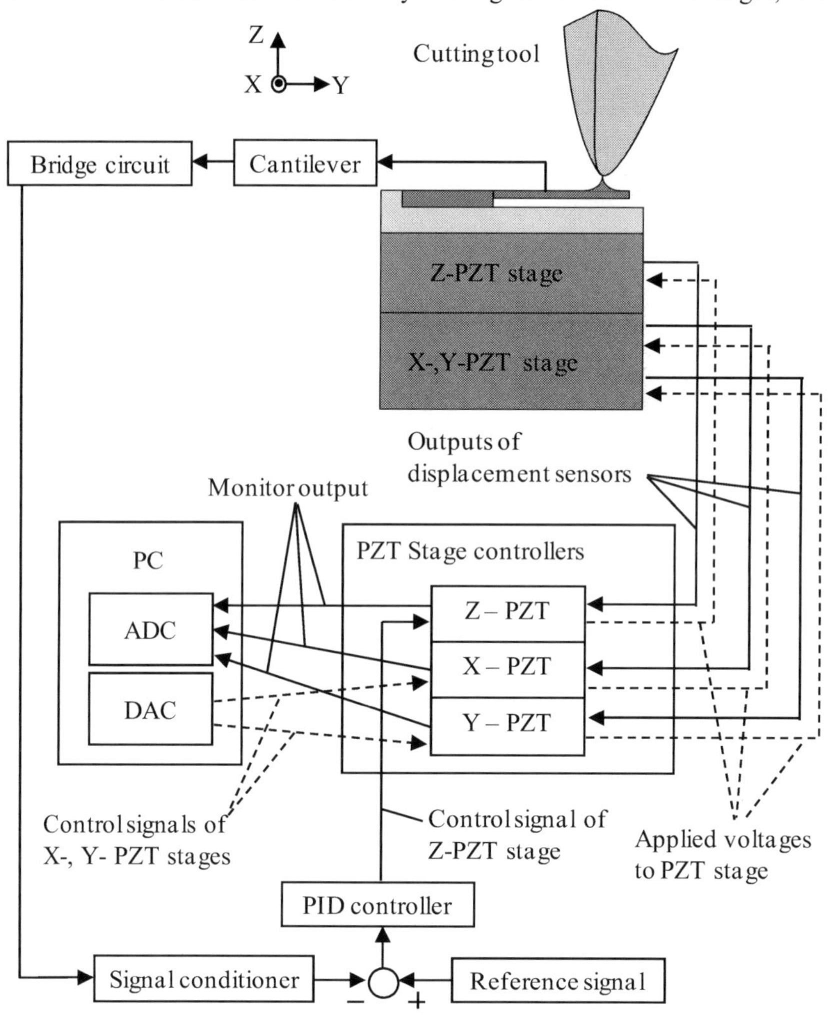

Figure 9.23. Electronics for the AFM probe unit

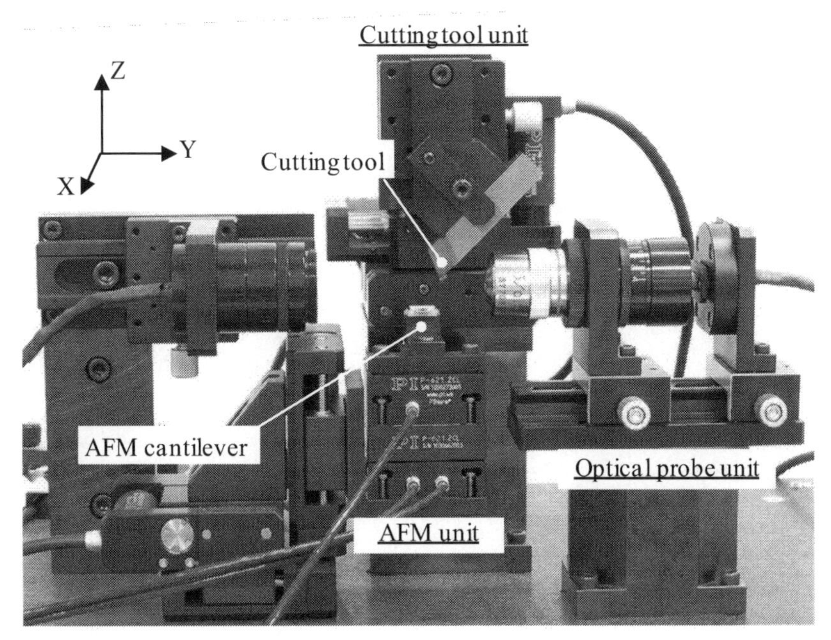

Figure 9.24. Photograph of the measuring instrument for 3D cutting edge profiles

are controlled by the PC via a digital to analog converter. The voltage outputs of the capacitive displacement sensors embedded in the *X*- and *Y*-PZT stages are input to the PC via the analog to digital convertor. Figure 9.24 shows a photograph of the designed and constructed measuring instrument.

9.3.2 Instrument Performance

The ability of the instrument to align the cutting tool and the AFM cantilever is first tested. The specimen was a single-crystal diamond cutting tool with a nose radius of 0.2 mm. The tool had a rake angle of 0° and a clearance angle of 7°. The AFM cantilever had a tip radius on the order of 20 nm and had not been used before the experiment.

Figures 9.25 and 9.26 show the results of positioning the tool edge along the *X*- and *Y*-directions, respectively. The error bar in the figure shows the full band of the deviation of the output data at each position. The data in each figure is fitted with a least mean square curve. In the experiment of Figure 9.25, the cutting tool was moved toward the center of the laser beam in steps of 1 μm by the *X*-directional servo motor-driven stage of the cutting tool unit. The curve of the lock-in amplifier output, which corresponds to the AC amplitude of the photodiode output, versus the *X*-displacement was in good agreement with the simulation result. The lock-in amplifier output attained its minimum value at $x = 0$ μm, which was the center of the laser beam along the *X*-direction. The positioning resolution was comparable to the 1-μm step. The resolution can be improved by utilizing the fitting curve.

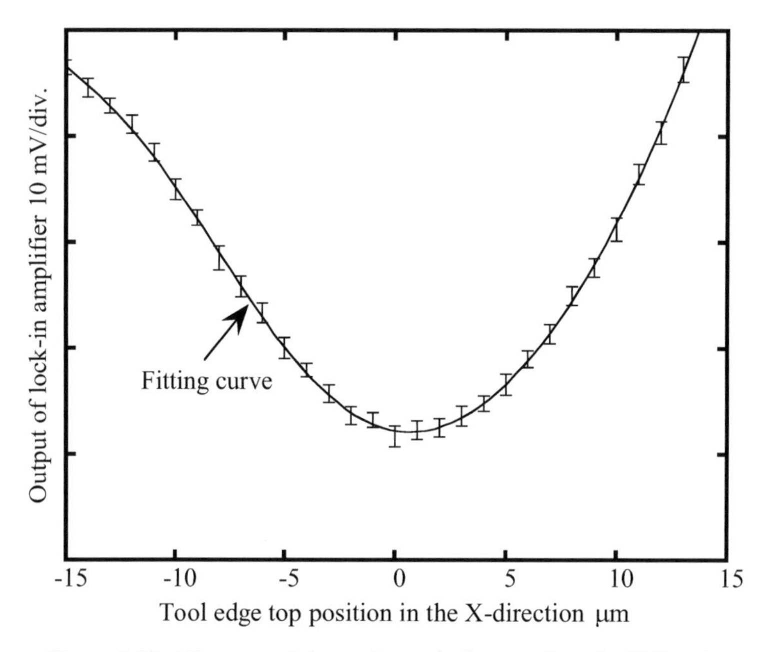

Figure 9.25. Alignment of the cutting tool edge top along the *X*-direction

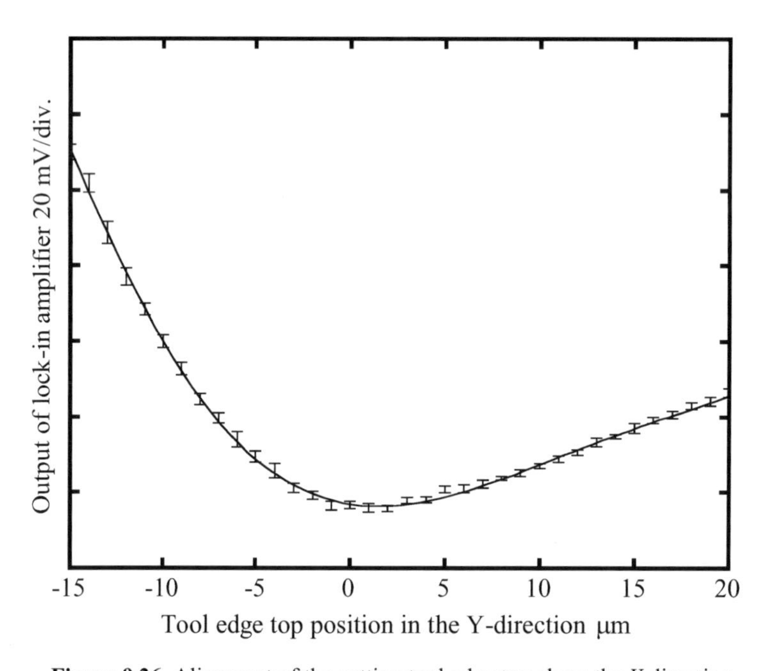

Figure 9.26. Alignment of the cutting tool edge top along the *Y*-direction

Similarly, the result in Figure 9.26 indicated that the tool edge top approached the center of the laser beam waist along the *Y*-direction at $y = 0$ μm. In the experiment (Figure 9.26), the cutting tool was moved toward the center of the laser beam in steps of 1 μm by the *Y*-directional servo motor-driven stage of the cutting tool unit. The *Y*-directional resolution of positioning was also on the order of 1 μm. Figure 9.27 shows the results obtained by scanning the tool edge top along the *Z*-direction in steps of 0.1 μm by the *Z*-directional servo motor-driven stage of the cutting tool unit. The derivative of the lock-in amplifier output was calculated from the average of the lock-in amplifier output at each position. As can be seen in the figure, the tool edge top approached the center of the beam waist at approximately $z = 0$ μm, where the derivative of the lock-in amplifier reached its maximum value. The positioning resolution along the *Z*-direction was approximately 0.5 μm.

At this point, the tool edge top had been aligned to the center of the laser beam waist in the *X*-, *Y*- and *Z*-directions. After the coordinates of the tool edge top at the beam waist center were recorded, the tool edge was pulled out from the laser beam, and the AFM probe was brought into the laser beam by the *X*- and *Y*-directional servo motor-driven stages with steps of 1 μm, respectively. Figures 9.28 (a) and 9.28 (b) show the results of positioning the AFM tip along the *X*- and *Y*-directions, respectively. It can be seen in Figure 9.28 (a) that the AFM tip was aligned to the center of the laser beam waist along the *X*-direction at $x = 0$ μm. The positioning resolution in the *X*-direction was approximately 1 μm. However, when aligning the AFM tip in the *Y*-direction, there are multiple peaks within the scanning range of 80 μm in the outputs shown in Figure 9.28 (b), which make the alignment difficult in this direction.

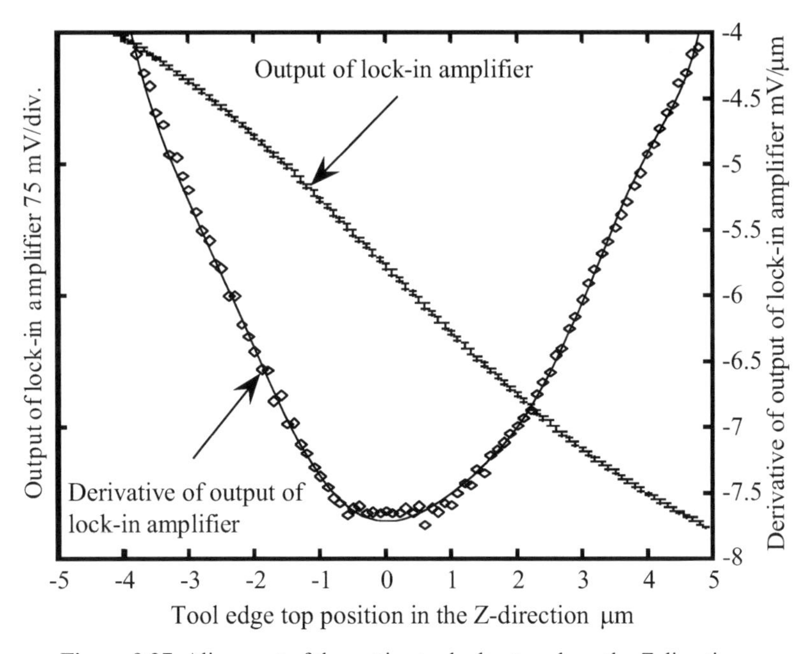

Figure 9.27. Alignment of the cutting tool edge top along the *Z*-direction

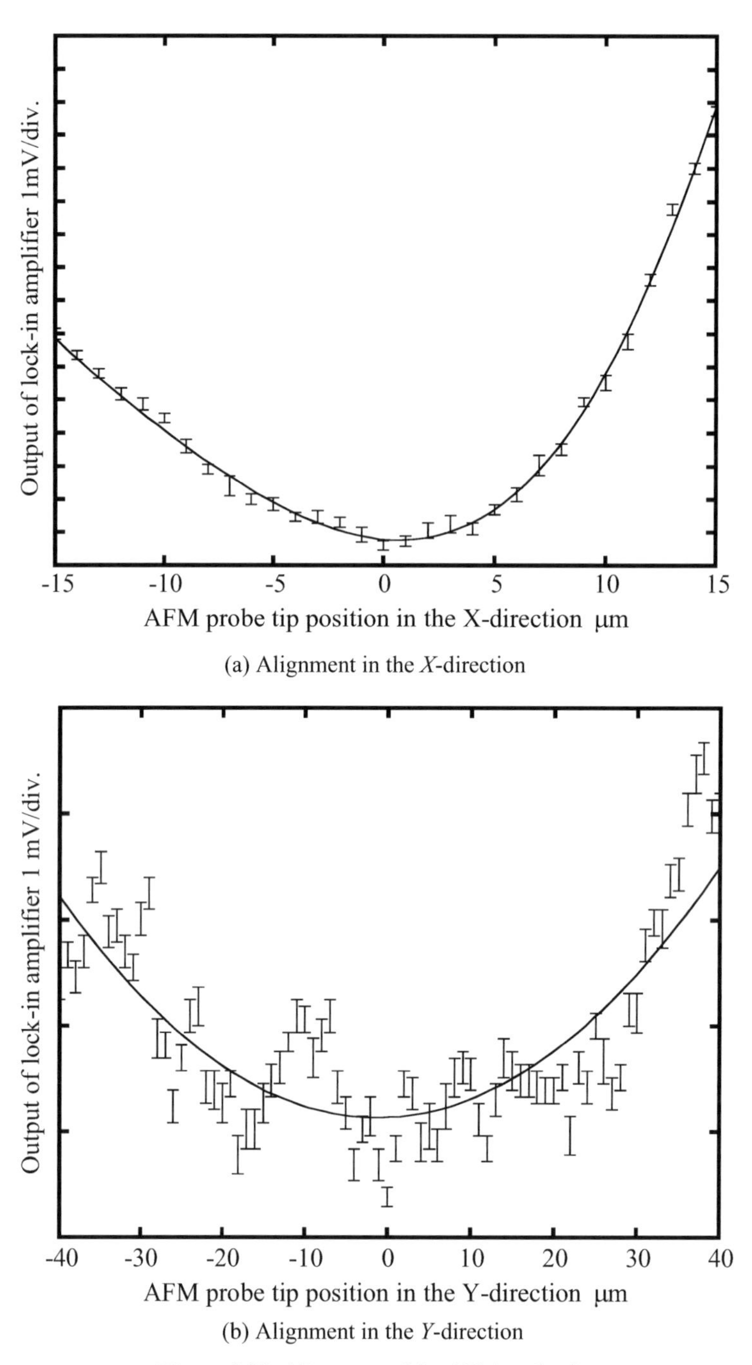

(a) Alignment in the *X*-direction

(b) Alignment in the *Y*-direction

Figure 9.28. Alignment of the AFM probe tip

The multiple peaks shown in the *Y*-directional output were caused by the light diffraction. Figure 9.29 shows an image of the laser light spot. The diffraction pattern can be observed from the image. To solve this problem, the alignment procedure is improved as shown in Figure 9.30.

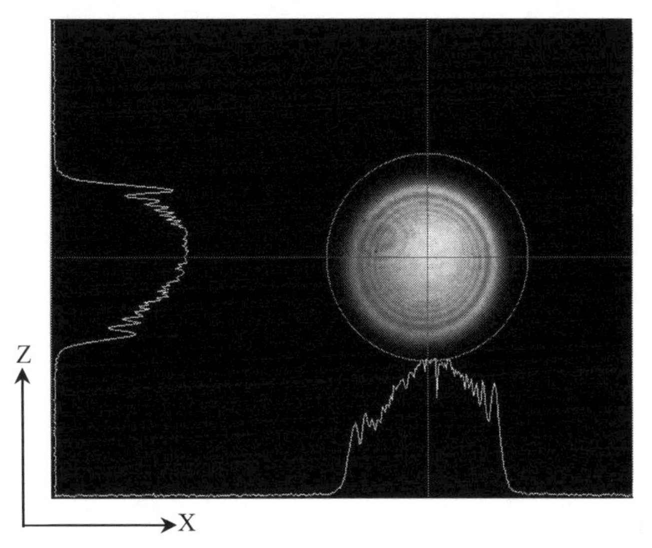

Figure 9.29. Light diffraction caused by the imperfection of the optics

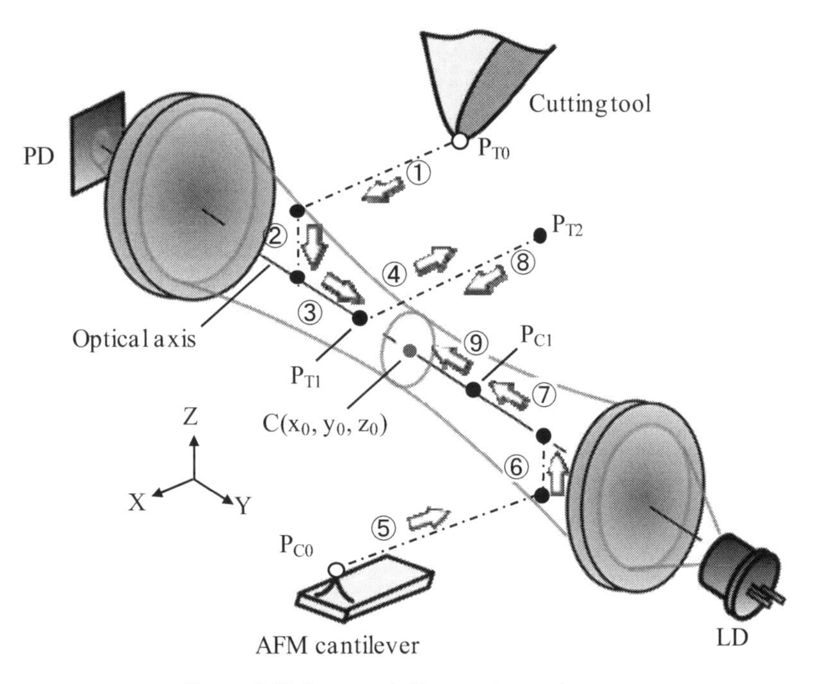

Figure 9.30. Improved alignment procedure

Assume that the cantilever tip and the cutting edge top are at P_{C0} and P_{T0} with unknown coordinates before the alignment, respectively. The coordinates of point C, which are also unknown, are assumed to be (x_0, y_0, z_0). The alignment steps are as follows:

Step 1: Move the tool alignment table along the X-direction until the X-coordinate of the cutting edge top of the tool becomes x_0 by tracking the minimum of the output of the PD.

Step 2: Move the tool alignment table along the Z-direction until the Z-coordinate of the cutting edge top of the tool becomes z_0 by tracking the maximum of the derivative of the PD output.

Step 3: Move the tool alignment table along the Y-direction until the Y-coordinate of the cutting edge top of the tool gets close to y_0 by tracking the minimum of the PD output. Record the position of the alignment table at P_{T1}.

Step 4: Move the tool alignment table until the tool gets to a point P_{T2}, which is out of the laser beam.

Step 5: Move the AFM alignment table along the X-direction until the X-coordination of the AFM cantilever tip becomes x_0 by tracking the minimum of the PD output.

Step 6: Move the AFM alignment table along the Z-direction until the Z-coordinate of the AFM cantilever tip becomes z_0 by tracking the minimum of the PD output.

Step 7: Move the AFM alignment table along the Y-direction until the Y-coordinate of the AFM cantilever tip gets close to y_0 by tracking the minimum of the PD output.

Step 8: Move the tool alignment table until the tool returns back to P_{T1}.

Step 9: Move the AFM alignment table along Y-direction until the AFM cantilever tip touches the cutting edge top of the micro-tool.

Once the alignment was accomplished in the experiment, the AFM probe tip was made to contact the tool edge top for 3D edge profile measurement. The tool edge top was kept stationary and the AFM probe tip was moved by the Z-directional PZT stage of the AFM unit toward the tool edge top along the Z-direction to obtain the force curve, which was the relationship between the contact force and the displacement of the AFM probe. The contact force was calculated from the piezo-resistive sensor of the AFM cantilever. The result is shown in Figure 9.31. As can be seen in the figure, the AFM probe tip was made to contact with the tool edge top when the displacement was 4.3 μm. The reference contact force was determined to be from 0.1 to 0.4 μN for the profile measurement with the AFM unit.

Figure 9.32 shows the 3D edge profile measurement of the cutting tool with a nose radius of 0.2 mm. The tool edge was scanned in a Y-directional raster pattern with equal X-steps. The scanning range in the X-direction was 40 μm and the number of scanning lines was 200. The range in the Y-direction was 1.65 μm and the number of lines was 270. The measurement time was 200 s, which was mainly limited by the servo control electronics of the AFM unit. It can be seen that the 3D edge profile of the cutting edge was measured successfully.

The alignment and measurement ability of the instrument for micro-cutting tools with nose radius less than 100 μm was also tested.

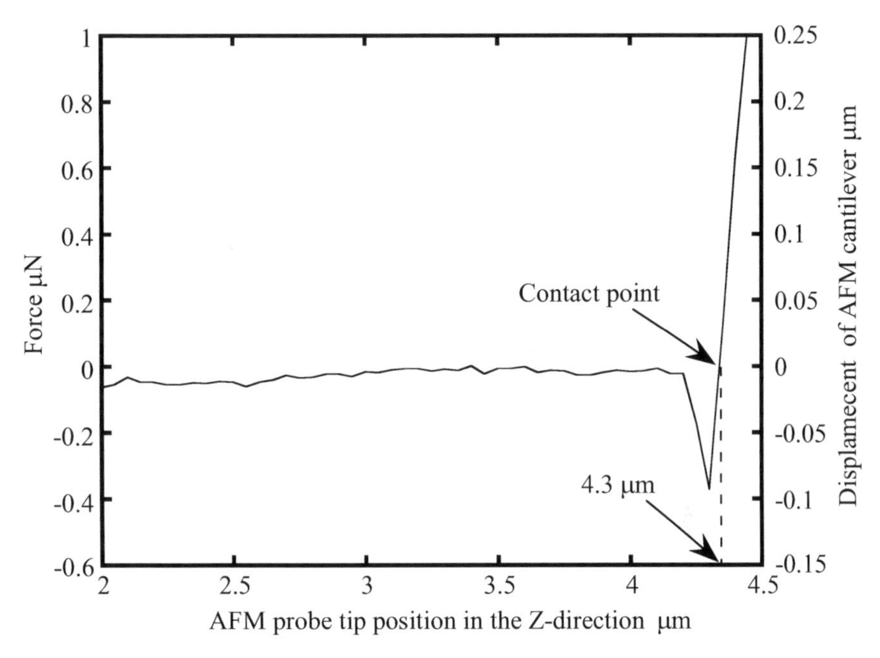

Figure 9.31. Force curve and the reference force for profile measurement

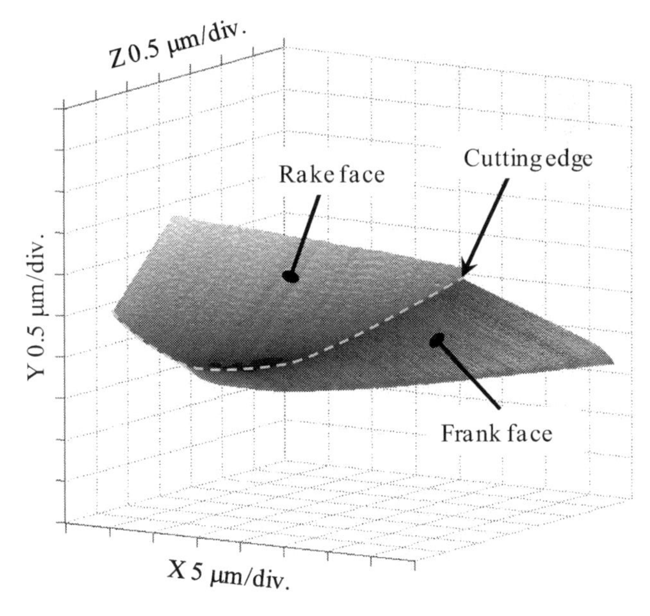

Figure 9.32. 3D profile measurement result of a round nose shape cutting tool with a nose radius of 0.2 mm

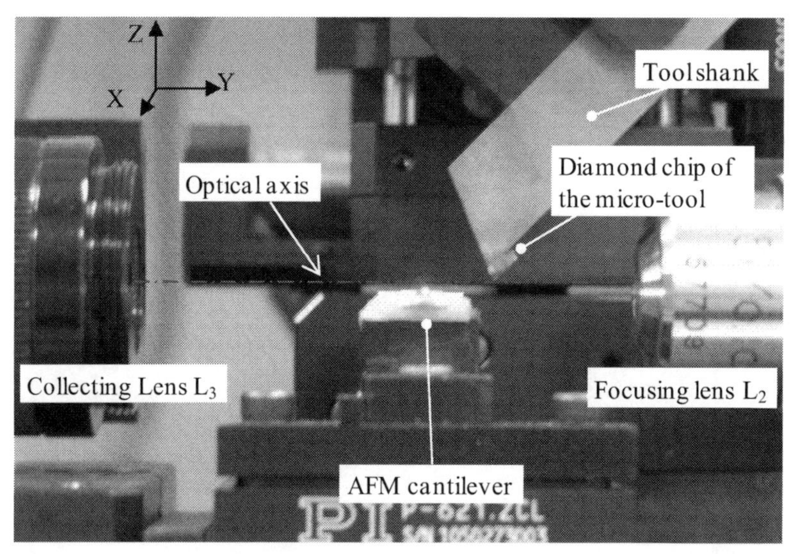

Figure 9.33. Alignment and measurement of a micro-tool with a nominal nose radius of 1.5 μm in the measuring instrument

Figure 9.33 shows a photograph of the AFM cantilever and a micro-tool with a nominal nose radius of 1.5 μm in the measuring instrument under alignment. As can be seen in the figure, the single-crystal diamond chip of the micro-tool was bonded on a stainless shank. The rake angle and the clearance angle were 0° and 7°, respectively.

Figures 9.34 to 9.36 show the alignment results of the micro-tool in the *X*-, *Z*- and *Y*-directions, respectively. During the alignment in the *X*-direction, the tool was moved toward the optical axis of the laser beam by the *X*-stage of the tool alignment table with a step of 0.7 μm and a velocity of 1 step/s. As can be seen in Figure 9.34, the output of the PD reached the minimum value of 3.8 V when the displacement of the *X*-stage was 36 μm, which means that the cutting edge top of the micro-tool was aligned to the optical axis along the *X*-direction. Then the tool was moved by the *Z*-stage of the tool alignment table with a step of 0.7 μm and a velocity of 1 step/s for alignment in the *Z*-direction. As can be seen in Figure 9.35, the derivative of the PD output reached the maximum of 0.29 V/μm when the cutting edge top of the tool was positioned at the optical axis. The displacement of the *Z*-stage was –45 μm. Finally, the tool was moved by the *Y*-stage of the tool alignment table with a step of 7 μm and a velocity of 1 step/s for the alignment in the *Y*-direction. Figure 9.36 shows the result. The tool was positioned around the laser beam waist when the PD output reached the minimum of 2.0 V and the displacement of the *Y*-stage was 630 μm. The tool was then moved out from the laser beam after the positions of the *X*-, *Y*- and *Z*-stages of the tool alignment table had been recorded so that the alignment of the AFM cantilever tip could be carried out. The sensitivity of the alignment in the *Y*-direction was lower than those in the *X*- and *Z*-directions. The alignment resolutions in the *X*- and *Z*-directions were approximately 1 μm and that in the *Y*-direction was approximately 20 μm.

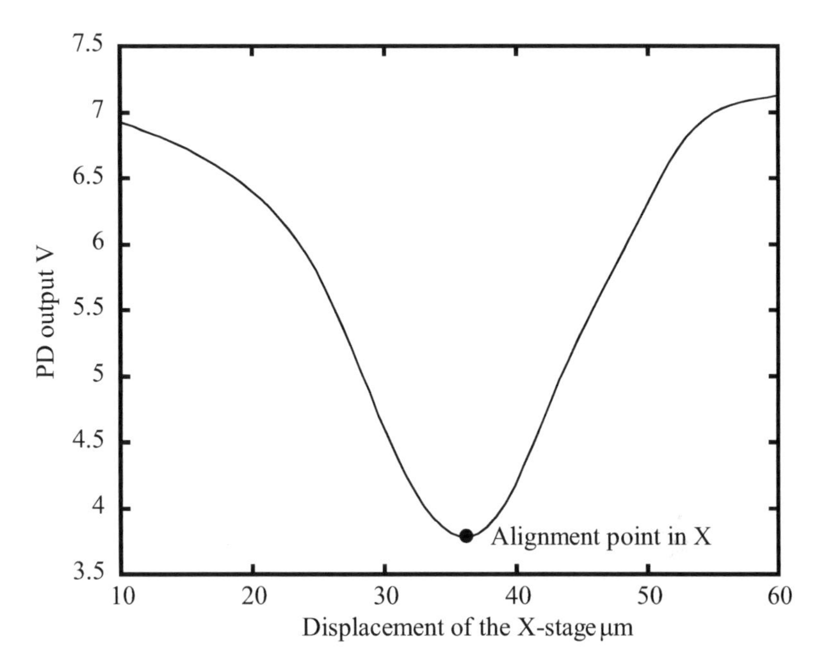

Figure 9.34. Alignment result of the micro-tool with a nominal nose radius of 1.5 μm in the *X*-direction

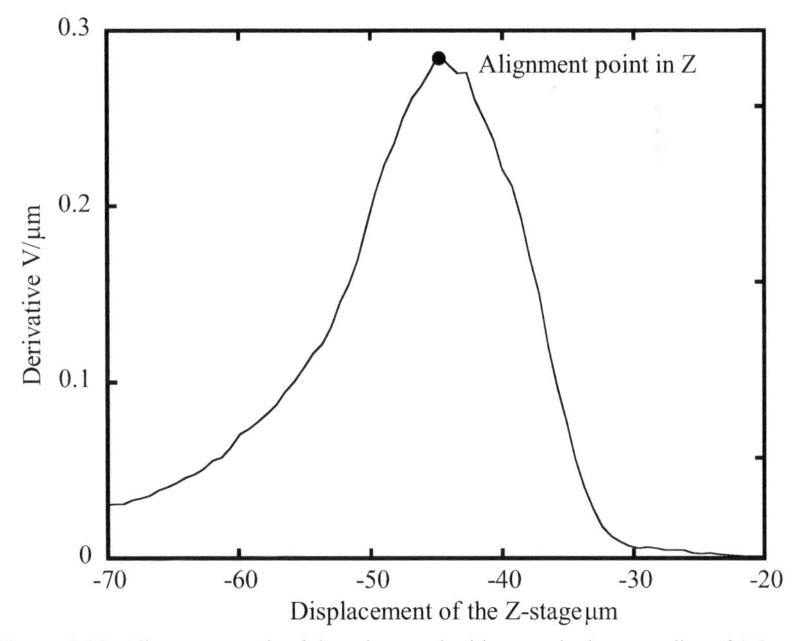

Figure 9.35. Alignment result of the micro-tool with a nominal nose radius of 1.5 μm in the *Z*-direction

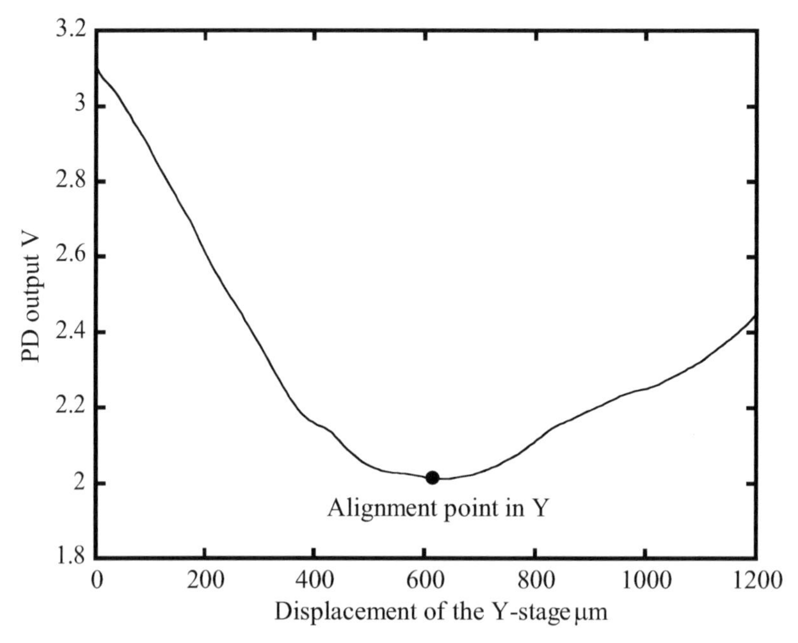

Figure 9.36. Alignment result of the micro-tool with a nominal nose radius of 1.5 μm in the *Y*-direction

To solve the problem of misalignment in the *Y*-direction, the AFM cantilever was moved by the *Y*-stage of the AFM alignment table to make contact to the tool cutting edge for profile measurement by the AFM. During this process, the deflection of the AFM cantilever was monitored through measuring the resistance output of the cantilever. Once the contact was established, the *Z*-directional position of the cantilever was servo controlled by the *Z*-PZT actuator in such a way that the deflection as well as the resistance output of the piezo-resistive cantilever was kept stationary. The measuring force of the AFM, which was proportional to the deflection of the AFM cantilever, was set to be 0.1 μN.

Figure 9.37 shows the result of the 3D cutting edge profile of the micro-tool. The scanning of the AFM tip was carried out in a *Y*-directional raster pattern with equal *X*-steps by using the *X*- and *Y*-PZT actuators. The scanning ranges were 5 μm (*X*) × 2.3 μm (*Y*). The number of scanning points in the *Y*-direction was 500 and that of scanning lines in the *X*-direction was 100. The scanning time was approximately 3.5 min. The measurements were repeated three times. Sectional profiles in the *YZ*-plane were extracted for evaluation of the cutting edge sharpness (ρ), the nose radius (*r*) and the edge contour out-of-roundness as shown in Figure 9.38. The three repeated results are plotted in the figure. The edge sharpness was evaluated to be approximately 40 nm with a repeatability of 5 nm. Figure 9.39 shows the evaluated results of out-of-roundness. The out-of-roundness was evaluated to be 42 nm with a repeatability of 8 nm. The nose radius was evaluated to be 1.520 μm with a repeatability of 10 nm. Figure 9.40 shows a photograph of the instrument for inspection of the cutting tools. Figure 9.41 shows the measurement results of micro-cutting tools by the instrument.

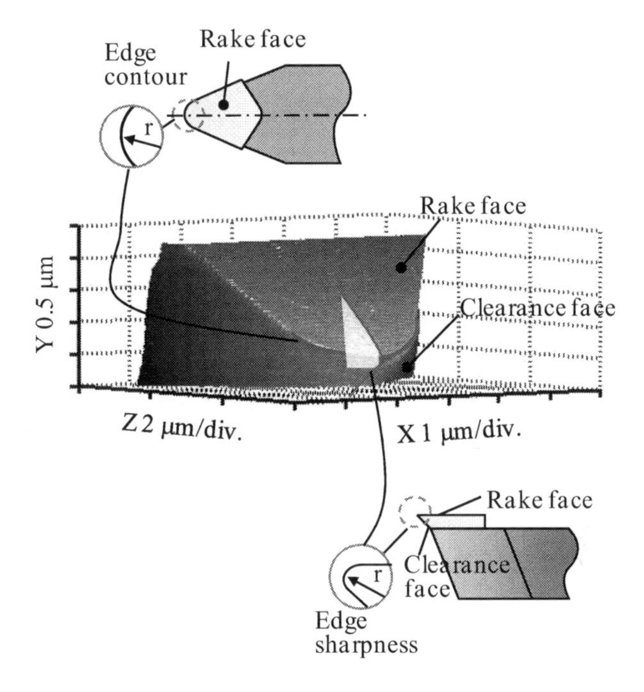

Figure 9.37. Measurement result of the 3D edge profile of the micro-cutting tool with a nominal nose radius of 1.5 μm

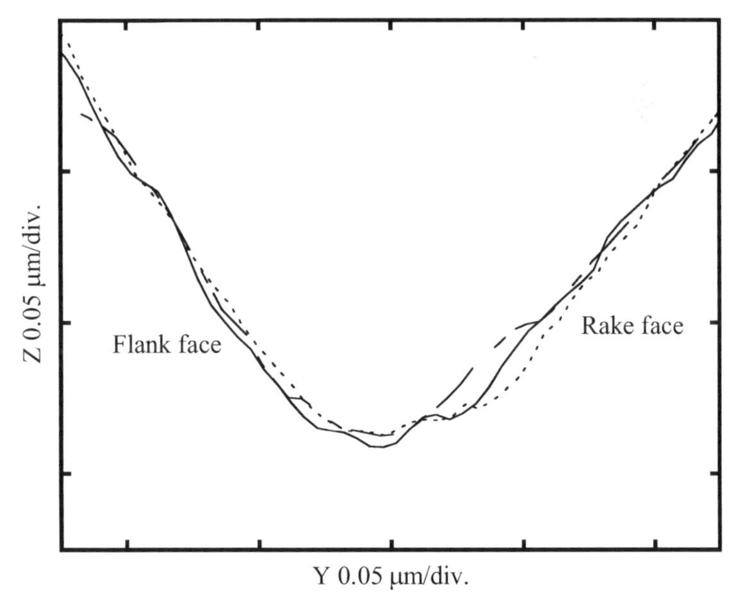

Figure 9.38. Measurement result of the edge sharpness of the micro-cutting tool with a nominal nose radius of 1.5 μm

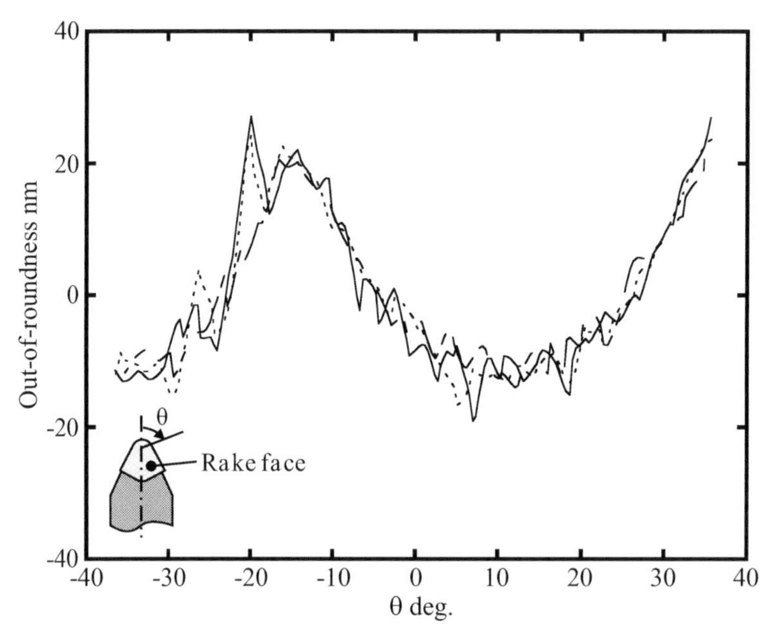

Figure 9.39. Measurement result of out-of-roundness of the micro-cutting tool with a nominal nose radius of 1.5 μm

Figure 9.40. Photograph of the measuring instrument developed for inspection of diamond cutting tools

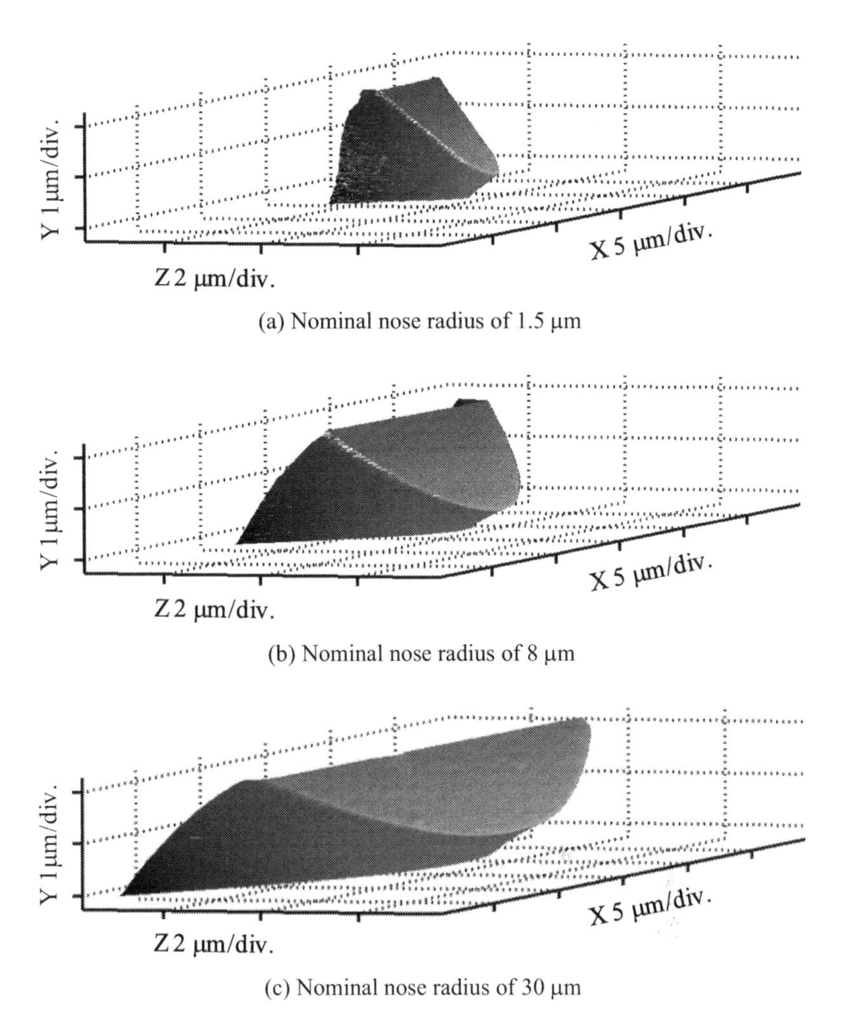

(a) Nominal nose radius of 1.5 μm

(b) Nominal nose radius of 8 μm

(c) Nominal nose radius of 30 μm

Figure 9.41. 3D measurement results of micro-tools

9.4 Summary

An optical alignment sensor has been proposed for automatic alignment of the probe tip of an AFM cantilever and the cutting edge top of a single-point diamond tool. The laser beam with its optical axis along the *Y*-direction from the laser source of the optical sensor is focused to form a beam waist near the AFM probe tip and the top of the cutting tool edge. The laser beam is then collected by a lens and is directed to a photodiode. The output of the photodiode, which is verified to be a function of the *X*-, *Y*- and *Z*-positions of the AFM tip and/or the tool edge top, is utilized for the alignment of the AFM tip with the tool edge top. It has also been verified that it is necessary to align the tool edge top along the *X*-direction before aligning it in the *Z*-direction.

A measuring instrument has been designed and constructed based on the proposed optical alignment probe. The instrument consists of an AFM unit for precision profile measurement of the cutting edge in addition to the optical probe. The AFM unit employs closed-loop controlled PZT scanners with travel strokes of 100 μm. The linearity of the scanning motion, which directly influences the measurement accuracy of the AFM, is 0.02% of the scanning range. The optical probe is designed to have enough working space for the AFM unit and the tool. The AFM unit and the micro-tool are mounted on alignment tables driven by *X*-, *Y*- and *Z*-directional linear stages for fast alignment of the AFM cantilever and the micro-tool cutting edge.

Round nose tools with different nominal nose radii have been measured to demonstrate the feasibility of the measuring instrument. The smallest tool had a nominal nose radius of 1.5 μm. The time for mounting and aligning the AFM cantilever and the 1.5 μm tool was approximately 20 min and that for measuring the 3D cutting edge profile of the tool was approximately 3.5 min. The cutting edge sharpness, the nose radius and the edge contour out-of-roundness of the tool have been measured to be 40 nm, 1.520 μm and 42 nm, respectively.

References

[1] Krauskopf B (1984) Reflecting demands for precision. Manuf Eng 92(5):90–100
[2] Moriwaki T, Okuda K (1989) Machinability of copper in ultra-precision micro diamond cutting. Ann CIRP 38(1):115–118
[3] Ikais N, Shimada S, Tanaka H (1992) Minimum thickness of cut in micromachining. Nanotechnology 3:6–9
[4] Bruzzone AAG, Costa HL, Lonardo PM, Lucca DA (2008) Advances in engineered surfaces for functional performance. Ann CIRP 57(2):750–769
[5] Ohmori H, Katahira K, Naruse T, Uehara Y, Nakao A, Mizutani M (2007) Microscopic grinding effects on fabrication of ultra-fine micro tools. Ann CIRP 56(1):569–572
[6] Paul E, Evans CJ, Mangamelli A, McGlauflin ML, Polvani RS (1996) Chemical aspects of tool wear in single point diamond turning. Precis Eng 18:4–19
[7] Gao W, Araki T, Kiyono S, Okazaki Y, Yamanaka M (2003) Precision nano-fabrication and evaluation of a large area sinusoidal grid surface for a surface encoder. Precis Eng 27(3):289–298
[8] Maeda Y, Uchida H, Yamamoto A (1989) Measurement of the geometric features of a cutting tool edge with the aid of a digital image processing technique. Precis Eng 11(3):165–171
[9] Doiron TD (1989) Computer vision based station for tool setting and tool form measurement. Precis Eng 11(4):231–238
[10] Oomen JM, Eisses J (1992) Wear of monocrystalline diamond tools during ultraprecision machining of nonferrous metals. Precis Eng 14(4):206–218
[11] Soãres S (2003) Nanometer edge and surface imaging using optical scatter. Precis Eng 27(1):99–102
[12] Drescher J (1993) Scanning electron microscopic technique for imaging a diamond tool edge. Precis Eng 15(2):112–114

[13] Yousefi R, Ichida Y (2000) A study on ultra–high-speed cutting of aluminum alloy: formation of welded metal on the secondary cutting edge of the tool and its effects on the quality of finished surface. Precis Eng 24(4):371–376

[14] Malz R, Brinksmeier E, Preuß W, Kohlscheen J, Stock HR, Mayr P (2000) Investigation of the diamond machinability of newly developed hard coatings. Precis Eng 24(2):146–152

[15] Born DK, Goodman WA (2001) An empirical survey on the influence of machining parameters on tool wear in diamond turning of large single-crystal silicon optics. Precis Eng 25(4):247–257

[16] Asai S, Taguchi Y, Horio K, Kasai T, Kobayashi A (1990) Measuring the very small cutting-edge radius for a diamond tool using a new kind of SEM having two detectors. Ann CIRP 39(1):85–88

[17] Binning G, Quate CF, Gerber CH (1986) Atomic force microscope. Phys Rev Lett 56:930–933

[18] Danzebrink HU, Koenders LK, Wilkening G, Yacoot A, Kunzmann H (2006) Advances in scanning force microscopy for dimensional metrology. Ann CIRP 55(2):1–38

[19] Lucca DA, Seo YW (1993) Effect of tool edge geometry on energy dissipation in ultra-precision machining. Ann CIRP 42(1):83–86

[20] O' Shea DC (1984) Elements of modern optical design. Wiley, New York

10

Scanning Image-sensor System for Measurement of Micro-dimensions

10.1 Introduction

Measurement of micro-dimensions such as length, depth, width and radius of the nanomanufactured micro-structures is an important task for precision nanometrology. Some of the micro-structures are made on the workpiece over a long stroke and the measurement of the micro-dimension must be made over the entire stroke. In such cases, it requires the measurement system to have fast enough measuring speed.

An image-sensor system, which consists of an objective lens and an image sensor, is a good candidate for measurement of micro-dimensions [1–5]. Image sensors, including area-image sensors and line-image sensors, have the advantages of non-contact, high speed, as well as high resolution when an objective lens with a high magnification is employed. Image-sensor systems always have the disadvantage of a small field of view when they are used for the measurement of micro-dimensions over a long stroke. This can be overcome by scanning the system over the micro-dimension. However, there are a number of challenges for the scanning image sensor systems. One is the influence of scanning motion errors. Because the depth of focus of a high-magnification objective lens is small, typically on the order of several micrometers, scanning error motions larger than the depth of focus will introduce measurement errors. This is especially a challenge for a micro-dimension measurement over a long stroke. Another challenge is how to provide three-dimensional information of the micro-dimension in a short measurement time because the image provided by the image-sensor system is only a two-dimensional projection of the three-dimensional structure. To make a three-dimensional measurement possible, it is necessary to move the image-sensor system along the optical axis to take multiple images, which makes the measurement of the micro-dimension over a long stroke time-consuming.

In this chapter, two scanning image-sensor systems are presented for fast and accurate measurement of the micro-width of a long slit and the micro-radius of a long edge, respectively.

10.2 Micro-width Measurement by Scanning Area-image Sensor

10.2.1 Micro-width of Long Tool Slit

Measurement of the micro-width of a tool for slot die coating is discussed in Section 10.2. The slot die coating is for applying liquid coatings, such as adhesives and low-viscosity liquids [6]. A slot die is employed as the coating tool in this technology [7]. The main function of the coating tool is to spread liquids into the substrate film as shown in Figure 10.1. The tool has a slit length of several meters and a width on the order of 100 μm, from which the applied liquids come out (Figure 10.2). The slit is formed by two pieces of precision elements, which are machined by a precision grinding machine. Each of the precision parts has a flat area with a dimension of up to several meters in the X-direction and several hundred millimeters in the Y-direction. The deviations of the slit width, which greatly influence the performance of the coating tool, are required to be less than several micrometers over the entire slit. Measurement of the slit width deviation is thus essential for the process control in the manufacturing of the coating tool [8].

Conventionally, the slit width is measured by using a mechanical thickness gauge point by point along the X-direction of the slit as shown in Figure 10.2. The measurement, which relies on the skill of the operator, is time-consuming and lacks accuracy. There also exits a possibility of damaging the edge of the slit. Although it is possible to carry out the slit width measurement by scanning a displacement

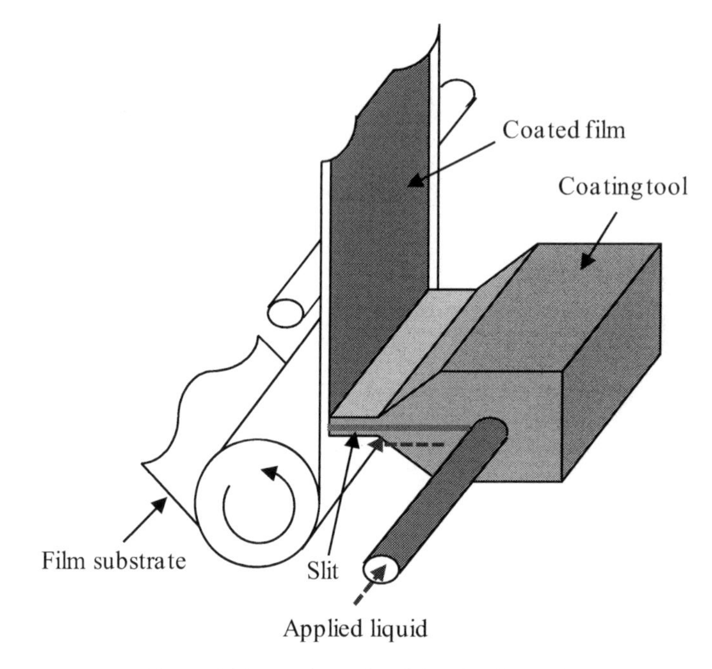

Figure 10.1. A schematic of the slot die coating

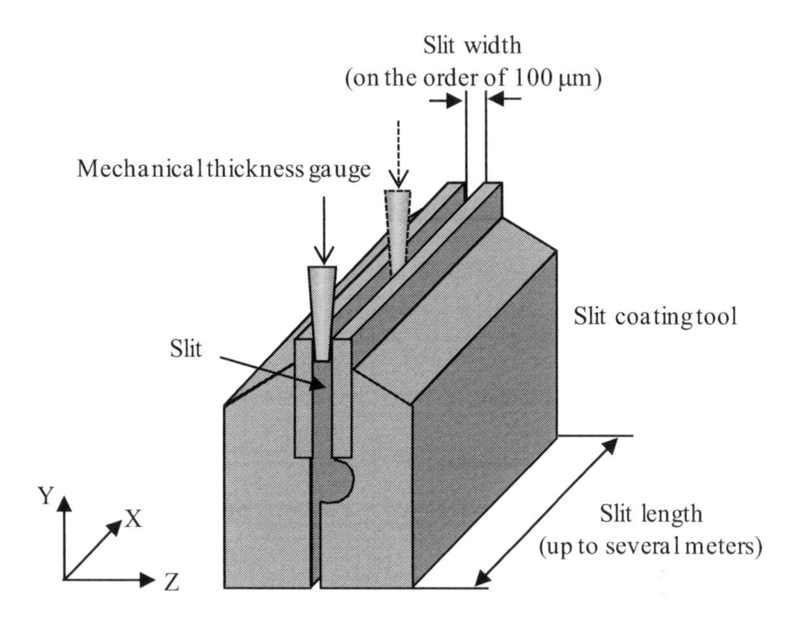

Figure 10.2. The coating tool with a long slit and the conventional measurement method using a mechanical thickness gauge

sensor across the slit along the *Z*-direction, the measurement accuracy is influenced by the edge affect of the sensor and the error motion of the scanning mechanism. The measurement time is also too long.

In Section 10.2, a scanning area-image-sensor method is presented for micro-slit measurement of the coating tool. This method, in which the slit width on the top surface of the slit is scanned by an area-image sensor along the *X*-direction, can be expected to provide non-contact, fast and accurate measurements.

10.2.2 Evaluation of Slit Width

Figure 10.3 shows a schematic of the slit width measurement by using an area-image sensor [9]. An area-image sensor with an objective lens is mounted on the top of the slit to take the image of the slit. The numbers of pixels of the area-image sensor are 720 and 480 in the *Z*-direction and the *X*-direction, respectively. The magnification of the objective lens is set to be 450. The field of view of the image is 680 μm (Z) × 510 μm (X). Figure 10.4 shows a part of the original image of the slit area taken by the image sensor. The dark area of the image is the slit part since no lights were reflected back to the image sensor. The white area is the top surface of the precision element forming the slit.

Figure 10.5 shows a close-up view of the edge part (part A) in Figure 10.4. It can be seen that the boundary in the image between the areas of the slit and the top surface of the precision element is not clear because of the defocus caused by the slope of the top surface of the precision element. To identify the boundary of the slit, the binary image [10] shown in Figure 10.6 is evaluated by making the image processing as follows:

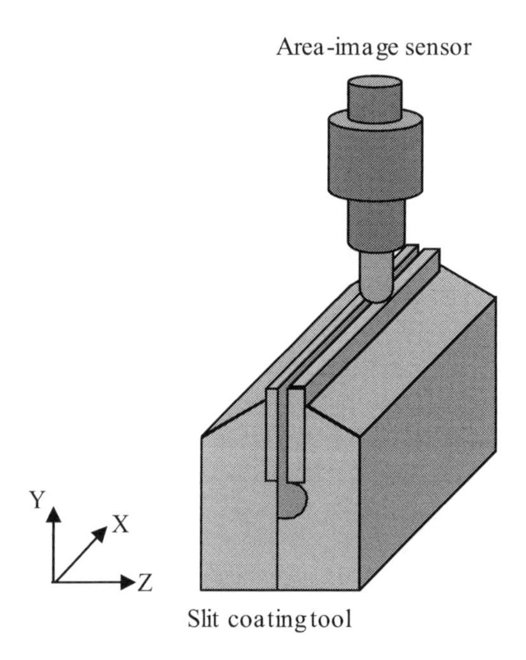

Figure 10.3. The slit width measurement method by using an array image sensor

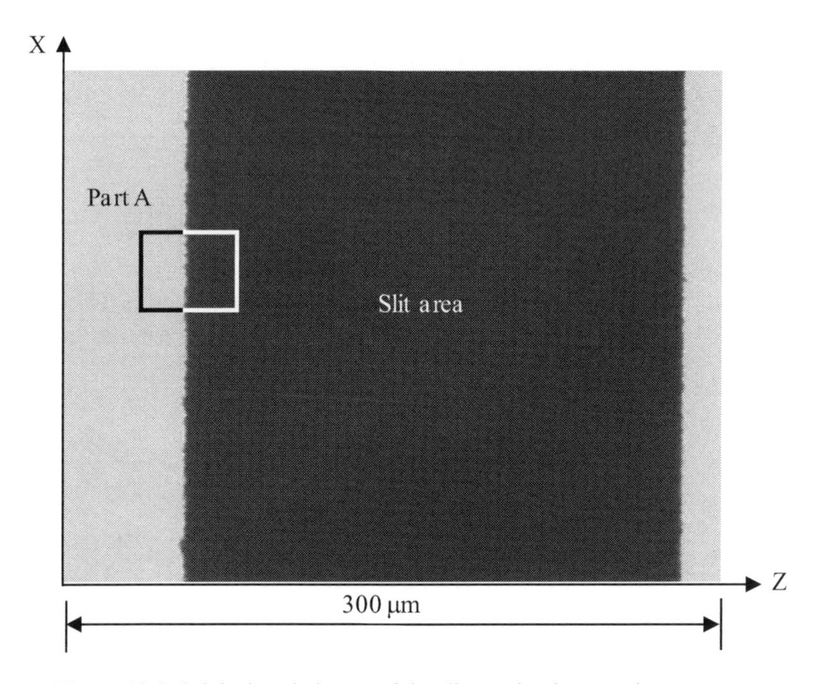

Figure 10.4. Original static image of the slit area by the array image sensor

$$I_B(x,z)=\begin{cases}255 & \left(I(x,z)\geq T\right)\\ 0 & \left(I(x,z)<T\right)\end{cases}, \tag{10.1}$$

where $I(x, z)$ is the original intensity of the pixel of the area-image sensor at the coordinate of (x, z). $I(x, z)$ ranges from 0 to 255. $I_B(x, z)$ is the intensity of the binary image, which is either 0 or 255. T is the threshold value. Figure 10.7 shows the binary image of Figure 10.4. Figure 10.8 shows the intensities of the line BB' in Figure 10.7 before and after the binary processing. As seen in the figures, the boundary of the edge becomes clear after the binary processing.

Figure 10.7 also shows the range for evaluation of the slit width at a position x_i along the X-direction. To reduce the influence of the random errors, the binary data of N line pixels along the X-direction are employed. The slit width is evaluated by averaging the data over this range.

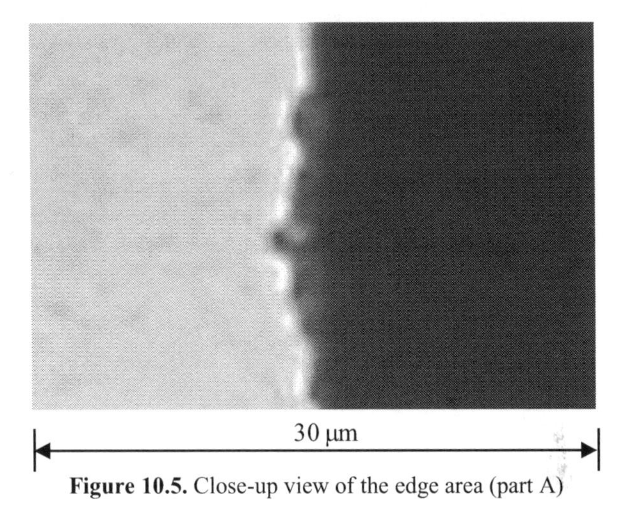

Figure 10.5. Close-up view of the edge area (part A)

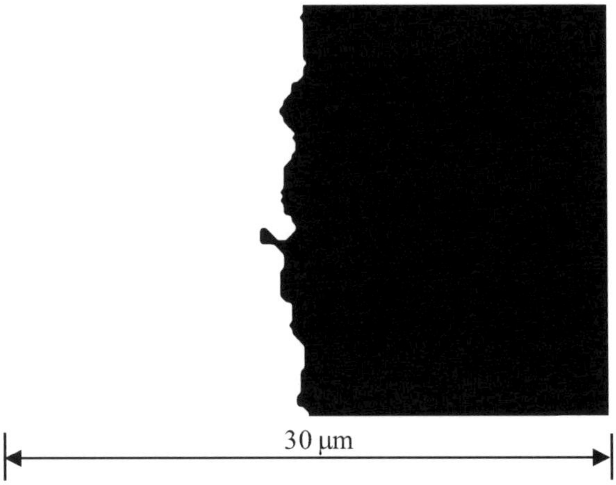

30 μm

Figure 10.6. The binary image of Figure 10.5

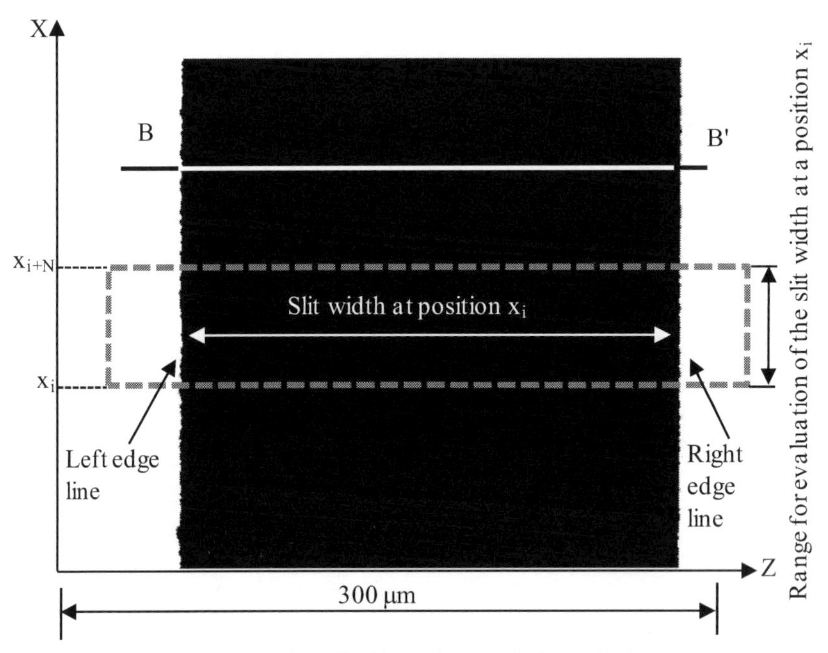

Figure 10.7. The binary image of Figure 10.4

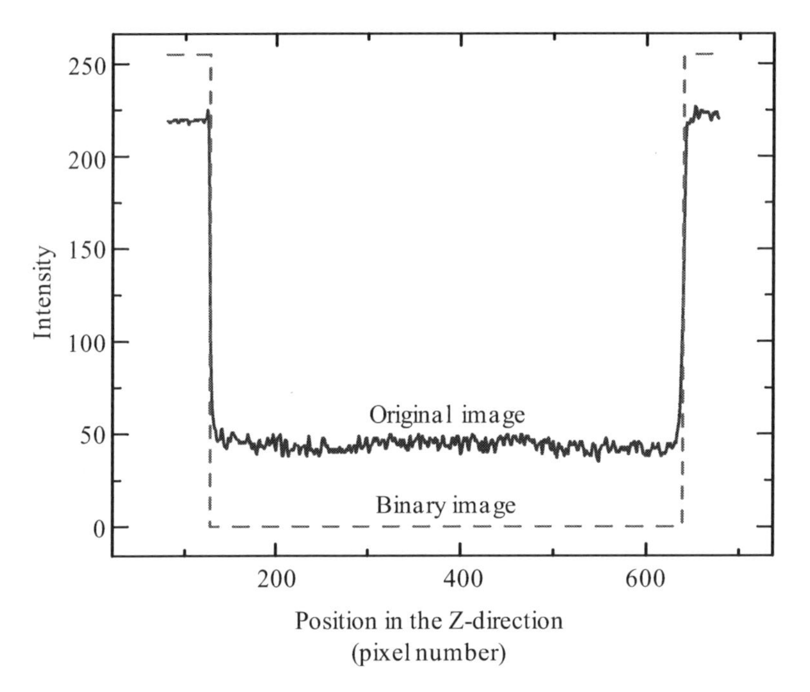

Figure 10.8. Intensity distribution of the cross-section *BB'* in Figure 10.7

10.2.3 Scanning Area-image Sensor

The slit width over the entire slit can be measured by scanning the area-image sensor with a moving stage as shown in Figure 10.9. There are two scanning modes for the measurement. In mode 1, the area-image sensor moves step by step along the *X*-direction. At each step, the area-image sensor takes a static image of the slit. This type of measurement is simple but time-consuming. In mode 2, the image sensor is moved along the *X*-direction at a constant speed to take a video of the slit. The measurement time can be greatly shortened in this scanning mode. Assume that the time to take one frame of image is T seconds and the scanning velocity is v mm/s. Each frame of image is an average over a range of Tv mm along the *X*-direction. In the following, modes 1 and 2 are referred to as the static measurement mode and the continuous measurement mode, respectively.

Figure 10.10 shows the error motions associated with the scanning. As can be seen in Figure 10.10 (a), the error motion along the *Z*-direction does not influence the measurement because the images of the two edges of the slit are taken by the area-image sensor simultaneously. On the other hand, however, the error motion along the *Y*-direction shown in Figure 10.10 (b) directly influences the focus position of the image sensor. Because the depth of focus of a high-magnification objective is only several micrometers, an error motion larger than this amount will introduce errors to the slit width measurement.

Figures 10.11 and 10.12 show the images taken by the image sensor mounted at different *Y*-positions. The image shown in Figure 10.11 was taken at the focus position ($y = 0$ μm) and that shown in Figure 10.12 was taken at a defocus position of 50 μm ($y = -50$ μm) by moving the camera along the *Y*-direction.

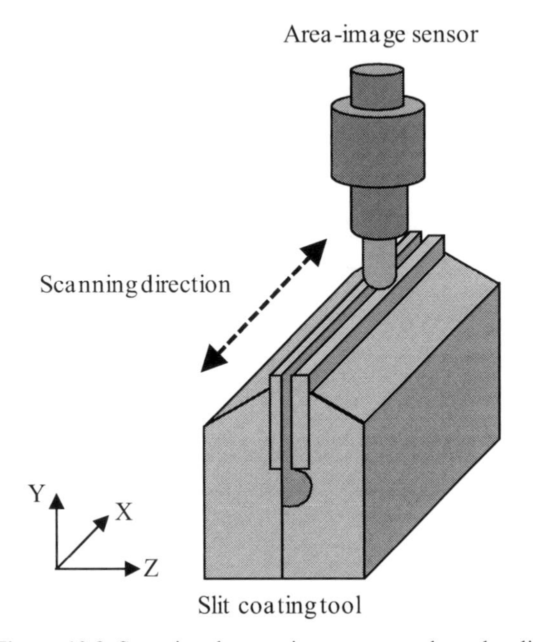

Figure 10.9. Scanning the area-image sensor along the slit

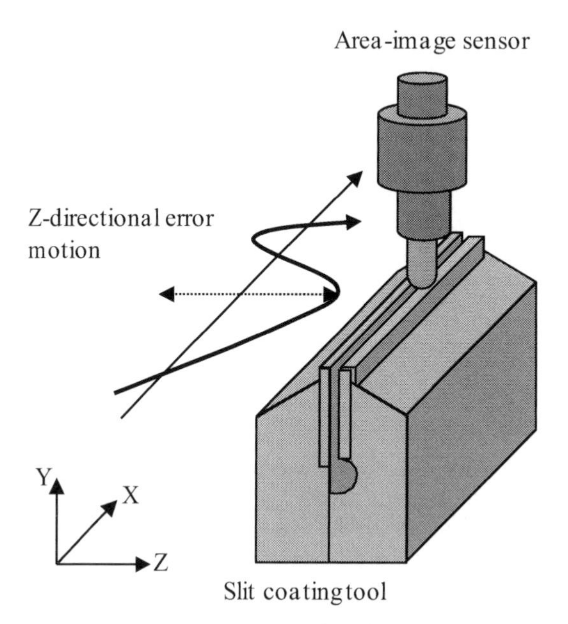

(a) Error motion along the *Z*-direction

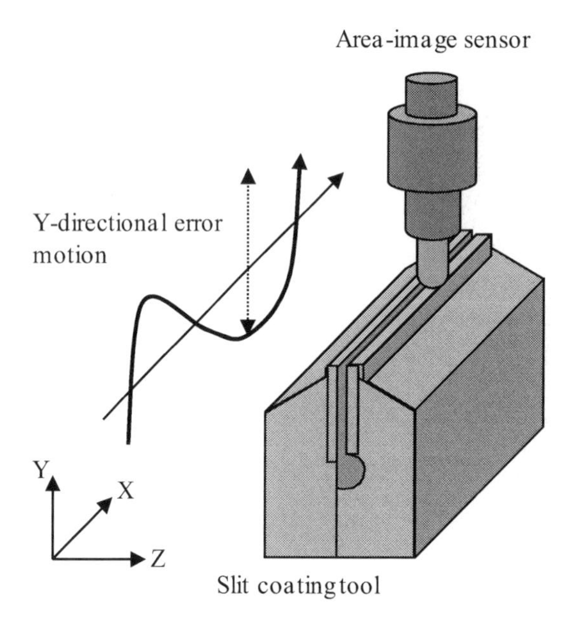

(b) Error motion along the *Y*-direction

Figure 10.10. Scanning error motions

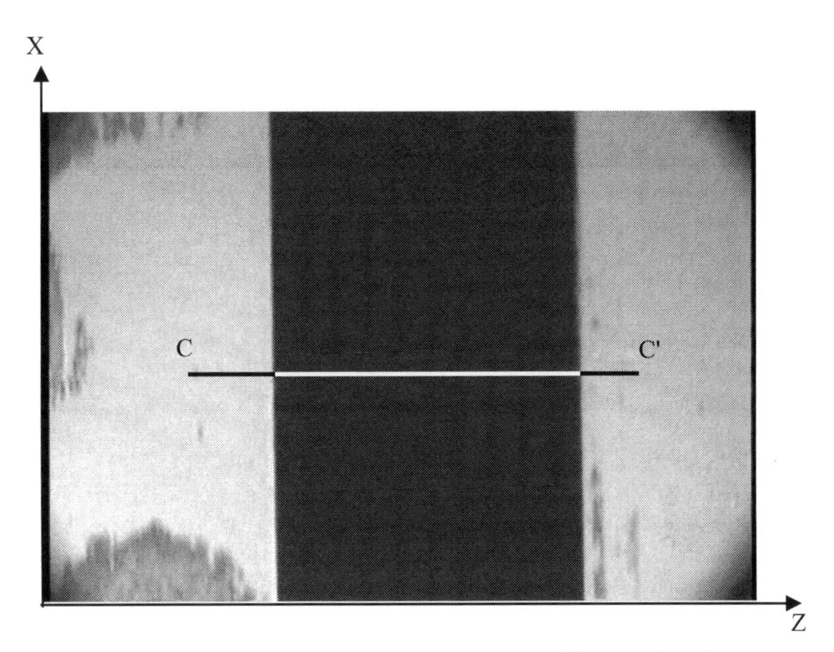

Figure 10.11. An image taken at the focus position ($y = 0$ μm)

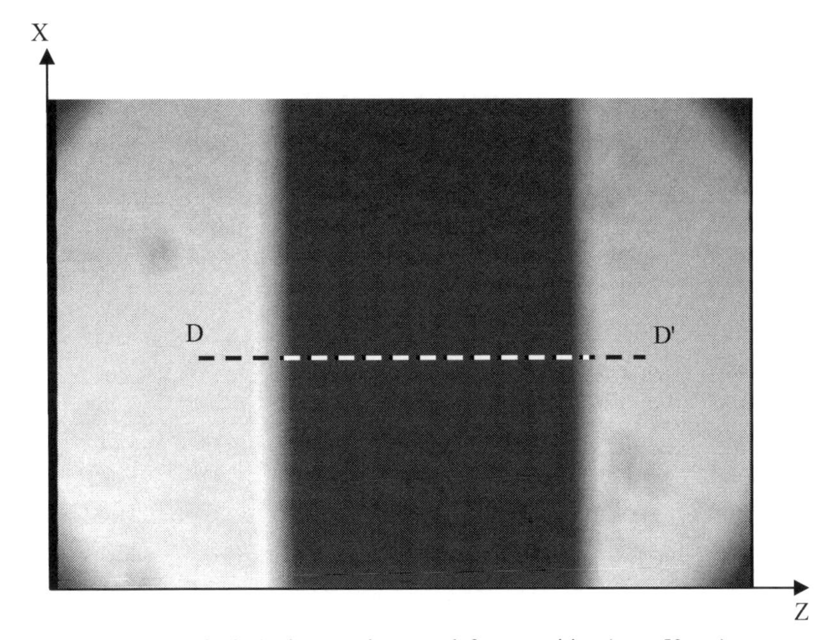

Figure 10.12. An image taken at a defocus position ($y = -50$ μm)

The influence of the defocus on the image can be observed by comparing the images. The out-of-flatness errors of the top surfaces of the precision elements or tilt of the coating tool also have the same influence. In the static measurement mode, the focus can be re-adjusted at each sampling position to avoid the influence of the *Y*-direction error motion. It is also possible to employ a servo control system for the *Y*-position of the image sensor in the continuous measurement mode. However, this will make the measurement system complicated.

A simple method is proposed to reduce the influence of the *Y*-directional error motion, especially in the continuous measurement mode. In this method, a proper threshold value *T* shown in Equation 10.1 is chosen for the binary processing. Figure 10.13 shows the intensity distributions of the binary images at the line *CC'* in Figure 10.11 and the line *DD'* in Figure 10.12. The binary images were obtained with different threshold values of 127 and 145, respectively. The numbers of pixels of the area-image sensor were 720 and 480 in the *Z*-direction and the *X*-direction, respectively. The corresponding field of view of the image was 310 μm (*Z*) × 230 μm (*X*). As can be seen in the figure, both the left and right edge lines of the intensity distribution of the image taken at $y = -50$ μm shift to the left when compared to that of the image taken at $y = 0$ μm.

The evaluated slit width, which is the distance between the left edge line and the right edge line, changes considerably when the threshold value *T* is set to be 127. On the other hand, the evaluated slit width changes little when *T* is set to be 145. The result indicates that the influence of the *Y*-directional error motion can be reduced by choosing a proper threshold value *T*.

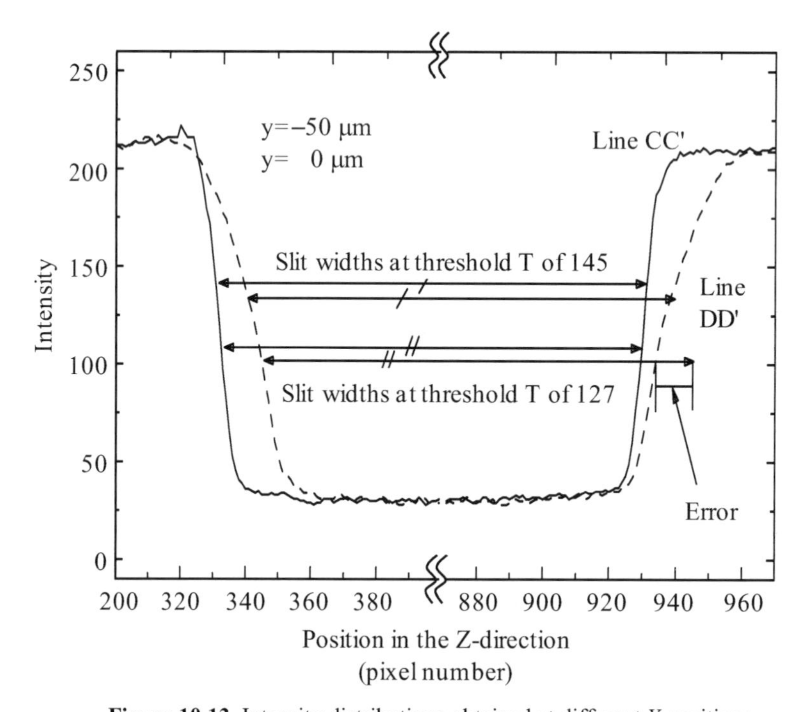

Figure 10.13. Intensity distributions obtained at different *Y*-positions

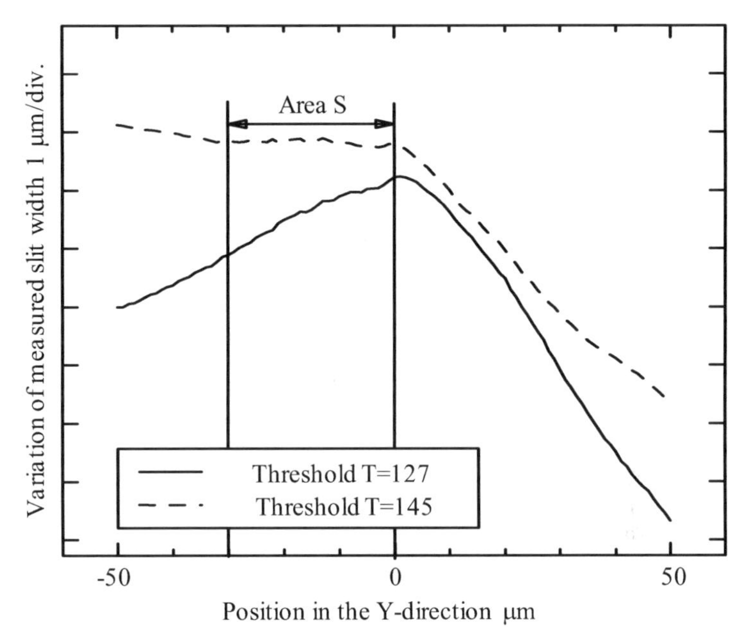

Figure 10.14. Variation of the evaluated slit width with respect to the *Y*-position of the image sensor

Figure 10.14 shows the variation of the evaluated slit width with respect to the *Y*-position of the image sensor. The magnification of the objective lens was set to be 450, corresponding to a field view of 680 μm (*Z*) × 510 μm (*X*). It can be seen that the variation of the evaluated slit width was less than 0.1 μm within the area of *S* (-30 μm $< y < 0$ μm) when the threshold value *T* was set to be 145. It means that the measurement of slit width will not be influenced if the sum of the *Y*-directional error motion and the out-of-flatness of the top surfaces of the coating tool is less than 30 μm.

10.2.4 Slit Width Measurement in Production Line

Slit width measurement of a precision coating tool was also carried out in the production line. Figure 10.15 shows the measurement setup. The precision coating tool was mounted on the worktable of a planer-type surface grinding machine on which the precision elements forming the coating tool had been machined. The area-image sensor with the objective lens was mounted on the grinding head. The image sensor was kept stationary and the coating tool was moved by the worktable along the *X*-direction so that the slit could be scanned by the image sensor.

The width measurement was carried out in the measurement modes of mode 1 (the static measurement mode) and mode 2 (the continuous measurement mode). Table 10.1 shows the measurement parameters.

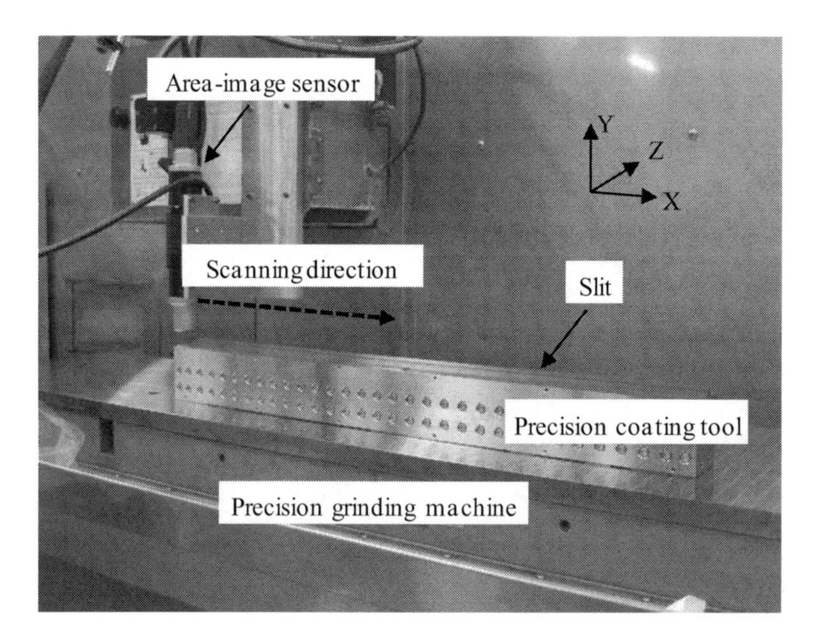

Figure 10.15. Slit width measurement of a precision coating tool on a machine tool

Table 10.1. Measurement parameters

	Mode 1	Mode 2
Length of coating tool	1400 mm	
Nominal width of slit	150 μm	
Numbers of pixels of the area-image sensor	1360 (*Z*) × 1024 (*X*)	720 (*Z*) × 480 (*X*)
Scanning velocity of the table		42.5 mm/s
Sampling points along the *X*-direction	14	1000
Measurement time	15 min	30 s

Figure 10.16 shows the measurement results of mode 1 and mode 2. The results of two repeated measurements of mode 1 are shown in Figure 10.16 (a). The results by the conventional thickness gauge are also plotted for comparison. As can be seen in Figure 10.16 (a), the measurement results of mode 1 were consistent with those by the thickness gauge. It can also be seen that the resolution of mode 1 was higher than that of the thickness gauge, which was 0.5 μm for the slit width measurement. Figure 10.16 (b) shows the measurement results of mode 2. The worktable was moved at a velocity of 42.5 mm/s. The measurement time was approximately 30 s, which was much shorter than that of mode 1.

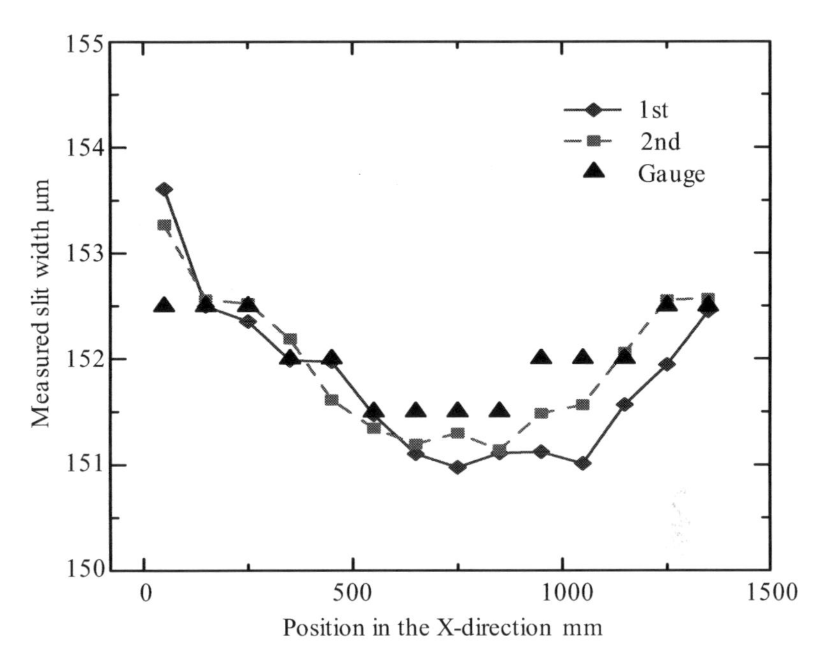

(a) Results by mode 1 and the conventional thickness gauge

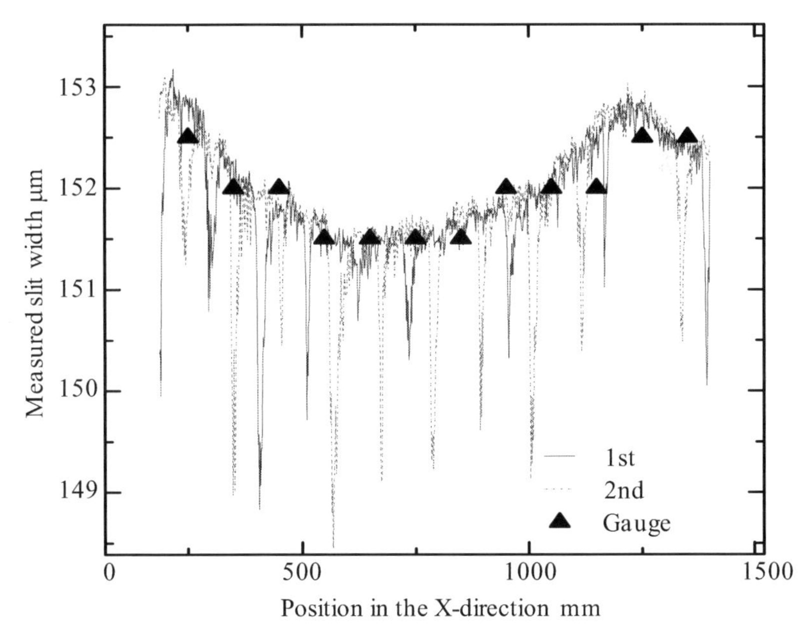

(b) Results by mode 2 and the conventional thickness gauge

Figure 10.16. Results of slit width measurement

The high-frequency components in the results shown in Figure 10.16 (b) were caused by the vibrations from other machine tools operating in the same machine shop. The results were fitted by six-degree polynomial curves to remove the influence of the vibration errors. The fitted results are shown in Figure 10.17. The difference between the two repeated measurement results is also shown in the figure. It can be seen that the repeatability of mode 2 for the slit width measurement was approximately 0.2 μm in Figure 10.17.

Figure 10.18 shows a slit width measuring instrument developed for in-line inspection of precision coating tools based on the technology described in Section 10.2.

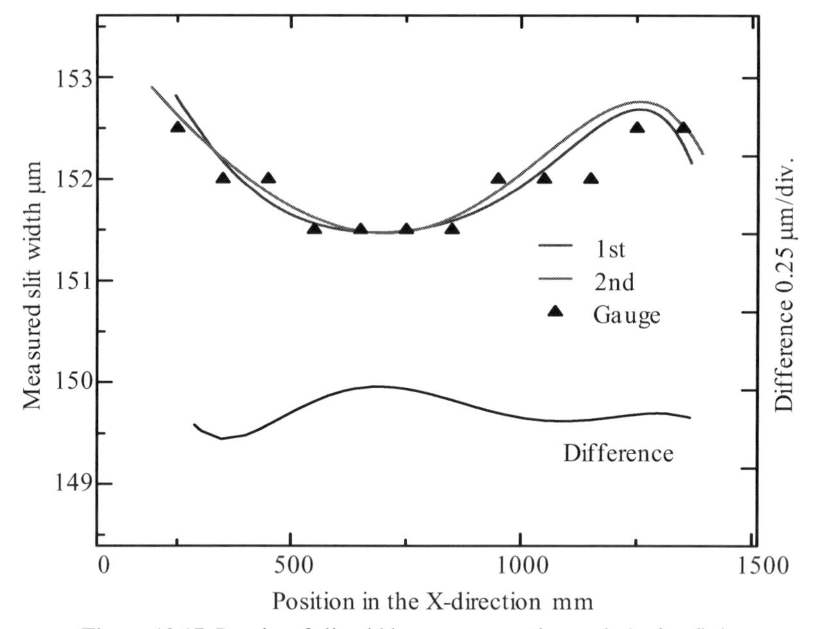

Figure 10.17. Results of slit width measurement by mode 2 after fitting

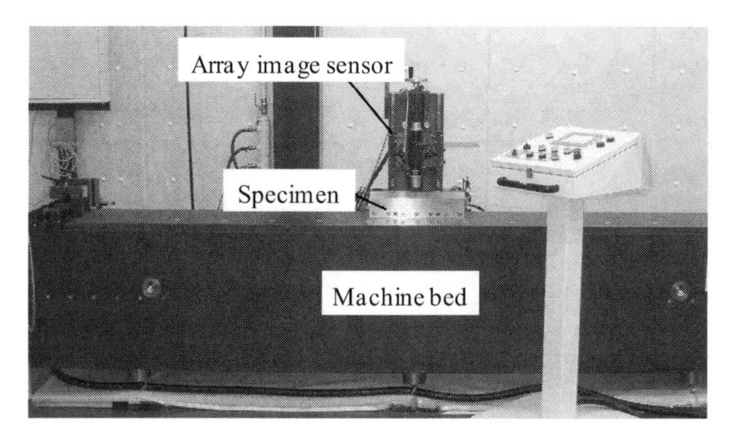

Figure 10.18. A slit width measuring instrument developed for in-line inspection of precision slit coating tools

10.3 Micro-radius Measurement by Scanning Line-image Sensor

10.3.1 Micro-radius of Long Tool Edge

Many tools made by or used in nanomanufacturing have long micro-edges. Figure 10.19 shows an example of this kind of tool. The tool, which is called the extrusion sheet die, is employed in the extrusion of plastic films. Similar to the slot die described in Section 10.2, the extrusion sheet die is configured by combining two precision elements. The length of the tool is up to several meters, and the edge radius of the precision element is less than 100 μm [11]. The function of the extrusion die is to form the desired film shape and dimension. Because the die edges are the final components in the extrusion process, the profiles of the die edges affect most strongly the quality of the extrudate, such as the uniformity, the thickness and the surface roughness. Measurement of the micro-edges, especially the micro-radius, over the entire tool length is thus necessary for quality control of the extrusion process.

Because the size of the tool is too large, it is difficult to mount it on a profile measuring machine for measurement of the edge. Conventionally, the measurement of the edge radius is indirectly measured through replication of the tool edge. Figure 10.20 shows a photograph of taking the replica by using a resin. The shape of the tool edge is transferred to the resin. The replicated edge shape is then measured by using a profile measuring machine. The replication process is time-consuming and the accuracy of replication is also not good enough. Moreover, it is impossible to measure the edge over the entire tool.

In this section, a fast and precision method is described for the measurement of the micro-radius by scanning a line-image sensor.

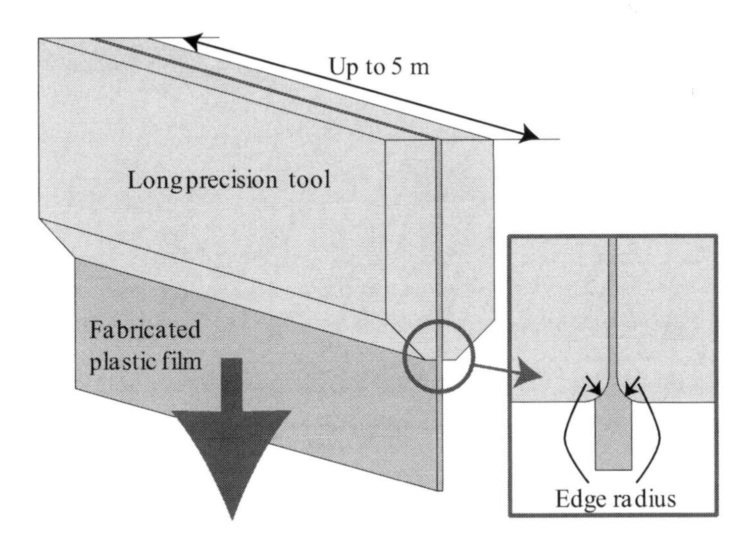

Figure 10.19. A schematic of a long precision tool with micro-edges

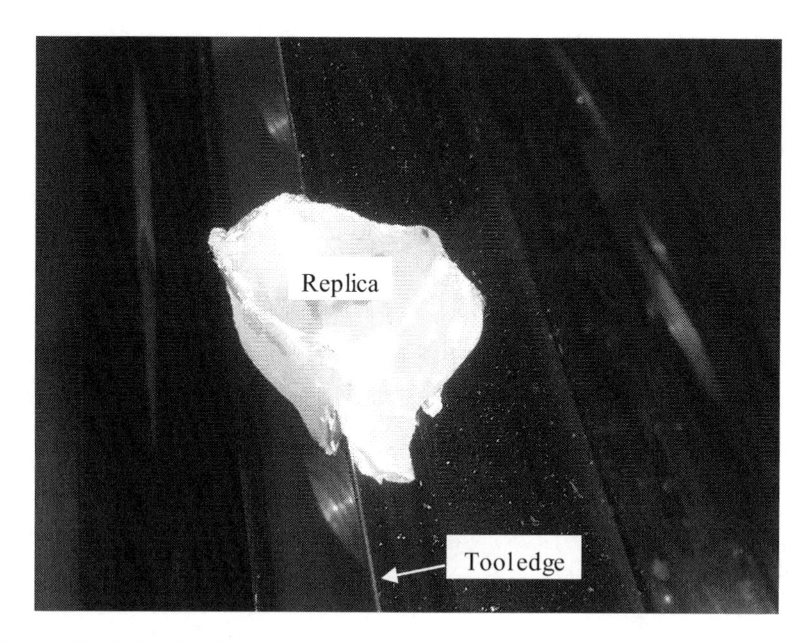

Figure 10.20. Replication of the tool edge in the conventional measurement of the micro-radius

10.3.2 Evaluation of Edge Radius

Figure 10.21 shows a schematic of the measurement system [12]. A line-image sensor with an objective lens [13] is employed to scan the tool edge along the *X*-direction. Figure 10.22 shows a sectional profile of the tool edge in the *YZ*-plane. The edge is assumed to be formed by an arc and two straight lines as shown in the figure. The edge radius is let to be *R*. An illumination light is projected onto the edge along the *Y*-axis. The angle between the incident light and the reflected light is let to be θ. The reflected light intensity is basically a function of the angle θ. The intensity reaches the maximum value when θ is equal to zero and the intensity gets smaller as θ increases. The line-image sensor can only receive the reflected light within a certain range of θ, which is determined by the numerical aperture of the objective lens, the size and sensitivity of the line-image sensor, as well as the reflectivity of the edge surface. Assume the range of the reflected lights, which can be received by the line-image sensor, is from $-\theta_0$ to θ_0. The corresponding edge width can be expressed by:

$$W = 2R\sin\theta_0 . \qquad (10.2)$$

Because the edge width *W* can be obtained from the image of the line-image sensor, the edge radius can thus be obtained as:

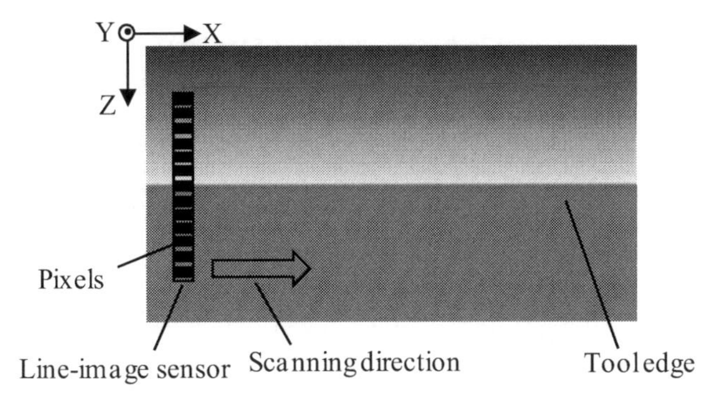

Figure 10.21. Scanning a line-image sensor along the tool edge

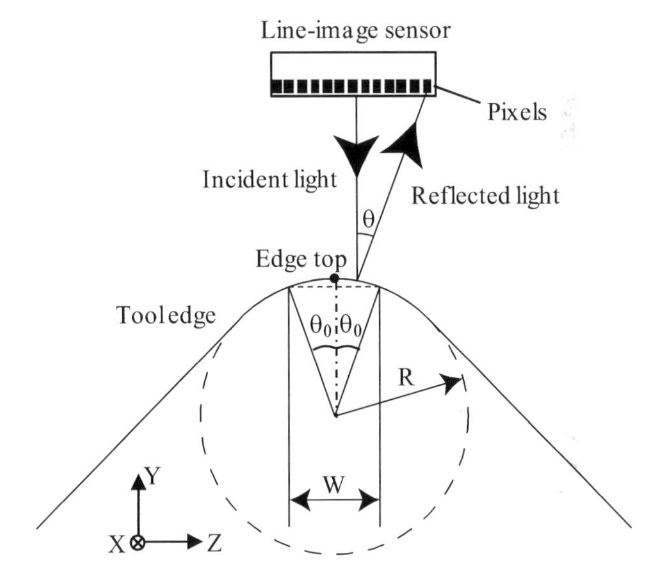

Figure 10.22. Relationship between the edge radius *R* and the edge width *W*

$$R = \frac{W}{2\sin\theta_0} = k \cdot W \,, \tag{10.3}$$

where

$$k = \frac{1}{2\sin\theta_0} \,. \tag{10.4}$$

Figure 10.23 shows an image of the tool edge taken by the line-image sensor. Because the line-image sensor can only take one line image along the *Z*-direction,

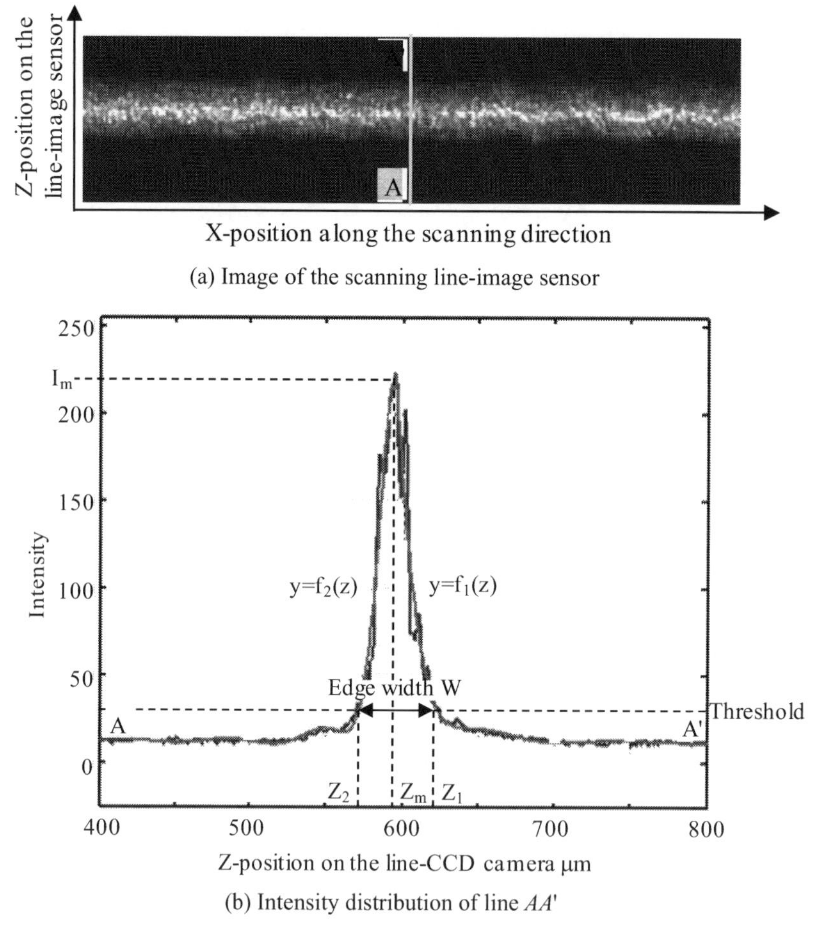

Figure 10.23. An example of the scanned image of the line-image sensor and evaluation of the edge width

it is necessary to scan the sensor along the *X*-direction to get the image of the edge over a certain range of the tool. The intensity output of the sensor at a position of x_i is also shown in the figure. The edge width is evaluated from the sensor output with the following steps:

Step 1: Determine the maximum intensity value I_m and the corresponding pixel position z_m.

Step 2: Fit the intensity curve in the range of $z>z_m$ by a function of $f_1(z)$ and that in the range of $z<z_m$ by a function of $f_2(z)$.

Step 3: Evaluate the pixel positions z_1 and z_2 where $f_1(z_1)$ and $f_2(z_2)$ are equal to a threshold value.

Step 4: Evaluate the edge width as follows:

$$W = z_1 - z_2 . \tag{10.5}$$

Because it is difficult to directly calculate the edge radius from the evaluated edge width based on Equation 10.3, a calibration process was introduced to determine the relationship between the edge width and the edge radius by using pin gauges with known radii. The line-image sensor with the objective lens and the lighting unit were mounted on a stepping motor-driven stage with its axis of motion along the *Y*-direction as shown in Figure 10.24. The pin gauge was kept stationary under the sensor. The line-image sensor had 5150 pixels aligned along the *Z*-direction over a length of 36.05 mm. The pixel pitch was 7 μm. The data rate was 40 MHz with a 10-bit video output. The objective lens had a magnification of 10 and a working distance of 50.86 mm. The field of view of the lens was 3.5 mm along the *Z*-direction and the resolution was 0.7 μm. The stepping motor stage has a resolution of 0.02 μm. Table 10.2 shows the specification of the pin gauges used in the calibration. The material of the pin gauge was carbide, which was the same as the tool. Figure 10.25 shows the measured widths of the pin gauges.

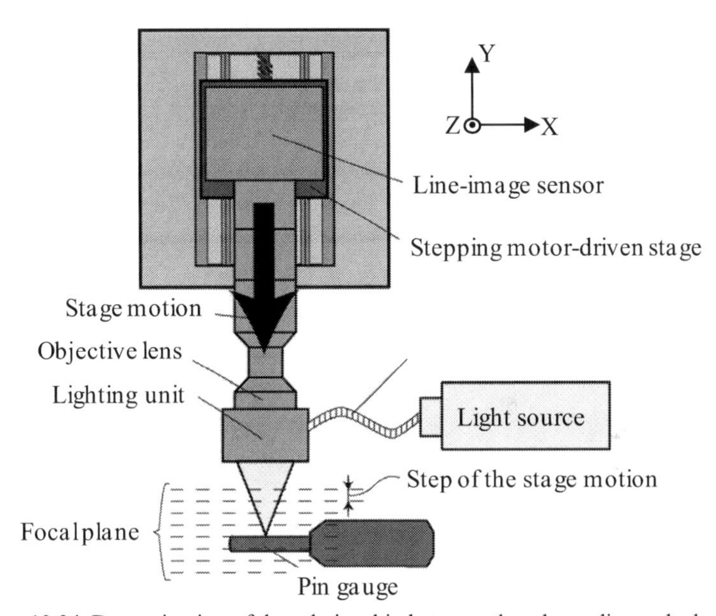

Figure 10.24. Determination of the relationship between the edge radius and edge width

Table 10.2. Specification of pin gauges

Diameter μm	Accuracy μm		Diameter μm	Accuracy μm
50	±0.3		300	±0.1
75	±0.1		400	±0.1
100	±0.1		500	±0.3
150	±0.1		800	±0.3
200	±0.1		1000	±0.4

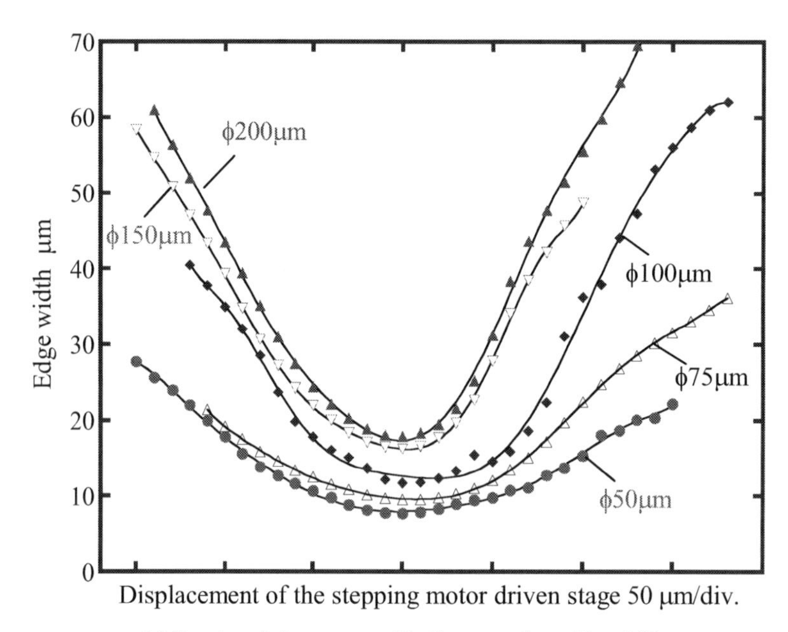

(a) Results of pin gauges with diameters from 50 to 200 μm

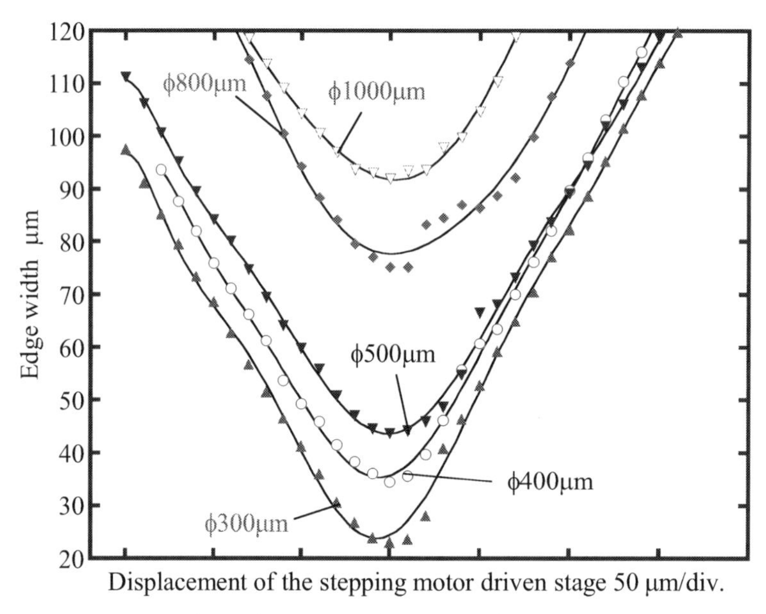

(b) Results of pin gauges with diameters from 300 to 1000 μm

Figure 10.25. Measured widths of pin gauges

In the experiment, the line-image sensor was moved by the stage along the *Y*-direction with a step of 10 μm over a range of 300 μm. At each step, the image of the edge was taken by the sensor to evaluate the edge width. As can be seen in Figure 10.25, the measured edge widths changed with the position of the stepping motor. The minimum was employed as edge width value of the pin gauge, which was obtained at the focus position of the lens. Figure 10.26 shows the relationship between the edge width value and the pin gauge radius. The edge width had a linear relationship with the edge radius. The relationship was employed in the following measurement of the die tool, which is called the width-based method.

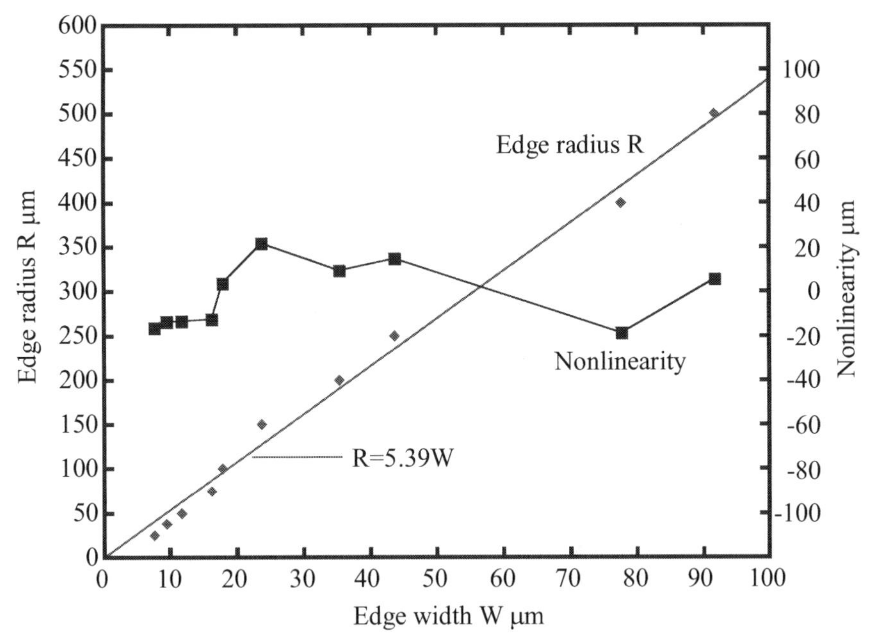

Figure 10.26. Relationship between the edge width and the edge radius

10.3.3 Edge Radius Measurement in Production Line

Figure 10.27 shows a schematic of the measurement system for edge radius of a long extrusion tool in the production line. The tool is kept stationary. The line-image sensor is moved by an *X*-stage to scan the tool edge along the *X*-direction. Because the length of the tool is on the order of several meters, large defocus errors could occur during the scanning, which are caused by the motion error of the scanning stage as well as the tilt and out-of-straightness of the tool edge with respect to the axis of the *X*-stage. As can be seen from Figure 10.25, a defocus error of the image sensor will introduce errors to the edge width measurement. Although it is possible to employ an autofocus mechanism for solving this problem, this will make the measurement longer and the instrument more complicated. As shown in Figure 10.28, multiple *X*-scans are carried out by stepping the image sensor along the *Y*-direction with a *Y*-stage.

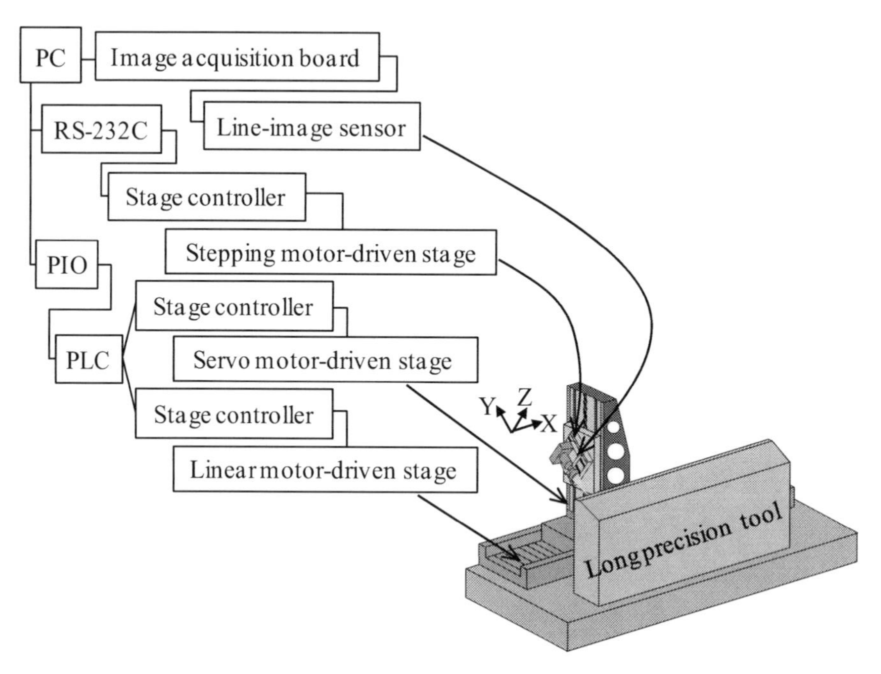

Figure 10.27. Edge radius measurement system for a long extrusion tool

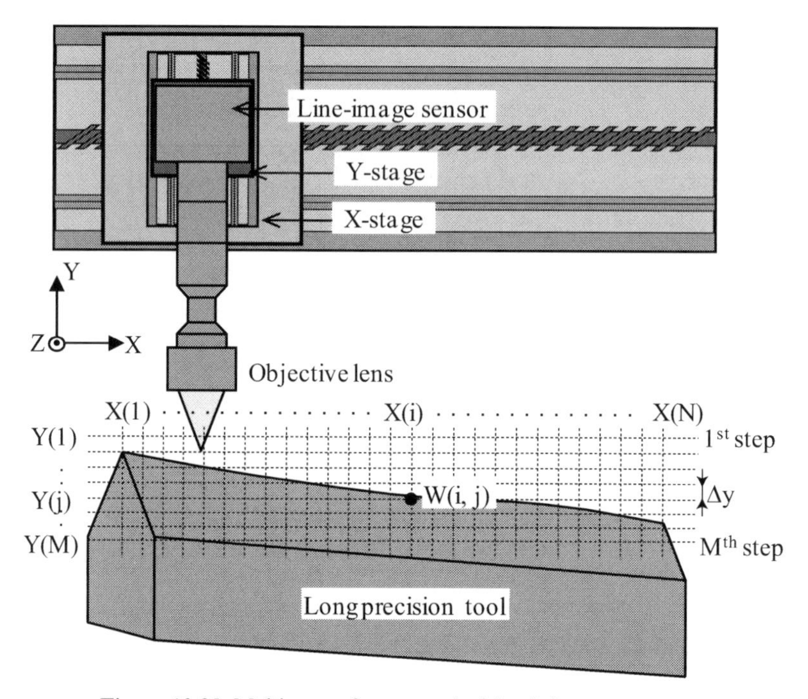

Figure 10.28. Multi-scans for removal of the defocus error

The steps for the scanning of the edge and evaluation of the edge widths are as follows:

Step 1: Set the measurement position along the *Y*-direction is $Y(1)$. Scan the image sensor along the *X*-direction to calculate the edge width values $W(i,1)$ ($i = 1, \ldots, N$).

Step 2: Move the *Y*-stage with a step of Δy so that the measurement position along the *Y*-direction becomes $Y(2)$. Scan the image sensor along the *X*-direction to calculate the edge width values $W(i, 2)$ ($i = 1, \ldots, N$).

Step 3: Repeat steps 1 and 2 to get all the edge width values $W(i, j)$ ($i = 1, \ldots, N, j = 1, \ldots, M$).

Step 4: Re-arrange the edge width data to get the curve of $W(i, j)$ with respect to $Y(j)$ ($j = 1, \ldots, M$) at $X(i)$ as shown in Figure 10.29.

Step 5: Determine the edge width at $X(i)$ from the minimum value W_{min} of $W(i, j)$ in Figure 10.29. The corresponding Y_{min} is the focus position at $X(i)$.

Experiments were carried out on a tool with a length of 5 m. The scanning velocity of the *X*-stage was 100 mm/s. The step Δy in the *Y*-direction was set to be 10 μm, which was the same as the depth of focus of the objective lens. The number of scans (M) was 14. The total time for taking the images was 30 min. The image shown in Figure 10.30 was called the unit image. There were 5000 lines of the

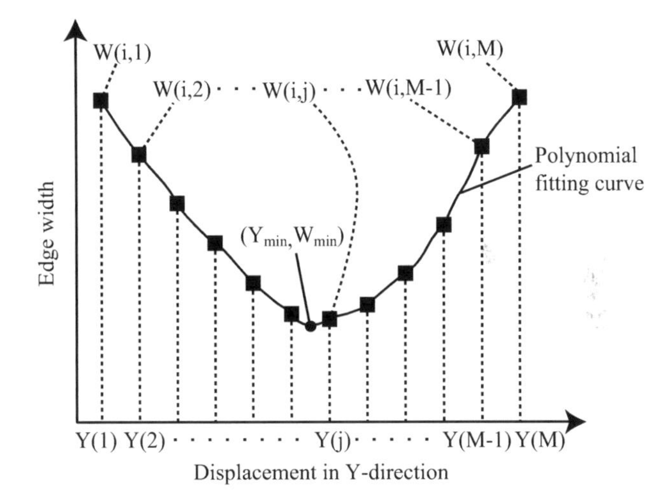

Figure 10.29. Determination of the edge width at $X(i)$

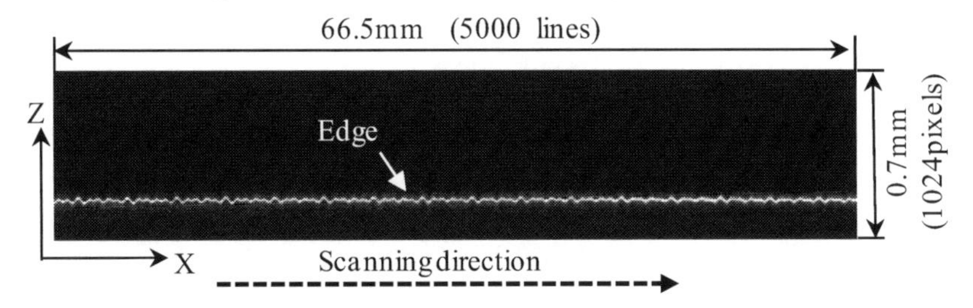

Figure 10.30. Unit image of a part of the tool edge obtained by scanning the line-image sensor ($j = 7$, $X = 2460.5$–2527 mm)

line-image sensor data in the unit image covering a length range of 66.5 mm along the tool edge. Each unit image was stored as an individual data file in the hard disk of the personal computer. There were 78 unit images for the entire tool edge. Figure 10.31 shows the measurement result of the edge radii of the tool. Figure 10.32 shows the corresponding focus position, which was the Y_{min} in Figure 10.29.

The image data obtained in Figure 10.28 can also be employed to evaluate the 3D edge profile by using the method of shape-from-focus (SFF) [14]. As can be seen in Figure 10.33, there are M unit images along the Y-direction. Figure 10.34 shows a schematic of the principle of the SFF method. In this method, the 3D profile of the target surface is measured by taking multiple images at different Y-positions. The Y-directional height position of the surface at the coordinate of (x, z) is determined from the most in-focus image. Figure 10.35 shows the 3D edge profile of the tool by the SFF method, and Figure 10.36 shows the edge radii evaluated from the 3D edge profile data. The results of three repeated measurements are shown in the figure. Comparing the results shown in Figures 10.31 and 10.36, it can be seen that the edge radii evaluated by the two different methods are consistent with each other.

On the other hand, both the width-based method and the SFF method are influenced by the repeatability of the out-of-straightness error motion of the X-directional scanning stage, which was evaluated by using a laser triangular displacement sensor instead of the line-image sensor to scan the side surface of the long tool as shown in Figure 10.37. Figure 10.38 shows a stability of the displacement sensor on the condition of keeping the scanning stage stationary. It can be seen that the sensor output was stable within a range of 0.3 μm. Figure 10.39 (a) shows the sensor output when the sensor was scanning the tool surface. The sensor output $m(x)$ changed approximately 250 μm over the scanning range of 5100 mm, which included the stage out-of-straightness error motion $e(x)$, the out-of-straightness profile error $f(x)$ of the tool surface and the term caused by the tool inclination angle θ. The defocus error caused by $m(x)$ to the image sensor can be removed by the method shown in Figures 10.28 and 10.33 on the condition that the repeatability of $m(x)$ is less than the depth of focus of the objective lens. The scanning was repeated sixteen times to investigate the repeatability of $m(x)$. The results are shown in Figure 10.39 (b). The repeatability of $m(x)$, which was mainly the repeatability of the stage motion, was approximately 1.5 μm. It can be seen that the repeatability, which was much smaller than the 10 μm depth of focus of the objective lens, was good enough for the measurement of the edge radius.

Figure 10.40 shows a photograph of an edge radius measuring instrument developed for in-line inspection of long extrusion tools based on the technology described in Section 10.3.

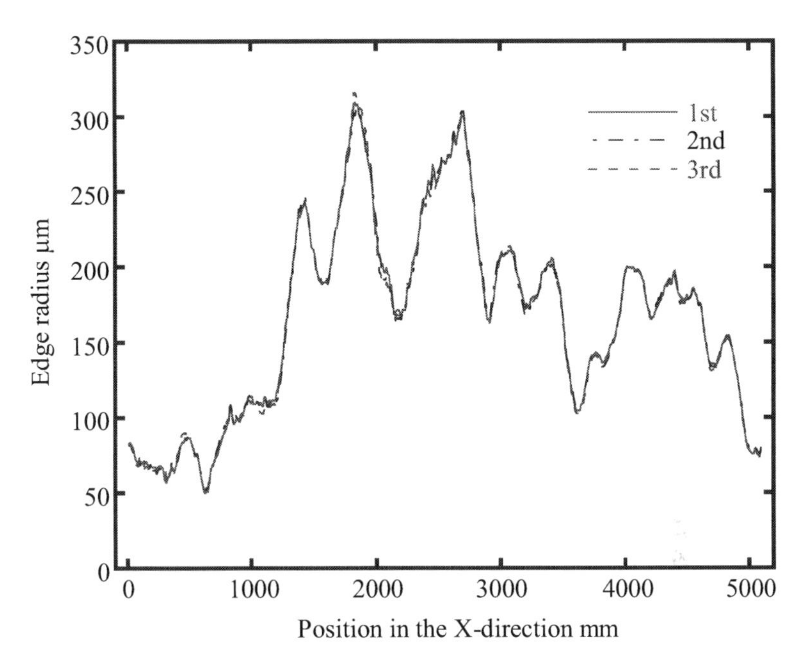

Figure 10.31. Results of three repeated measurements of the edge radii of the long tool by the width-based method

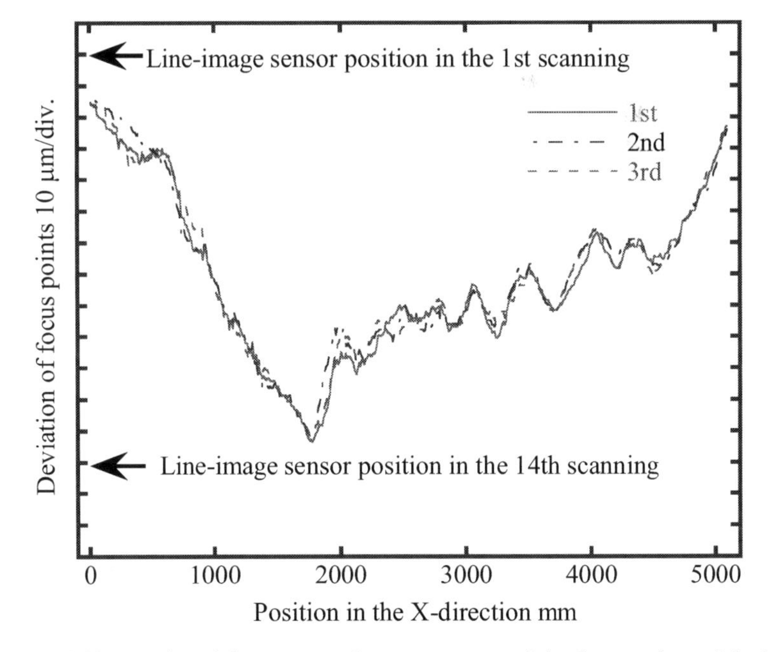

Figure 10.32. Results of three repeated measurements of the focus points of the line-image sensor in the *Y*-direction

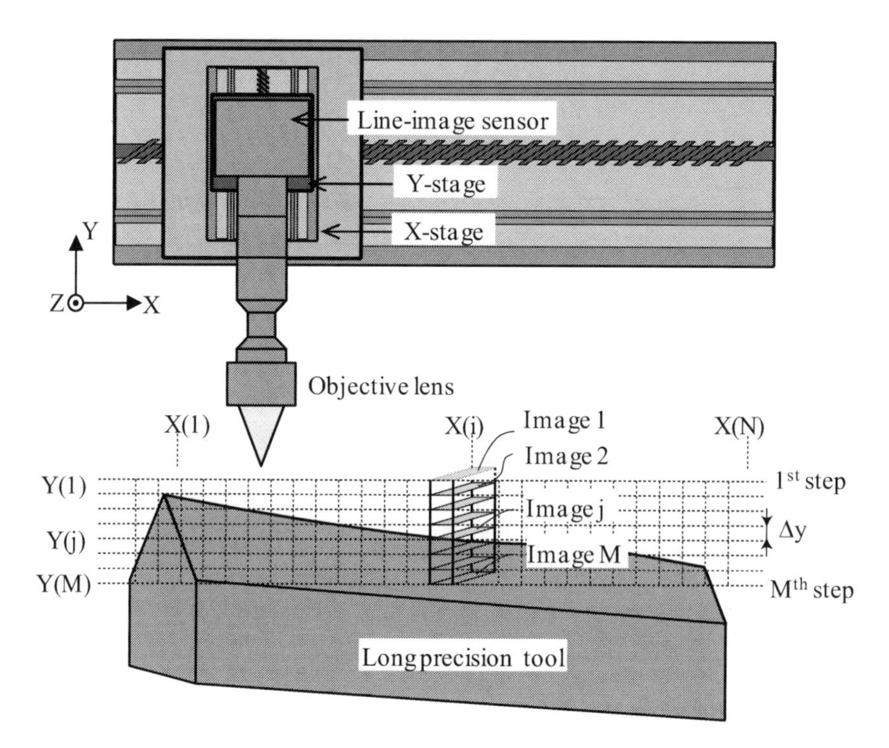

Figure 10.33. Results of three repeated measurements of the edge radii of the long tool

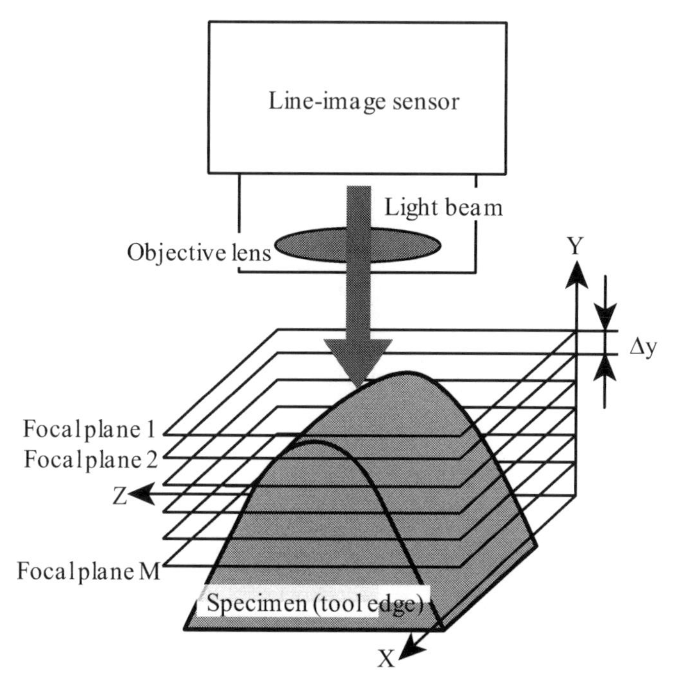

Figure 10.34. Principle of the SFF method

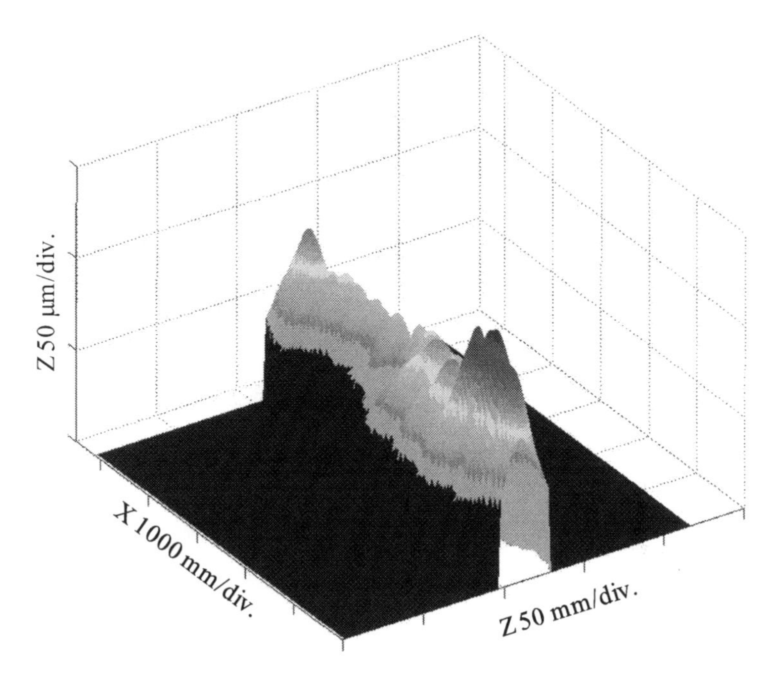

Figure 10.35. 3D edge profile of the long tool by the SFF method

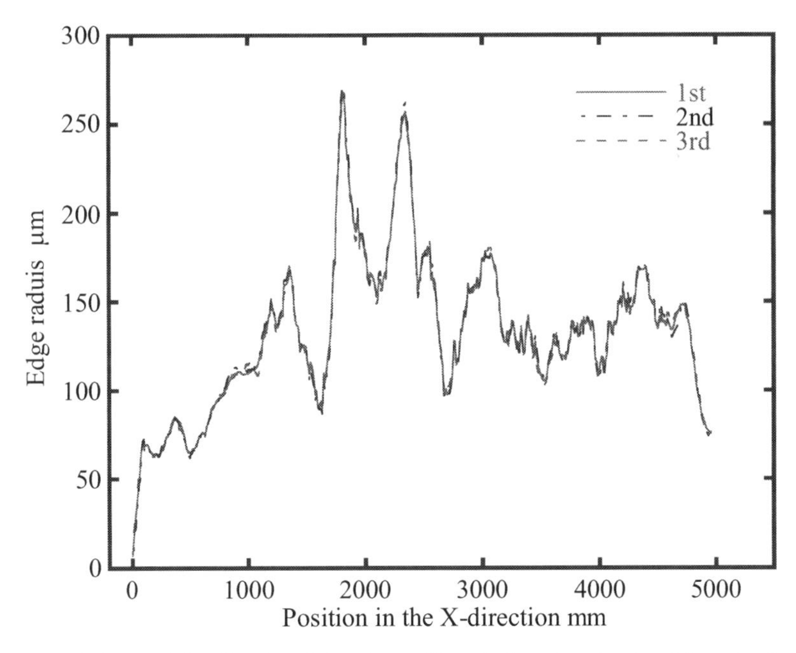

Figure 10.36. Results of the edge radii of the long tool by the SFF method

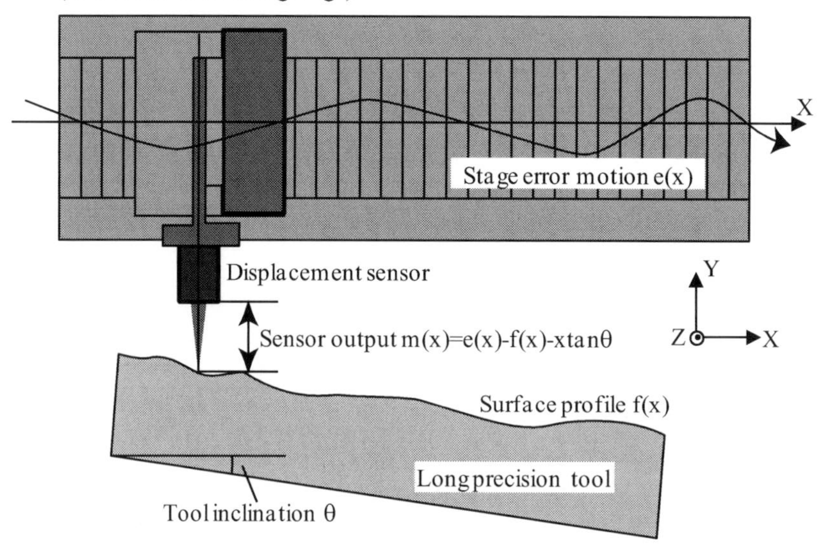

Figure 10.37. Setup for investigating the repeatability of the out-of-straightness error motion of the *X*-directional scanning stage

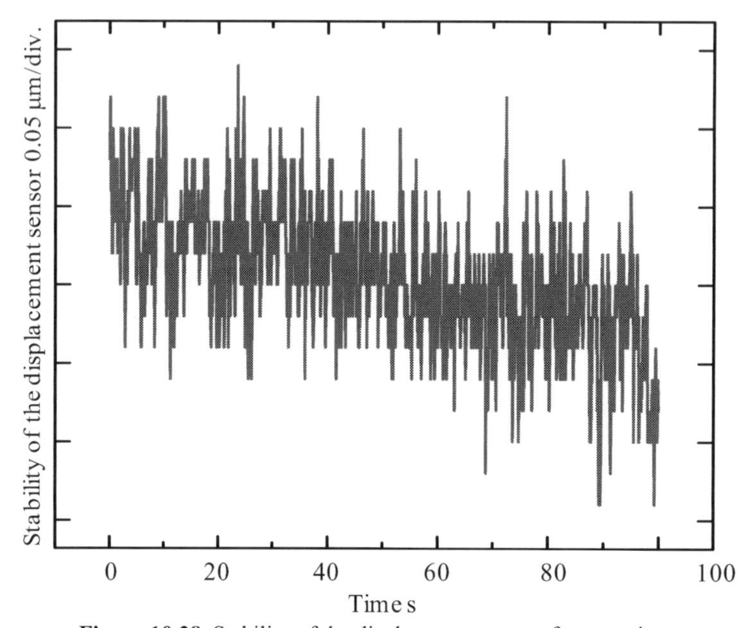

Figure 10.38. Stability of the displacement sensor for scanning

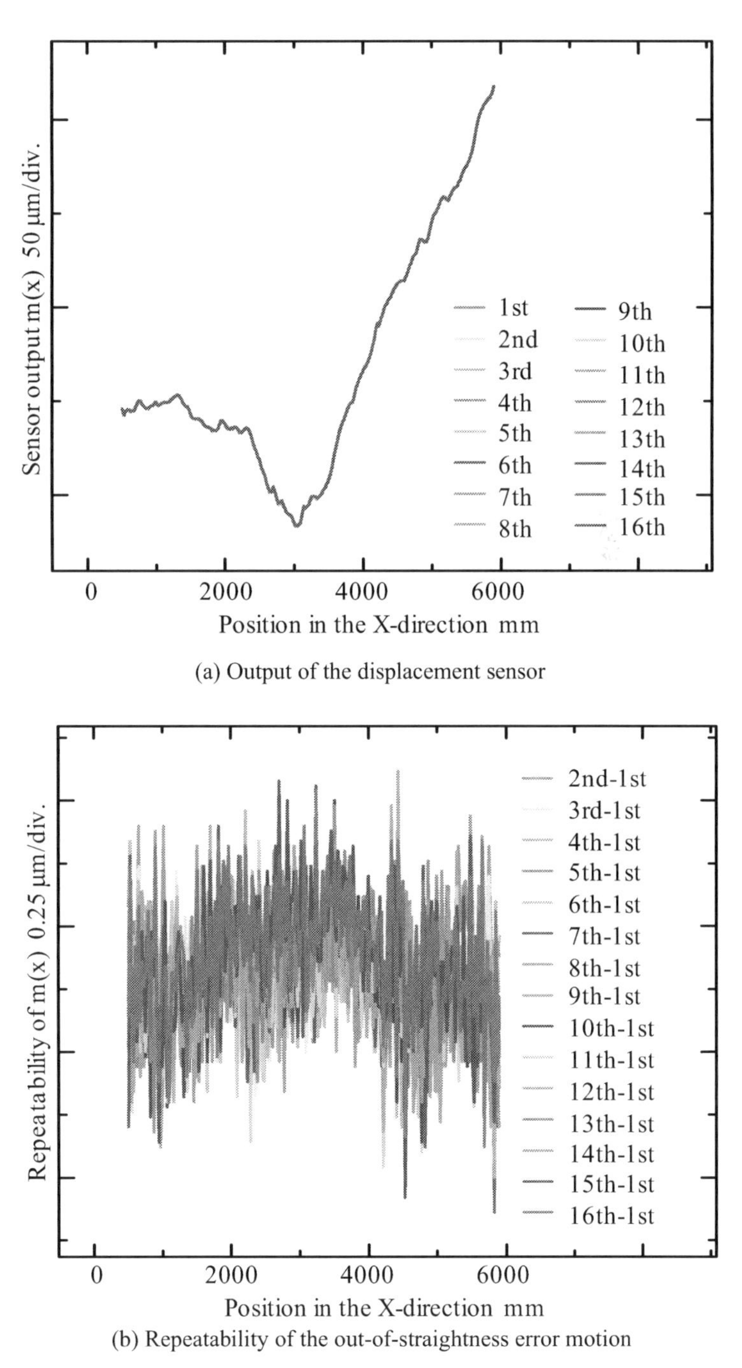

(a) Output of the displacement sensor

(b) Repeatability of the out-of-straightness error motion

Figure 10.39. Investigation of the repeatability of the out-of-straightness error motion of the *X*-directional scanning stage

Figure 10.40. An edge measuring instrument developed for in-line inspection of long extrusion tools

10.4 Summary

Scanning image-sensor systems are effective for fast measurement of micro-dimensions over long ranges. Two systems have been presented for measurement of micro-dimensions of precision long tools.

The first is a scanning area-image-sensor system employed for measurement of micro slit width of a long coating tool. The image of the two slit edges is captured by an area-image sensor. The image is processed using binarization to detect quickly the edge positions of the slit. The slit width is determined through a least square linear fitting of the edge data from binary image. The influence of the Z-directional straightness of the scanning stage can be reduced by choosing a proper threshold for binarization of the image. The slit width over the entire length of a 1.4 m long coating tool was measured in 33 s.

The second system is for a micro-edge radius measurement of a long extrusion tool by scanning a line-image sensor. In the width-based method, the edge width of the tool at each scanning point is first evaluated from the image of the line-image sensor. The edge radius is then calculated from the edge width based on the calibration results using pin gauges. A multi-line scanning method has been employed to reduce the influence of the defocus error caused by the out-of-straightness error motion of the scanning stage. The multi-line scanned images have also been applied to the shape-from-focus (SFF) method. This method uses different focus levels to obtain a sequence of the images of the target surface in the same field of view of the image sensor. The edge radii of a 5100 mm long precision tool was measured within a measurement time of 30 min. Consistent results of edge radii have been obtained by the width-based method and the SFF method.

References

[1] Barrett RC, Quate CF (1991) Optical scan-correction system applied to atomic force microscopy. Rev Sci Instrum 62(6):1393–1399

[2] Asada N, Fujiwara H, Matsuyama T (1998) Edge and depth from focus. Int J Comput Vis 26(2):153–163

[3] Wang CC (1994) A low-cost calibration method for automated optical measurement using a video camera. Mach Vis Appl 7:259–266

[4] Baba M, Ohtani K (2001) A novel subpixel edge detection system for dimension measurement and object localization using an analogue-based approach. J Opt A Pure Appl Opt 3:276–283

[5] Fan KC, Lee MZ, Mou JI (2002) On-line non-contact system for grinding wheel wear measurement. Int J Adv Manuf Technol 19:14–22

[6] Ferron Magnetic Inks (2010) http://www.ferron-magnetic.co.uk/coatings/index.html. Accessed 1 Jan 2010

[7] Ryotec (2009) http://www.ryotec.co.jp/english/index.html. Accessed 1 Jan 2010

[8] Furukawa M, Gao W, Shimizu H, Kiyono S, Yasutake M, Takahashi K (2003) Slit width measurement of a long precision slot die. J JSPE 69(7):1013–1017 (in Japanese)

[9] Keyence Corporation (2010) CCD camera, http://www.keyence.com. Accessed 1 Jan 2010

[10] Gonzalez RC, Woods RE (2002) Digital image processing. Prentice Hall, Upper Saddle River, NJ

[11] Kostic MM, Reifschneider LG (2006) Encyclopedia of chemical processing. Taylor and Francis, London

[12] Motoki T, Gao W, Furukawa M, Kiyono S (2007) Development of a high-speed and high-accuracy measurement system for micro edge radius of long precision tool. J JSPE 73(7):823–827 (in Japanese)

[13] Nippon Electro-Sensory Devices Corporation (2010) http://www.ned-sensor.co.jp. Accessed 1 Jan 2010

[14] Nayer SK, Nakagawa Y (1994) Shape from focus. IEEE Trans PAMI 16(8):824–831

Index